THE GENETICS AND BIOLOGY OF SEX DETERMINATION

The Novartis Foundation is an international scientific and educational charity (UK Registered Charity No. 313574). Known until September 1997 as the Ciba Foundation, it was established in 1947 by the CIBA company of Basle, which merged with Sandoz in 1996, to form Novartis. The Foundation operates independently in London under English trust law. It was formally opened on 22 June 1949.

The Foundation promotes the study and general knowledge of science and in particular encourages international co-operation in scientific research. To this end, it organizes internationally acclaimed meetings (typically eight symposia and allied open meetings and 15–20 discussion meetings each year) and publishes eight books per year featuring the presented papers and discussions from the symposia. Although primarily an operational rather than a grant-making foundation, it awards bursaries to young scientists to attend the symposia and afterwards work with one of the other participants.

The Foundation's headquarters at 41 Portland Place, London W1B 1BN, provide library facilities, open to graduates in science and allied disciplines. Media relations are fostered by regular press conferences and by articles prepared by the Foundation's Science Writer in Residence. The Foundation offers accommodation and meeting facilities to visiting scientists and their societies.

Information on all Foundation activities can be found at http://www.novartisfound.org.uk

Novartis Foundation Symposium 244

THE GENETICS AND BIOLOGY OF SEX DETERMINATION

2002

JOHN WILEY & SONS, LTD

Published in 2002 by John Wiley & Sons Ltd,
Baffins Lane, Chichester,
West Sussex PO19 1UD, England

National 01243 779777
International (+44) 1243 779777
e-mail (for orders and customer service enquiries): cs-books@wiley.co.uk
Visit our Home Page on http://www.wiley.co.uk
or http://www.wiley.com

Other Wiley Editorial Offices

John Wiley & Sons, Inc., 605 Third Avenue,
New York, NY 10158-0012, USA

WILEY-VCH Verlag GmbH, Pappelallee 3,
D-69469 Weinheim, Germany

Jacaranda Wiley Ltd, 33 Park Road, Milton,
Queensland 4064, Australia

John Wiley & Sons (Asia) Pte Ltd, 2 Clementi Loop #02-01,
Jin Xing Distripark, Singapore 129809

John Wiley & Sons (Canada) Ltd, 22 Worcester Road,
Rexdale, Ontario M9W 1L1, Canada

Novartis Foundation Symposium 244
ix+266 pages, 40 figures, 13 tables

Library of Congress Cataloging-in-Publication Data

The genetics and biology of sex determination / [editors, Derek Chadwick, Jamie Goode].
p. cm. – (Novartis Foundation symposium ; 244)
Includes bibliographical references and indexes.
ISBN 0-470-84346-2 (alk. paper)
1. Sex determination, Genetic–Congresses. I. Chadwick, Derek. II. Goode, Jamie. III. Series.
QP278.5 .G466 2002
571.8′1–dc21 2001057371

British Library Cataloguing in Publication Data

A catalogue record for this book is available from the British Library

ISBN 0 470 84346 2

Typeset in 10½ on 12½ pt Garamond by Dobbie Typesetting Limited, Tavistock, Devon.
Printed and bound by Antony Rowe Ltd, Eastbourne
This book is printed on acid-free paper responsibly manufactured from sustainable forestry, in which at least two trees are planted for each one used for paper production.

Contents

Symposium on The genetics and biology of sex determination, held at the Novartis Foundation, London, 1–3 May 2001

Editors: Derek Chadwick (Organizer) and Jamie Goode

This symposium is based on a proposal made by Peter Koopman

Participants

Richard R. Behringer Department of Molecular Genetics, University of Texas M D Anderson Cancer Center, 1515 Holcombe Blvd, Houston, TX 77030, USA

Philippe Berta Human Molecular Genetics Group, Institut de Génétique Humaine, UPR CNRS 1142, 141 rue de la Cardonille, 34396 Montpellier Cedex 5, France

Monica Bullejos (*Novartis Foundation Bursar*) Departmento de Biología Experimental, Facultad de Ciencias Experimentales, Universidad de Jaen, Paraje de las Lagunillas S/N, E-23071 Jaen, Spain

Paul Burgoyne Laboratory of Developmental Genetics, National Institute for Medical Research, Mill Hill, London NW7 1AA, UK

Giovanna Camerino Biologia Generale e Genetica Medica, Dipartimento di Patologia Umana ed Ereditaria, Università di Pavia, Via Forlanini 14, Pavia 27100, Italy

Blanche Capel Department of Cell Biology, Box 3709, Duke University Medical Center, Durham, NC 27710, USA

Brian Charlesworth Institute of Cell, Animal and Population Biology, University of Edinburgh, Edinburgh EH9 3JT, UK

Russell D. Fernald Program in Neuroscience , Department of Psychology, Jordan Hall/Building 420, Stanford University, Stanford, CA 94305-2130, USA

Peter N. Goodfellow Discovery Research, GlaxoSmithKline, Gunnels Wood Road, Stevenage, Herts SG1 2NY, UK

Jennifer A. Marshall Graves Department of Zoology, Comparative Genomics Research Unit, Research School of Biological Sciences, The Australian National University, Canberra, ACT 2601, Australia

Andrew Greenfield MRC Mammalian Genetics Unit, Harwell, Oxon OX11 0RD, UK

Vincent R. Harley Prince Henry's Institute of Medical Research, Level 4, Block E, Monash Medical Centre, 246 Clayton Road, Melbourne, VIC 3168, Australia

Nathalie Josso Unité de Recherches sur l'Endocrinologie du Développement, INSERM U493 Ecole Normale Supérieure, 1 rue Maurice-Arnoux, 92120 Montrouge, France

Peter Koopman Institute for Molecular Bioscience, The University of Queensland, Brisbane, QLD 4072, Australia

Robin Lovell-Badge MRC National Institute for Medical Research, The Ridgeway, Mill Hill, London NW7 1AA, UK

Anne McLaren Wellcome/CRC Institute, Tennis Court Road, Cambridge CB2 1QR, UK

Ursula Mittwoch The Galton Laboratory, Department of Biology, University College London, Wolfson House, 4 Stephenson Way, London NW1 2HE, UK

Ken-ichirou Morohashi Department of Developmental Biology, National Institute for Basic Biology, Myodaiji 38, Okazaki 444-8585, Japan

Francis Poulat Institut de Genetique Humaine, UPR CNRS 1142, 141 rue de la Cardonille, 34396 Montpellier Cedex 5, France

Marilyn Renfree Department of Zoology, The University of Melbourne, VIC 3010, Australia

Andreas Schedl Human Molecular Genetics Unit, University of Newcastle upon Tyne, Ridley Building, Newcastle upon Tyne NE1 7RU, UK

Gerd Scherer Institute of Human Genetics and Anthropology, University of Freiburg, Breisacherstrasse 33, D-79106 Freiburg, Germany

Roger V. Short (*Chair*) Royal Women's Hospital, 132 Grattan Street, Carlton, Melbourne, VIC 3053, Australia

Andrew H. Sinclair Department of Paediatrics, University of Melbourne, and Murdoch Children's Research Institute, Royal Children's Hospital, Melbourne, VIC 3052, Australia

Amanda Swain Section of Gene Function and Regulation, Chester Beatty Laboratories, 237 Fulham Road, London SW3 6JB, UK

Eric Vilain Department of Human Genetics, UCLA School of Medicine, 6335 Gonda Center, 695 Charles Young Drive, Los Angeles, CA 90095-7088, USA

Adam Wilkins BioEssays, Editorial Office, 10/11 Tredgold Lane, Napier Street, Cambridge CB4 3PP, UK

David Zarkower Department of Genetics, Cell Biology and Development, University of Minnesota, 6-160 Jackson Hall, 321 Church Street, SE, Minneapolis, MN 55455, USA

An introduction to the genetics and biology of sex determination

Roger V. Short

Department of Obstetrics and Gynaecology, Royal Women's Hospital, 132 Grattan Street, Melbourne, Victoria 3053, Australia

The Ciba/Novartis Foundation meetings are amazing. I remember the first one I attended, back in 1958. Last week I was in the University of California in Berkeley, talking to Professor Howard Bern, the distinguished comparative biologist. He said, 'Do you know how my scientific career began? It was when, as a young graduate student, I was invited to a Ciba Foundation meeting in 1952, on germ cells' (Ciba Foundation 1953). I hope that in another 40 years' time, some of you will be saying something similar about this meeting. It is the discussions that we have at these meetings that are so exciting.

I would like to set the scene. I should probably start with a word of explanation. The first question that many of you will be asking is, why are there so many Australians in the room? You might think that it is because Peter Koopman proposed the meeting, but that isn't the reason. Sex 'down under' is done rather differently, so we have much to learn from Gondwanaland about the evolution of sex.

We are going to hear a great deal at this meeting about the evolution of sex determination, which is currently a very exciting topic. But let me remind all of you how we define sex. If you produce many small highly motile gametes, you are male. If you produce fewer, large, sessile gametes, you are female. Although we are going to be discussing sex determination, almost all of the papers will be dealing not with the type of gametes that are ultimately produced, but with the morphology of the gonadal soma. I think we need to remember that the somatic sex of the gonad is a secondary issue; it is germ cell sex that ultimately determines maleness or femaleness. Although we know much about the genetic control of gonadal somatic differentiation, we are largely ignorant of the genetic control of the germ cells.

Let me say a few words about the gametes. The biggest single cell that has ever existed is the egg of *Aepyornis*, the giant elephant bird from Madagascar. One egg could contain around five gallons of liquid! This may have been the species' undoing, because when humans first landed on Madagascar about 2000 years ago, they found that Aepyornis eggs made wonderful water containers, and so they raided the nests, leaving 'holy' eggs as testimony of their activity.

Why are eggs so big? Why are sperm so small? Anisogamy is at the very heart of sexual differentiation. One of the reasons for the large size of the female gamete is that mitochondrial DNA is exclusively maternally inherited, hence the oocyte at ovulation has to contain all the mitochondrial DNA for the new individual. In contrast, the male gamete is designed as a highly condensed nuclear DNA warhead that can traverse great distances before penetrating the egg. Following blastoff at orgasm the male gamete is propelled by rocket boosters in the form of the mitochondrial DNA in the midpiece sheath, which drives the beating of the sperm's tail. Although the midpiece sheath actually enters the egg at fertilization, all this paternal mitochondrial DNA is subsequently destroyed by the cytoplasm of the oocyte. So here we are, sexually reproducing organisms, parasitized by mitochondrial DNA which is reproducing vegetatively within us and is exclusively inherited from our mothers. It may be this asymmetrical inheritance of our mitochondrial DNA that has necessitated the sexual dimorphism of the gametes, and hence the major sex differences in the gonads.

Study of the germ cells has an illustrious history. Charles Darwin could not understand how it was that the gametes could transmit information across the generations. He thought that there must be particles, which he called 'gemmules', that were pieces of information from within every somatic cell that was handed over to the gametes. However, he had only a vague understanding of fertilization, and did not appreciate that a single spermatozoon was required to fertilize the egg. August Weizmann then proposed an alternative view, the continuity of germplasm. He envisaged an immortal germline which budded off a mortal soma at each generation, and morphologists imagined that they could see the sequestered germplasm in the newly fertilized egg prior to the first cell division.

Thanks to the cloning of Dolly the sheep, Cumulina the mouse, and many others, we now know that almost any somatic cell nucleus in the body, if inserted into an enucleated oocyte, can produce a new individual that is fully fertile. Thus there is something magical in the cytoplasm of the oocyte that can restore totipotency to a differentiated somatic cell nucleus, and transform soma into sex, somatic cell into germ cell. Each of us in this room therefore has the potential to restore our germ cells from our own somatic cells by nuclear transplantation cloning. This technology, coupled with recent advances in germ cell transplantation, will ensure that germ cell creation, manipulation and repair will be a fruitful area for future research.

One fascinating aspect of sex determination only recently occurred to me, when I was thinking about the way in which mitochondrial DNA is transmitted from one generation to the next. Since males only possess their mother's mitochondrial DNA, it is somewhat ironic that a man's fertility is determined by the motility of his spermatozoa, which is controlled by his mother's mitochondrial DNA in the midpiece sheath of his sperm. So sexual inequality reigns supreme, and the female of the species is more deadly than the male. Maybe it was prophetic foresight that led William Harvey, in the frontispiece of his 1651 volume *De Generatione Animalium*, to have Zeus holding apart the two halves of an egg inscribed with those prophetic words, '*Ex ovo omnia*'.

So in conclusion, I would like to plead for more attention to be paid to the germ cells as not just the arbiters of sex, but also the determinants of sex. After all, the sex-determining gene *Sry* may turn the somatic tissue of the gonad of a female mouse into a testis, but it is incapable of transforming the oogonia into spermatogonia. And in the female, it needs an oocyte to induce the gonadal stroma to develop into hormone-secreting follicular cells, so the somatic tissue of the ovary is at the mercy of the germ cells.

With those thoughts, I would like to introduce the first paper.

Reference

Ciba Foundation 1953 Mammalian germ cells. Churchill, London (Ciba Found Symp 16)

Sex-determining genes in mice: building pathways

Robin Lovell-Badge, Clare Canning and Ryohei Sekido

Division of Developmental Genetics, MRC National Institute for Medical Research, The Ridgeway, Mill Hill, London NW7 1AA, UK

Abstract. Sry is active in the mouse for a very brief period in somatic cells of the genital ridge to initiate Sertoli cell differentiation. SRY protein must act within the context of other gene products required for gonadal development and must itself act on one or more target genes that will ensure the further differentiation and maintenance of Sertoli cells. Over the last few years several genes have been found that have important roles in gonadal development and sex determination. These include genes encoding transcription factors such as *Lhx9*, *Wt1*, *Sf1*, *Dax1*, *Gata4*, *Dmrt1* and *Sox9*, and some involved in cell–cell signalling, including *Amh*, *Wnt4* and *Dhh*. While more await discovery, it is now possible to start putting some of the known genes into pathways or networks. *Sox9* probably occupies a critical role in mammals for both the initiation and maintenance of Sertoli cell differentiation. Data will be presented that are consistent with SRY acting directly on *Sox9* to ensure its up-regulation. SF1 is also central to gonadal differentiation. Our results imply that it contributes to transcriptional activation of several relevant genes, not just those required for male development, including *Sox9* and *Amh*, but also those that can have an antagonistic effect on Sertoli cell differentiation, such as *Dax1*. Progress in establishing other regulatory interactions will also be discussed.

2002 The genetics and biology of sex determination. Wiley, Chichester (Novartis Foundation Symposium 244) p 4–22

Sry was discovered in 1990. Over the following year it was proven to be the Y-linked testis determining gene in both mice and humans through a combination of mutation studies and transgenic experiments (Sinclair et al 1990, Gubbay et al 1990, 1992, Berta et al 1990, Koopman et al 1991). At this time, life seemed simple. *Sry* was the only gene so far identified that was known to be involved in diverting the pathway of gonadal development to make a testis rather than an ovary. We also knew two of the factors that effectively exported the male signal to the rest of the developing embryo. These were testosterone (and other androgens) made in Leydig cells by a series of P450 gene products, and anti-Müllerian hormone (AMH, otherwise known as Müllerian inhibiting substance, MIS), a transforming growth factor (TGF)β-like protein made by Sertoli cells, two factors predicted by Jost through his experiments conducted over 50 years ago (Jost 1953, Munsterberg & Lovell-Badge 1991, Josso & Picard 1986). Of course, we knew things would not stay simple for long. There had to be many other genes involved; in early gonadal development, in the sex-determination step itself and for the differentiation of all the various cell lineages making up the developing gonad along the male or female pathway.

Current models of the pathway or more accurately the network of genes involved look at first sight very complex. However, this can be simplified by breaking the various components into separate, albeit interacting, parts.

Cell lineages

First, we can consider the different cell lineages that make up the developing gonads. *Sry* acts within the supporting cell lineage, between 10.5 and 12.0 days post coitum (dpc) in the mouse, triggering the differentiation of Sertoli cells rather than follicle cells (Palmer & Burgoyne 1991). Cell marking and BrdU-labelling experiments have shown that cells of this lineage originate, at least in part and conceivably entirely, from the coelomic epithelium prior to 11.5 dpc (Karl & Capel 1998, Schmal et al 2000). A proportion of the cells entering the XY genital ridge end up in an interstitial location where they form an undefined cell type. The remainder give rise to Sertoli cells. These rapidly begin to influence all the other bipotential lineages within the gonad. The germ cells, which have migrated into the genital ridge via the mesonephros, become arrested in mitosis rather than entering meiosis, which is characteristic of germ cells within the ovary. The latter seems to be the default pathway as germ cells that have failed to migrate into the gonad of either sex enter meiosis at about the same time (McLaren & Southee 1997). Steroidogenic cells, which are also likely to be within the genital ridge by 11.5 dpc, but whose origin is uncertain, will differentiate relatively early in the testis, where they become Leydig cells (Morohashi 1997). These cells are already beginning to produce testosterone by 12.5 dpc, as well as insulin-like growth factor 3 (INSL3), a third factor essentially predicted by Jost's experiments, but only recently discovered, which is responsible for the transabdominal phase of testicular descent (Nef & Parada 1999, Zimmermann et al 1999). The ovarian theca cells are not obvious and seem to have little functional role until much later. Finally, but critically, subsequent to SRY action there is a reorganization of connective tissue cells into the testicular pattern. This includes the migration of cells from the mesonephros into the developing testis (Martineau et al 1997, Tilmann & Capel 1999). These cells give rise to peritubular myoid cells and endothelial cells. The myoid cells, which are perhaps the only cell lineage unique to testis, have an important role in the morphological differentiation of the testis as they participate with the Sertoli cells to form the epithelial testis cords. The endothelial cells contribute to the characteristic vasculature of the testis, which is likely to be important to support the more rapid growth of the testis, compared to the ovary, and to allow efficient export of testosterone, INSL3 and AMH, the three factors that masculinize the remainder of the embryo.

For each of these lineages there is a decision of cell fate. Any such decision requires at least two processes. Firstly, an initiation step, which can involve extrinsic factors such as growth factors or intrinsic 'switches' such as SRY. This is then followed by a process that reinforces this initial decision, leading to maintenance of the pattern of gene expression required for the cell phenotype, where regulatory loops and/or long term changes in chromatin organization are required. The regulatory loops can be cell autonomous or involve crosstalk with another cell type. In this respect, the myoid cells may also have a critical role in helping to maintain Sertoli cell differentiation. Indeed it is likely that the continued differentiation of each cell type depends on interactions with all the others. But if we first restrict ourselves to the supporting cell lineage it is easier to understand how SRY might work.

Genetic pathways

The molecular events occurring within the supporting cell lineage can also be simplified by separating the network of genes and their protein products into three main themes. This is illustrated in Fig. 1, but it must be stressed that this is only a model. Many interactions remain to be established and it is highly likely that additional critical genes will be found.

We can place a linear pathway in the centre, beginning with *Sry*. If *Sry* is expressed, the related gene *Sox9*, which is switched on at a low level beforehand, becomes expressed at high levels (Morais da Silva et al 1996, Kent et al 1996). *Sox9* then stays at a high level throughout Sertoli cell development and is likely to be involved in the initiation and maintenance of Sertoli cell-specific gene expression. *SOX9* is known to be important for testis differentiation in humans as heterozygous mutations of the gene, which are responsible for the severe dwarfism syndrome, campomelic dysplasia, also lead to XY female sex reversal in about 75% of cases (Foster et al 1994, Wagner et al 1994, Kwok et al 1995, Sudbeck et al 1996, Meyer et al 1997, Wunderle et al 1998, Pfeifer et al 1999). The mutations can be regulatory or inactivating mutations within the coding region. Therefore, heterozygous levels of SOX9 are insufficient for normal cartilage development and close to a threshold for gonadal development, below which Sertoli cells either do not begin to differentiate or they fail to be maintained as such. In the mouse, a heterozygous null mutation does not seem to compromise Sertoli cell differentiation, but this may simply reflect a lower threshold (Bi et al 1999). Unfortunately, homozygous null embryos do not survive long enough to assess

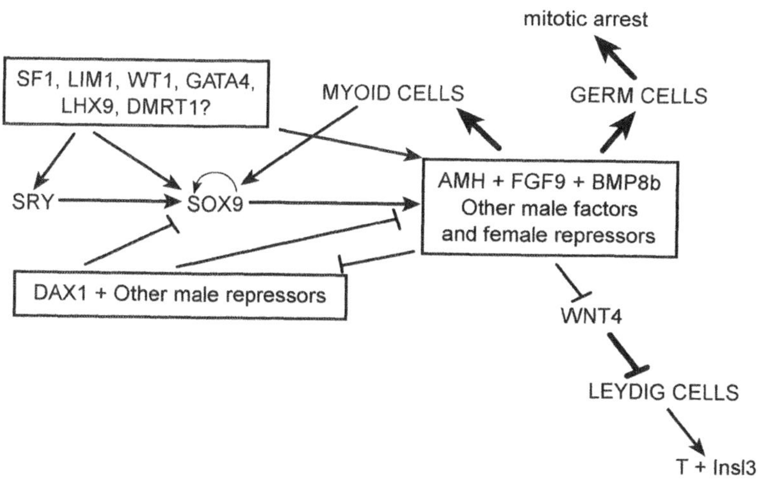

FIG. 1. Model of the genetic interactions during sex determination in the mouse. The central pathway (right-centre box) is essential for male development. Factors indicated in the lower left box are required as anti-testis genes to ensure that the central pathway does not operate in the XX gonad. Factors above in the upper left box are required for gonadal development, and act as positive factors for the central pathway but also for the repressive, anti-testis genes. All these factors act within the supporting cell lineage, but also signal to the other lineages within and outside the developing gonad. See text and relevant chapters in this volume for further details of the pathway and genes. T, testosterone; Insl3, insulin-like growth factor 3.

the precise role of *Sox9*. However, gain-of-function experiments reveal the central importance of SOX9 for Sertoli cell differentiation and sex determination in the mouse (see below). So far, the only known direct target gene for SOX9 in the gonad is *Amh*, but a number of other genes begin to be expressed within early Sertoli cells at the same time, including *Dhh* and *Fgf9* (De Santa Barbara et al 1998, Arango et al 1999, Bitgood et al 1996). Moreover, it seems likely that there will be a substantial number of genes dependent on SOX9 for their expression later on in Sertoli cells. Several genes are also down-regulated shortly after SOX9 expression has increased. These include *Sry*, *Dax1* and *Wnt4* (Swain et al 1998, Vainio et al 1999). SOX9 is thought to function as both an architectural protein in a similar way to SRY (by virtue of its HMG box DNA binding domain; Pontiggia et al 1994), and a transcriptional activator (it has a strong activation domain at its C-terminus; Sudbeck et al 1996). So it seems likely that an as yet unidentified repressor mediates the down-regulation of these genes, possibly itself activated by SOX9. However, perhaps in certain contexts SOX9 can mediate repression itself, simply by acting as an architectural factor through bending of DNA via its HMG box domain.

There are then two opposing forces acting on this central pathway. There is a set of factors that are required for gonadal development, including LIM1, LHX9, WT1, GATA4 and SF1 (see Swain & Lovell-Badge 2001 for review and elsewhere in this volume). Many of these factors act at several stages, or continuously, and can be considered to have a positive role with respect to gonadal development and in particular Sertoli cell differentiation. Null mutations in each of these genes are known to lead to a failure of gonadal development in both sexes. The exception to this is *Gata4*, where its role in gonadal development is unknown because the null mutation is an early embryonic lethal (Viger et al 1998). *Lhx9*, *Wt1* and *Sf1* homozygous mutants all show a similar phenotype with respect to the genital ridge, which begins to develop but the cells die through apoptosis at about 11.5 dpc (Birk et al 2000, Kreidberg et al 1993, Luo et al 1994). The similar phenotype suggests that there may be epistatic relationships among them, and there is evidence that the expression of *Sf1* depends on LHX9 (Birk et al 2000). Both of these are relatively specific to the gonad, although *Sf1* is also expressed in the adrenals and pituitary and hypothalamus. *Wt1* expression is much more widespread, being in the metanephros, coelomic epithelium, heart, etc. The gonads are, however, the only place where all three are expressed, so together they could be responsible for gonad-specific expression of other genes.

All these genes may serve as transcriptional activators of genes in the central pathway. There is strong evidence that SF1, WT1 and GATA4 participate along with SOX9 for *Amh* transcription (De Santa Barbara et al 1998, Arango et al 1999, Viger et al 1998). In this case SOX9 is the limiting factor as all the others are expressed from the beginning of genital ridge development, whereas *Amh* only begins to be expressed once SOX9 levels become significantly higher at 11.5 dpc. Studies where the binding sites for SF1 and SOX9 in the minimal regulatory region of *Amh* were mutated *in vivo* would also fit with this (Arango et al 1999). All the other factors could bind to their target sequences but cannot initiate transcription until SOX9 is able to initiate formation of the appropriate complex through its ability to bend DNA, via its HMG box. There are suggestions that *Sry* may depend on WT1 and we have some evidence that expression of *Sox9* in the genital ridge is dependent on SF1, as *Sox9* transcript levels are absent in homozygous *Sf1* mutant embryos at about 11 days (Hossain & Saunders 2001, A. Swain & R. Lovell-Badge, unpublished data).

A heterozygous mutation in *SF1* and partial loss of function mutations in *WT1* (notably in Frasier syndrome) can lead to XY female sex reversal in humans (Achermann et al 1999, Barbaux et al 1997). This suggests that these factors act positively to encourage Sertoli cell differentiation, but it is not clear whether this is at the level of initiation or maintenance. Moreover, as both genes seem to be required for cell survival and perhaps proliferation, they may have a more critical role in the development of testes than ovaries, as increased cell proliferation is a characteristic of the former. The sex reversal seen with these partial loss-of-function mutations could also be explained by an effect on the central pathway as both *Sry* and *Sox9* need to be expressed above a critical threshold to induce testis formation.

Finally, there is a set of factors that act negatively on this central pathway. These can be considered antitestis genes, but may also include ovarian determining genes. The role of these genes is to ensure that an ovary develops in the absence of *Sry*. Unfortunately, to date we only know of one such factor, DAX1. This is most likely to be responsible for the dosage-sensitive sex reversal syndrome in humans, which involves duplication of the region of the X chromosome containing the gene *XP21* (see Swain et al 1998, and references therein). Transgenic mice carrying extra copies of the *Dax1* gene can also show XY female sex reversal in some circumstances. However, a loss of function mutation engineered in the mouse gene does not lead to male development in XX animals, suggesting that if it is an ovary-determining gene, it must be part of a redundant system, where other genes can compensate for its absence (Yu et al 1998). The gene encodes an unconventional member of the nuclear receptor superfamily, DAX1, which has a ligand-binding domain, but a novel N-terminal domain instead of a zinc finger DNA-binding domain. It is unclear whether DAX1 can bind DNA by itself, but there is substantial evidence that it interacts with SF1, a more typical orphan nuclear receptor, recruiting co-repressors and changing the activity of SF1 from that of transcriptional activator to repressor (e.g. Nachtigal et al 1998, Kawabe et al 1999). It is therefore simple to imagine that it can work as an antitestis gene, simply by antagonizing SF1. As *Sox9* expression probably depends on SF1, this is likely to be the critical point at which excess DAX1 leads to sex reversal. However, DAX1 has also been implicated as a repressor of *Amh* expression (Nachtigal et al 1998). While the two genes are hardly co-expressed — *Dax1* being down-regulated in the testis coincident with the up-regulation of *Amh* — it is possible that the persistent expression of DAX1 in the ovary serves to ensure that AMH is not made in the female embryo.

Interestingly, at least the initiation of expression of *Dax1* in the genital ridge depends on SF1 and perhaps some of the other 'positive factors'. We showed that an 11 kb 5′ fragment from *Dax1* is sufficient to drive expression of reporter genes within the developing gonad in a pattern identical to that of the endogenous gene (Swain et al 1998). Further characterization of this 11 kb has delineated an SF1 binding site that is essential to the initiation of this expression. Moreover the endogenous *Dax1* gene is not expressed in *Sf1* mutant genital ridges (Hoyle et al 2002). Therefore SF1 is directly responsible for the expression of its own antagonist, which makes for an intriguing regulatory loop as well as stressing the complexity of the network of interactions if viewed as a whole. It also reinforces the idea that SF1, and probably the other 'positively' acting factors grouped with it in Fig. 1, are largely neutral in the decision to follow the male or female pathway. It is just that the genes required for testis differentiation are sensitive flowers and those for the ovary are more robust.

From the above, it is clear that *Sox9* plays a central role in mammalian sex determination. It is a good candidate for a gene directly regulated by SRY. Moreover, there is now substantial evidence suggesting that it is the only critical gene downstream of SRY. These data include the following. Firstly, in transgenic mouse experiments where *Dax1* regulatory sequences were used to drive the expression of human SOX9 specifically in the genital ridge, only 1 out of more than 20 independent transgenic mice or lines showed sex reversal, but this one XX male looked identical to those made with mouse *Sry* as a transgene (A. Swain & R. Lovell-Badge, unpublished data). The reason for the low rate of sex reversal is probably due to the transient nature of *Dax1* expression in the male. In other words, if the transgene begins to induce Sertoli cell differentiation, then it will be turned off. Perhaps the one case that worked had a sufficiently high level of SOX9 expression, such that it was able to induce expression of the endogenous *Sox9* gene via a feedback loop. Secondly, a case of sex reversal in humans was reported where a duplication of 17q23-24 (the chromosomal region containing *SOX9*) led to XX male development (Huang et al 1999). Thirdly, the best evidence comes from a chance insertion of a transgene upstream of *Sox9* that has led to the constitutive activation of the gene in XX as well as XY gonads (Bishop et al 2000). Although there is some dependency on genetic background, this is sufficient to cause male development of all transgenic XX mice. The nature of the mutation, termed *Odsex*, is not understood, as it involves an insertion and deletion over 1 Mb upstream of *Sox9*. It could be due to the loss of a negative regulatory element, to a less-specific long-range position effect on chromatin or to a direct effect of enhancer elements contained within the transgene on *Sox9* transcription. See also the recent paper by Shedl and colleagues (Vidal et al 2001).

SRY action

It then becomes important to establish whether SRY directly regulates *Sox9* and if so, how this is achieved. An important question, still unanswered after 11 years, is how does the SRY protein work? Is it a transcriptional activator or does it just exert its effects by altering chromatin structure, and how does it interact with any protein partners? These questions have been difficult to answer, partly because SRY has evolved so rapidly, such that the only part of the protein showing any conservation is the HMG box DNA binding domain (Whitfield et al 1993, Tucker & Lundrigan 1993, Hacker et al 1995). Indeed, if the mouse and human genes are compared there is no homology outside the box, including the rest of the open reading frame, 5′ and 3′ untranslated regions, and flanking DNA. This implies that the only functional part of the gene is the HMG box itself. This seems to be borne out by mutation studies in cases of XY female sex reversal in humans, where almost all point mutations are located within the box. If the N and C-terminal domains were important then mutations affecting these would also have been frequent. This is seen for SOX9, where mutations leading to campomelic dysplasia can affect either the HMG box or the C-terminal activation domain.

On the other hand, the extent of non-synonymous versus synonymous changes in the non-box regions of SRY, as well as the non-uniform rate of change seen when comparing groups of related species, implies that there is selection for change, and therefore some function to these regions (Whitfield et al 1993). *In vitro* assays have demonstrated that the C-terminal glutamine-rich region of the mouse SRY protein can function as an activation domain, although only weakly, whereas the human protein has no demonstrable activation properties (Dubin & Ostrer 1994). Moreover, in recent experiments, Bowles et al (1999) showed that translational stop codons engineered into the mouse *Sry* open reading frame (ORF), either just C-terminal to the HMG box or just before the glutamine rich region, prevented the ability of an *Sry* transgene to give XX male sex reversal. This implied that the glutamine rich region was essential to mouse SRY function, although with the caveat that the authors were unable to show the presence of stable SRY protein *in vivo* because of the lack of suitable antibodies.

Finally, while a 14 kb genomic fragment of the mouse *Sry* gene readily gives XX male sex reversal in transgenic mice, we had been unable to obtain sex reversal with a 25 kb clone carrying the human *SRY* gene. This was despite showing that transcripts were present in the genital ridge (Koopman et al 1991). This could be interpreted as evidence that the mouse and human proteins act differently, implying that the conserved HMG box is not sufficient and that the other domains of the protein are important, presumably through interactions with other proteins. Indeed, interactions with other proteins have been shown *in vitro* for both the mouse and human C-terminal domains, albeit with different proteins in each case (Poulat et al 1997, Zhang et al 1999).

However, an alternative explanation is simply that the human gene is not correctly expressed in mice. It could be a quantitative problem, where levels of expression from the human *SRY* transgene are insufficient to act in the mouse. Indeed, it is possible that regulatory regions may be missing from the 25 kb genomic region, or that the human and mouse genes could be regulated in substantially different ways. To address this question, we have engineered two transgene constructs that are hybrids between the mouse and human sequences (C. Canning, I. Bar, G. Penney and R. Lovell-Badge, unpublished data). In the first, the mouse HMG box was replaced with the human N-terminal domain and HMG box, in the context of the mouse regulatory sequences contained within the

14 kb genomic region. This functioned efficiently in transgenic mice, giving XX male sex reversal. This shows that the human and mouse HMG boxes are interchangeable and is in line with similar experiments by Eicher and colleagues (Bergstrom et al 2000), who showed that the mouse SRY HMG box could be replaced with that of either SOX3 or SOX9 and still function. However, in all these experiments the C-terminal glutamine-rich domain of mouse SRY was still present. We therefore engineered a second construct where the whole human *SRY* ORF was inserted in the context of the mouse regulatory sequences, including its own stop codon, so the only protein that could be made was that of human SRY. This was also able to give sex reversal in transgenic mice. The resulting XX males were identical in phenotype to those produced with the mouse *Sry* transgene and we could detect human SRY protein of the correct size within the genital ridge at 11.5 dpc. Therefore, despite the extensive sequence differences, both human and mouse SRY proteins can function in mice, and there is no requirement for the glutamine-rich region or, presumably, any transactivation domain. It is still possible that relevant factors that interact with the human SRY C-terminal domain are present in mice, but given that this is just one representative of the many different SRY sequences existing in mammals, each of which would have to have its own specific partner, the simplest explanation is that there is no requirement for the non-box domains in sex determination. However, it is conceivable that SRY could have additional (male-specific) functions for which the non-box regions are required. Such functions could include anything from spermatogenesis to male behaviour, for which there could be selection to account for the rapid evolution of the sequence.

It is likely then, that for the role of SRY in sex determination, all that is required is an HMG box of the right type, expressed in a stable form at the appropriate time during gonadal development. In which case, although the HMG box will almost certainly be involved in interactions with other proteins, SRY may be acting solely as an architectural factor altering local chromatin structure at its binding site in a critical enhancer region of its target gene(s) (Pontiggia et al 1994). To really prove this, however, such an enhancer has to be found.

The relationship between SRY and SOX9

As discussed above, *Sox9* is the best candidate we have for a direct target of SRY. A high level of *Sox9* expression correlates with the presence of *Sry*: it is seen in both XY and XX *Sry* transgenic genital ridges and is absent from genital ridges that will develop as ovaries, whether XX or carrying a Y chromosome deleted for *Sry*. These genetic arguments are therefore consistent with *Sox9* being a downstream gene, although they cannot prove it is a direct target. To further explore this possibility, we wanted to look in detail at how SRY and SOX9 are expressed

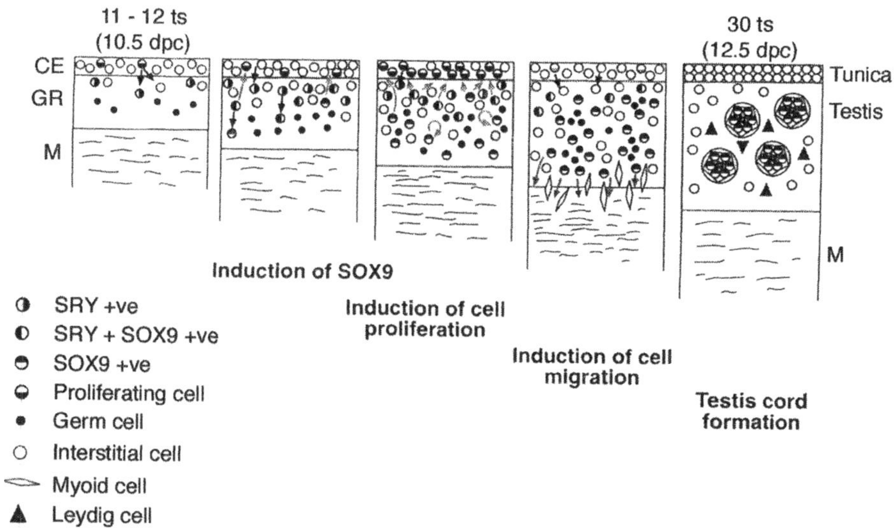

FIG. 2. Model of the cellular events relating to SRY and SOX9 expression. See text for details. Arrows indicate signalling between cells. CE, coelomic epithelium; GR, genital ridge; M, mesonephros.

during early testis development. As yet, we and others have been unable to derive good antibodies against mouse SRY, so we took an alternative strategy, inserting six copies of an epitope tag at the C-terminus of the *Sry* ORF, in the context of the mouse 14 kb genomic region. This was then used to derive transgenic mice. The tagged protein was functional, in that it caused XX male sex reversal, and could be detected by antibodies to the MYC epitope in the genital ridge. Co-localization experiments using antibodies against SOX9 allowed us to conclude that SRY is not expressed in cells of the coelomic epithelium, but is first found in cells just below this layer. SOX9 is induced shortly after the onset of SRY expression, perhaps within a few hours, but SRY is then rapidly lost as there are relatively few double-positive cells. We also made use of a second *Sry* transgenic construct, where a human placental alkaline phosphatase (*HPLAP*) reporter gene was inserted at the beginning of the ORF. This transgene does not allow expression of the SRY protein, so it does not cause sex reversal, but because HPLAP is a very stable enzyme, it acts as a short-term lineage label allowing us to tell which cells were expressing *Sry* at 12.5 dpc, at a time when transcripts for both the transgene and endogenous *Sry* are no longer present. When combined with the antibody data, we can conclude that all cells that have expressed *Sry* become Sertoli cells and, importantly no other cell type. Details of these experiments will be reported elsewhere (R. Sekido, I. Bar, V. Narvaez & R. Lovell-Badge, unpublished data), but the conclusions are summarized in Fig. 2.

Combining our data with those of Blanche Capel and co-workers (Martineau et al 1997, Karl & Capel 1998, Tilmann & Capel 1999, Schmahl et al 2000), we can propose a model that relates gene expression with the cell biology of the developing testis. At about 10.5 dpc, some SF1-positive cells within the coelomic epithelium divide, giving rise to daughter cells that enter the early genital ridge. These adopt two separate fates, one giving rise to an interstitial cell type of no known function, the other begins to express *Sry*. Once SRY protein accumulates above a critical threshold it induces a high level of SOX9 expression. These cells then signal back to the overlying coelomic epithelium to trigger an increase in proliferation of SF1-positive cells, the daughter cells of which then enter the genital ridge, giving rise to more interstitial and *Sry*-expressing cells. This cycle continues, with the coelomic epithelium acting as a factory generating more pre-Sertoli cells (although these also proliferate within the gonad), until shortly after 11.5 dpc when the process stops, coincident with the coelomic epithelium becoming SF1-negative. By this stage, *Sox9* expression will have also initiated the expression of other genes, such as *Amh*, and led to the repression of *Sry* and *Dax1*. The differentiating Sertoli cells also produce signals responsible for the migration of peritubular myoid and endothelial cells into the genital ridge from the mesonephros. Conceivably, the presence of these cells could be responsible for repressing further recruitment from the coelomic epithelium. It is possible that FGF9 is the signal responsible for proliferation or recruitment of cells from the coelomic epithelium and for the migration from the mesonephros (Colvin et al 2001).

The co-localization of SOX9 and SRY within the same cells and the rapid onset of SOX9 expression following the appearance of SRY is again entirely consistent with *Sox9* being a direct target of SRY. However, to prove this, it is still necessary to define the critical regulatory sequence responsible for the Sertoli cell-specific expression of *Sox9*. This poses a problem, however. *In vitro* cell transfection experiments suggested that a small 5′ region adjacent to the *Sox9* promoter could drive reporter gene expression in cells isolated from the early testis, but this same region did not work in transgenic mice to give any expression within the gonad (Kanai & Koopman 1999). In fact, human mutation studies, where translocation breakpoints leading to campomelic dysplasia and sex reversal were found to map up to a megabase 5′ to *SOX9*, and transgenic experiments using YACs containing up to 350 kb of *SOX9* genomic sequence, both suggested that the critical regulatory regions map a long way from the gene itself (Wunderle et al 1998, Pfeifer et al 1999). However, it is possible that *Sox9* is just particularly sensitive to long range position effects. We have therefore begun to readdress this problem, beginning with a mouse *Sox9* BAC clone including about 70 kb 5′ and 30 kb 3′ flanking DNA, into which a β-galactosidase reporter gene has been engineered (R. Sekido & R. Lovell-Badge, unpublished results). In preliminary experiments this can give robust Sertoli cell-specific expression within the gonads of transgenic mice. It does not reproduce all the other sites of *Sox9* expression, for example within developing cartilage, but this result does suggest that it will be possible to define the critical regulatory region that responds to SRY by further analysis of the sequences contained within this BAC.

Conclusions

Considerable progress has been made over the last 11 years, such that it is now possible at least to formulate reasonable models of how sex determination may work in mammals. An impressive number of genes have been discovered that clearly play an important role in the process. Moreover, from the model of the network of gene interactions outlined in Fig. 1, one can imagine how this can be altered in evolution, simply by changing the rate-limiting step. This can explain how sex determination can work in the few mammalian species that do not have *Sry* (Just et al 1995) and perhaps also in other vertebrates using a completely different switch, such as the ZZ/ZW system of birds or environmental mechanisms in reptiles. One could even choose a different cell lineage to be the critical one—for example, steroidogenesis seems to play a more leading role in sex determination in many lower vertebrates.

However, we are no doubt still missing many relevant genes, in particular for the female pathway, both those that can be considered antitestis genes and those that are actively required for the specification of the cell types characteristic of the ovary. We are also missing many of the details of gene, protein and cellular interactions, which are necessary for a true understanding of the process. All of this should keep us off the streets for at least the next 10 years.

Acknowledgements

We are very grateful to Amanda Swain, Blanche Capel and Paul Burgoyne and to other members of the laboratory for valuable discussions and for permission to refer to unpublished data.

References

Achermann JC, Ito M, Ito M, Hindmarsh PC, Jameson JL 1999 A mutation in the gene encoding steroidogenic factor-1 causes XY sex reversal and adrenal failure in humans. Nat Genet 22:125–126

Arango NA, Lovell-Badge R, Behringer RR 1999 Targeted mutagenesis of the endogenous mouse *Mis* gene promoter: in vivo definition of genetic pathways of vertebrate sexual development. Cell 99:409–419

Barbaux S, Niaudet P, Gubler MC et al 1997 Donor splice-site mutations in WT1 are responsible for Frasier syndrome. Nat Genet 17:467–470

Bergstrom DE, Young M, Albrecht KH, Eicher EM 2000 Related function of mouse SOX3, SOX9, and SRY HMG domains assayed by male sex determination. Genesis 28:111–124

Berta P, Hawkins JR, Sinclair AH et al 1990 Genetic evidence equating SRY and the testis-determining factor. Nature 348:448–450

Bi W, Deng JM, Zhang Z, Behringer RR, de Crombrugghe B 1999 Sox9 is required for cartilage formation. Nat Genet 22:85–89

Birk OS, Casiano DE, Wassif CA et al 2000 The LIM homeobox gene Lhx9 is essential for mouse gonad formation. Nature 403:909–913

Bishop CE, Whitworth DJ, Qin Y et al 2000 A transgenic insertion upstream of Sox9 is associated with dominant XX sex reversal in the mouse. Nat Genet 26:490–494

Bitgood MJ, Shen L, McMahon AP 1996 Sertoli cell signaling by Desert hedgehog regulates the male germline. Curr Biol 6:298–304

Bowles J, Cooper L, Berkman J, Koopman P 1999 Sry requires a CAG repeat domain for male sex determination in Mus musculus. Nat Genet 22:405–408

Colvin JS, Green RP, Schmahl J, Capel B, Ornitz DM 2001 Male-to-female sex reversal in mice lacking fibroblast growth factor 9. Cell 104:875–889

de Santa Barbara P, Bonneaud N, Boizet B et al 1998 Direct interaction of SRY-related protein SOX9 and steroidogenic factor 1 regulates transcription of the human anti-Mullerian hormone gene. Mol Cell Biol 18:6653–6665

Dubin RA, Ostrer H 1994 Sry is a transcriptional activator. Mol Endocrinol 8:1182–1192

Foster JW, Dominguez-Steglich MA, Guioli S et al 1994 Campomelic dysplasia and autosomal sex reversal caused by mutations in an SRY-related gene. Nature 372:525–530

Gubbay J, Collignon J, Koopman P et al 1990 A gene mapping to the sex-determining region of the mouse Y chromosome is a member of a novel family of embryonically expressed genes. Nature 346:245–250

Gubbay J, Vivian N, Economou A, Jackson D, Goodfellow P, Lovell-Badge R 1992 Inverted repeat structure of the Sry locus in mice. Proc Natl Acad Sci USA 89:7953–7957

Hacker A, Capel B, Goodfellow P, Lovell-Badge R 1995 Expression of Sry, the mouse sex determining gene. Development 121:1603–1614

Hoyle C, Narvaez V, Alldus G, Lovell-Badge R, Swain A 2002 Dax1 expression is dependent on SF1 in the developing gonad. Mol Endocrinol, in press

Hossain A, Saunders GF 2001 The human sex-determining gene *SRY* is a direct target of *WT1*. J Biol Chem 276:16817–16823

Huang B, Wang S, Ning Y, Lamb AN, Bartley J 1999 Autosomal XX sex reversal caused by duplication of SOX9. Am J Med Genet 87:349–353

Josso N, Picard JY 1986 Anti-Mullerian hormone. Physiol Rev 66:1038–1390

Jost A 1953 Problems in fetal endocrinology: the gonadal and hypophyseal hormones. Recent Prog Horm Res 8:379–418

Just W, Rau W, Vogel W et al 1995 Absence of Sry in species of the vole Ellobius. Nat Genet 11:117–118

Kanai Y, Koopman P 1999 Structural and functional characterization of the mouse Sox9 promoter: implications for campomelic dysplasia. Hum Mol Genet 8:691–696

Karl J, Capel B 1998 Sertoli cells of the mouse testis originate from the coelomic epithelium. Dev Biol 203:323–333

Kawabe K, Shikayama T, Tsuboi H et al 1999 Dax-1 as one of the target genes of Ad4BP/SF-1. Mol Endocrinol 13:1267–1284

Kent J, Wheatley SC, Andrews JE, Sinclair AH, Koopman P 1996 A male-specific role for SOX9 in vertebrate sex determination. Development 122:2813–2822

Koopman P, Gubbay J, Vivian N, Goodfellow P, Lovell-Badge R 1991 Male development of chromosomally female mice transgenic for Sry. Nature 351:117–121

Kreidberg JA, Sariola H, Loring JM et al 1993 WT-1 is required for early kidney development. Cell 74:679–691

Kwok C, Weller PA, Guioli S et al 1995 Mutations in SOX9, the gene responsible for Campomelic dysplasia and autosomal sex reversal. Am J Hum Genet 57:1028–1036

Luo X, Ikeda Y, Parker KL 1994 A cell-specific nuclear receptor is essential for adrenal and gonadal development and sexual differentiation. Cell 77:481–490

Martineau J, Nordqvist K, Tilmann C, Lovell-Badge R, Capel B 1997 Male-specific cell migration into the developing gonad. Curr Biol 7:958–968

McLaren A, Southee D 1997 Entry of mouse embryonic germ cells into meiosis. Dev Biol 187:107–113

Meyer J, Sudbeck P, Held M et al 1997 Mutational analysis of the SOX9 gene in campomelic dysplasia and autosomal sex reversal: lack of genotype/phenotype correlations. Hum Mol Genet 6:91–98

Morais da Silva S, Hacker A, Harley V, Goodfellow P, Swain A, Lovell-Badge R 1996 Sox9 expression during gonadal development implies a conserved role for the gene in testis differentiation in mammals and birds. Nat Genet 14:62–68

Morohashi K 1997 The ontogenesis of the steroidogenic tissues. Genes Cells 2:95–106

Munsterberg A, Lovell-Badge R 1991 Expression of the mouse anti-Müllerian hormone gene suggests a role in both male and female sexual differentiation. Development 113:613–624

Nachtigal MW, Hirokawa Y, Enyeart-VanHouten DL, Flanagan JN, Hammer GD, Ingraham HA 1998 Wilms' tumor 1 and Dax-1 modulate the orphan nuclear receptor SF-1 in sex-specific gene expression. Cell 93:445–454

Nef S, Parada LF 1999 Cryptorchidism in mice mutant for Insl3. Nat Genet 22:295–299

Palmer SJ, Burgoyne PS 1991 In situ analysis of fetal, prepubertal and adult XX–XY chimaeric mouse testes: sertoli cells are predominantly, but not exclusively, XY. Development 112:265–268

Pfeifer D, Kist R, Dewar K et al 1999 Campomelic dysplasia translocation breakpoints are scattered over 1 Mb proximal to SOX9: evidence for an extended control region. Am J Hum Genet 65:111–124

Pontiggia A, Rimini R, Harley VR, Goodfellow PN, Lovell-Badge R, Bianchi ME 1994 Sex-reversing mutations affect the architecture of SRY-DNA complexes. Embo J 13:6115–6124

Poulat F, Barbara PS, Desclozeaux M et al 1997 The human testis determining factor *SRY* binds a nuclear factor containing PDZ protein interaction domains. J Biol Chem 272: 7167–7172

Schmahl J, Eicher EM, Washburn LL, Capel B 2000 Sry induces cell proliferation in the mouse gonad. Development 127:65–73

Sinclair AH, Berta P, Palmer MS et al 1990 A gene from the human sex-determining region encodes a protein with homology to a conserved DNA-binding motif. Nature 346:240–244

Sudbeck P, Schmitz ML, Baeuerle PA, Scherer G 1996 Sex reversal by loss of the C-terminal transactivation domain of human SOX9. Nat Genet 13:230–232

Swain A, Lovell-Badge R 2001 Sex determination and differentiation. In: Rossant J, Tam P (eds) Mouse development. Academic Press, London, p 371–393

Swain A, Narvaez V, Burgoyne P, Camerino G, Lovell-Badge R 1998 Dax1 antagonizes Sry action in mammalian sex determination. Nature 391:761–767

Tilmann C, Capel B 1999 Mesonephric cell migration induces testis cord formation and Sertoli cell differentiation in the mammalian gonad. Development 126:2883–2890

Tucker PK, Lundrigan BL 1993 Rapid evolution of the sex determining locus in Old World mice and rats. Nature 364:715–717

Vainio S, Heikkila M, Kispert A, Chin N, McMahon AP 1999 Female development in mammals is regulated by Wnt-4 signalling. Nature 397:405–409

Vidal VP, Chaboissier MC, de Rooij DG, Schedl A 2001 Sox9 induces testis development in XX transgenic mice. Nat Genet 28:216–217

Viger RS, Mertineit C, Trasler JM, Nemer M 1998 Transcription factor GATA-4 is expressed in a sexually dimorphic pattern during mouse gonadal development and is a potent activator of the Mullerian inhibiting substance promoter. Development 125:2665–2675

Wagner T, Wirth J, Meyer J et al 1994 Autosomal sex reversal and campomelic dysplasia are caused by mutations in and around the SRY-related gene SOX9. Cell 79:1111–1120

Whitfield LS, Lovell-Badge R, Goodfellow PN 1993 Rapid sequence evolution of the mammalian sex-determining gene SRY. Nature 364:713–715

Wunderle VM, Critcher R, Hastie N, Goodfellow PN, Schedl A 1998 Deletion of long-range regulatory elements upstream of *SOX9* causes campomelic dysplasia. Proc Natl Acad Sci USA 95:10649–10654

Yu RN, Ito M, Saunders TL, Camper SA, Jameson JL 1998 Role of Ahch in gonadal development and gametogenesis. Nat Genet 20:353–357

Zhang J, Coward P, Xian M, Lau YF 1999 In vitro binding and expression studies demonstrate a role for the mouse Sry Q-rich domain in sex determination. Int J Dev Biol 43:219–227

Zimmermann S, Steding G, Emmen JM et al 1999 Targeted disruption of the Insl3 gene causes bilateral cryptorchidism. Mol Endocrinol 13:681–691

DISCUSSION

Wilkins: It is clear that there is a complex network of interactions taking place here. If there are evolutionary pressures to change the timing of expression of one component, this will have knock-on effects on other components. It is possible that the early *Sox9* expression in mammals is in some way a response to selective pressures. I would submit that in order to make sense of such shifts in expression, we have to understand the whole network (which is difficult) and compare it in all these organisms.

A specific question: it seemed to me that the *Dmrt* genes were conspicuous by their absence in your diagram. How do they fit into your scheme?

Lovell-Badge: I think they are important. But the experiments don't quite show this yet. This is probably because of functional redundancy.

Zarkower: We have some preliminary results that show that *Dmrt1* can cause some sex reversal if we sensitize the background. On the basis of the evolutionary conservation of early male-specific expression among a range of vertebrates, it seems likely that *Dmrt1* has early as well as late functions.

Burgoyne: Roger Short, I felt you threw down the gauntlet in your introduction in suggesting that the germ cells have a role in gonadal sex determination. One of my main research interests is in the genetic basis for germ cell sex differentiation; I nevertheless feel that I should support the soma view. You have to differentiate between the determination process—that is, the fate decision to go down the male or female pathways—and the differentiation process itself. If you take an XX *Sry* gonad or an XO *Sry* gonad, the soma imposes the fate decision for the germ cells to go down the male pathway, because they become prospermatogonia and not oocytes. Subsequently, XX germ cells with *Sry* don't make it very far down the process, because two X chromosomes become lethal.

XO germ cells make stem cell spermatogonia and then they arrest because they need genes on the Y chromosome. However, these are both requirements for the differentiation process; the fate decision is imposed by the gonad. I would say that sex determination of the germ cells is mediated by the supporting cells.

Short: I agree, but I would still like to know why two X chromosomes are lethal to a male germ cell. What is it about the second X that is inducing lethality?

Capel: Why can't it be different for the two sexes? In the female, the germ cells do control the pathway; in the male, *Sry* interferes with the ability of the germ cells to control the pathway. What I am suggesting is that in the absence of *Sry* the germ cells will enter meiosis and dictate the formation of an ovary.

Burgoyne: They interact back on the system and are involved in the differentiation process.

Capel: But in the presence of *Sry* their ability to enter meiosis is blocked. The soma is then imposing the male pathway, whereas in the absence of *Sry* the germ cells are imposing the female pathway.

McLaren: All the germ cells are probably pre-programmed cell-autonomously to enter meiosis and follow the female pathway, unless they are prevented from doing so by the testis (McLaren & Southee 1997). We don't know whether *Sry* or *Sox9* is needed in the testis for the inhibition of meiosis, but it is clearly something to do with Sertoli cells. In the testis, differentiation of the somatic component occurs even without germ cells, but in the differentiation of the ovary the female germ cells call the tune (McLaren 2000).

In XX↔XY chimeras, one gets a small number of XX Sertoli cells. They almost certainly express *Sox9*. There are only a few of them, so it is not like Blanche Capel's sandwich experiment in which she seems to see many XX *Sox9*-expressing cells induced. Do you think it is the other *Sry*- and *Sox9*-positive Sertoli cells who are inducing neighbours? In your first diagram you had SOX9 directly regulating its own expression: could there also be a paracrine effect on *Sox9* regulation?

Lovell-Badge: It might well be exactly the same thing that Blanche Capel was seeing with the migration. The XX Sertoli cells may differentiate slightly later. The initial Sertoli cells form because of *Sry*, and then if they are sufficient to induce the migration, there will be some XX cells induced to express *Sox9*, which become Sertoli cells.

Mittwoch: I have a question relating to the difference between human and mouse *Sry*. If *Sry* induces cell proliferation, one would expect the rate of proliferation to be different in humans and mice. Did I understand correctly that the effect of *Sry* does not depend on the protein, but on the regulatory sequences? Are they likely to specify the rate of proliferation?

Lovell-Badge: The proliferation is not directly due to *Sry*. It is due to the action of *Sox9*. The role of *Sry* is only to activate *Sox9* expression. The transgenic studies showing that the human SRY protein can work in mice tell us that the only part of the protein needed is the HMG box. Eva Eicher has done some experiments showing that the HMG box from other SOX proteins can be swapped. All that is needed is the expression of an HMG box of this type at the right time. This is sufficient to induce *Sox9* expression, and everything else follows on from that.

Harley: I would add that higher doses of SOX3 and SOX9 HMG box were required to replace SRY in Eicher's experiments, because different HMG boxes have different DNA sequence specificity. Can you comment on Harry Ostrer's experiments showing that the polyglutamine-rich region of mouse SRY can function as a transcriptional activation domain?

Lovell-Badge: There have been several reports about this. It is possible that it could work by making it a slightly stronger protein, bringing in its own transactivating domain. But this is clearly not necessary. There is no similar activation domain in the human SRY protein.

Schedl: Your data support the idea that *Sox9* can substitute for *Sry* function. We have done some experiments that also support this idea. I will report on these data in my paper (Guo et al 2002, this volume).

Graves: I have a question about the interaction between *Sry* and *Sox9*. Your co-expression studies are very nice, but do they show that there is a direct interaction?

Lovell-Badge: Unfortunately, not quite. The only way we will be able to prove this is by finding the regulatory region on *Sox9* where SRY binds. We assume that there is going to be a critical region where SRY can bind. It is possible that there is an autoregulatory feedback where SOX9 could also bind to this site. There may also be an SF1 binding site. However, looking in 70 kb of sequence we find a lot of potential sites for all the factors.

Behringer: Coming back to the expression of SOX9, have you looked at 10.5 days in the male and female? It should also be switched on in the female. Does the *Sox9* regulatory sequence have a switch element, or a gonad-specific element?

Lovell-Badge: We see a low level of expression in both sexes at early stages.

Behringer: The wholemount *in situ* suggests it should be more robust.

Lovell-Badge: It should be. Have you seen the early expression of *Sox9* in both sexes? Not everyone sees it. In some experiments we find it clearly; in others we don't. It really is at a low level.

Behringer: *Sox9* is expressed dimorphically in other tissues such as the Müllerian duct mesenchyme. Where else is the *lacZ* expressed?

Lovell-Badge: It is not expressed throughout much of the skeleton, for example. There is a bit of expression around the dorsal aorta.

Behringer: What about in the Müllerian duct?

Lovell-Badge: Not really.

Josso: I would like to return to the difference in chronology between *Sox9* and *Amh* expression in mammals and birds. It is not as different as all that. Before *Sox9* is expressed, *Amh* is expressed at a very low level. As soon as *Sox9* appears, *Amh* expression explodes and it is present at a high level. The only difference is that a little bit of *Amh* is expressed before.

Sinclair: However, in the alligator we see very strong expression of *Amh* before *Sox9* appears. This is also seen in the chicken.

Lovell-Badge: Chickens and alligators lack *Sry*. Perhaps there is not this early phase of turning on of *Sox9*, and it only really comes on in response to the migration of cells into the gonad.

Capel: In the alligator, one of the earliest indications of the male pathway is proliferation. Does this occur before or after *Sox9* appears?

Sinclair: Before.

Capel: So it is synchronous with the beginning of *Amh* expression. I think this is also true in chickens.

Lovell-Badge: I think other things are happening first, with *Sox9* being downstream of the critical sex-determining genes in the chick. For example, SF1 could be more important for *Amh* expression.

Capel: I don't know what to make of the timing differences between *Sox9* and *Amh*.

Sinclair: *Sox9* is clearly doing something later on, because it is being strongly up-regulated.

Lovell-Badge: It is probably important for the regulation of other genes such as *Dmrt1*.

Koopman: I would like to return to the structure–function data relating to *Sry*. Robin, it seems to me that your data suggest that either a mouse or human HMG box is needed, along with a mouse or human C-terminus, in any combination.

Lovell-Badge: That's true for the HMG box. Also, Eva Eicher's data show that a *Sox3* or *Sox9* HMG box would also work.

Koopman: Existing data suggest that some sort of C-terminus is needed also.

Lovell-Badge: Yes, but this could just be for stability. Or it could be that *Sry* has functions outside the gonad and that the reason why you have this rapid evolution of *Sry* is not for its role in sex determination, but for roles outside the genital ridge. For example, *Sry* could play a role in spermatogenesis, where it is known to be expressed in some species, or in the brain. This is very speculative, but could *Sry* be contributing to some aspects of behaviour that are sex specific, and is this the reason for its rapid evolution? The non-HMG box portion of the protein could be there partly to give stability, but it could also be doing other things.

Wilkins: It is possible that some of the changes in *Sry* are being driven by selective changes for these other functions, which would have required compensatory changes so that *Sry* could keep on doing its job with the molecules that it interacts with in sex determination.

Short: If I remember correctly, some time ago that you said you thought *Sry* had the fastest known rate of mutation of any gene. To what extent do you think the rapid mutation rate of *Sry* is because it is stuck out there on the Y chromosome and can't get any recombination repair?

Lovell-Badge: It is clearly evolving faster than some other genes on the Y chromosome, so that can't be the whole explanation. I wouldn't necessarily make the claim that it was the fastest-evolving gene; that's probably not the case. But it certainly does have a rapid rate of evolution. If you compare different primate species, the rate of *Sry* evolution isn't constant among them. The difference between some species is much greater than that between others. This implies that there may be selection.

Wilkins: I think we should avoid speaking of a rapid rate of mutation. There may be a rapid rate of evolution, but the mutation rate is likely to be the same for all genes.

References

Guo J-K, Hammes A, Chaboissier M-C et al 2002 Early gonadal development: exploring *Wt1* and *Sox9* function. In: The genetics and biology of sex determination. Wiley, Chichester (Novartis Found Symp 244), p 23–34

McLaren A 2000 Germ and somatic cell lineages in the developing gonad. Mol Cell Endocrinol 163:3–9

McLaren A, Southee D 1997 Entry of mouse embryonic germ cells into meiosis. Dev Biol 187:107–113

Early gonadal development: exploring *Wt1* and *Sox9* function

Jian-Kan Guo, Annette Hammes, Marie-Christine Chaboissier, Valerie Vidal, Yiming Xing, Frances Wong and Andreas Schedl[1]

Human Molecular Genetics Unit, University of Newcastle upon Tyne, Ridley Building, Newcastle upon Tyne, NE1 7RU, UK

Abstract. Prior to sex determination the gonadal anlage is formed as a bipotential primordium with the capacity to differentiate into either testes or ovaries depending on the presence or absence of the *Sry* gene. Knockout experiments have implicated five genes in the formation or survival of the gonadal primordium: *Wt1*, *Sf1*, *Lim1*, *Lhx9* and *Emx2*. We are particularly interested in the Wilms' tumour suppressor, WT1, which is characterized by complex posttranscriptional modifications. Here we will focus on published *in vitro* evidence suggesting distinct functions for the various isoforms and present our own results from *in vivo* experiments. Our data suggest that WT1 is an important regulator of the transcription or stability of the sex-determining gene *Sry*. One of the first genes expressed after the initial male sex-determining signal is the *Sox9* gene. Human *SOX9* has been implicated in male-to-female sex reversal. To analyse *Sox9* function in mouse development we have performed transgenic experiments and ectopically expressed this gene in XX gonads. Our data indicate that *Sox9* is sufficient to induce testis formation in mice. Here we will discuss our new data and present an updated model for *Wt1* and *Sox9* function in gonad formation and sex determination.

2002 The genetics and biology of sex determination. Wiley Chichester (Novartis Foundation Symposium 244) p 23–34

Genes involved in gonad formation and survival

The indifferent gonad in the mouse forms at embryonic day 10 as a swelling at the ventromedial side of the mesonephros. Proliferation of the coelomic epithelium results in the generation of the gonadal primordium, which due to the presence or absence of the SRY protein then differentiates along the male or female pathway. Despite intensive research over the last few decades very little is known about the molecular mechanism underlying the formation of the gonadal primordium. So far we know of five genes which seem to play an essential role

[1]This chapter was presented at the symposium by Andreas Schedl, to whom correspondence should be addressed.

during this process: The Wilms' tumour suppressor gene *Wt1* (Kreidberg et al 1993), the steroidogenic factor *Sf1* (Luo et al 1994), the Lim-type homeobox containing genes *Lim1* (Shawlot & Behringer 1995) and *Lhx9* (Birk et al 2000) and the evenskipped homologue *Emx2* (Miyamoto et al 1997). Knockout mutations in all of these genes result in mice lacking gonadal tissues, but the basis for this phenotype is different. Whereas the gonadal anlagen in *Wt1* and *Sf1* knockout mice seem to undergo apoptosis, gonads in *Lhx9* and *Emx2* knockout mice exhibit proliferative defects within the coelomic epithelium. The reason for the absence of gonadal tissue in *Lim1* knockout mice is still unclear, but recent evidence suggests that it is required for the differentiation of the intermediate mesoderm (Tsang et al 2000). We are particularly interested in *Wt1* and the role of its various isoforms in the formation and differentiation of the gonad.

Biochemical evidence for distinct functions of Wt1 isoforms

Wt1 is a complex gene. Through a combination of alternative splicing, RNA editing and three alternative translation start sites as many as 24 different isoforms are expressed from its locus (Fig. 1A). Of particular interest are isoforms produced by the usage of an alternative splice donor site at the end of exon 9 (Fig. 1), which leads to the insertion or omission of three amino acids (KTS) between zinc fingers 3 and 4. Because this insertion changes the spacing of the zinc fingers it has been proposed that it also changes the DNA binding specificity of this protein. Indeed, *in vitro* studies demonstrated distinct consensus sequences and affinities to DNA (Laity et al 2000) and the two isoforms differ in their potential to activate or repress the transcription from a variety of promoters (for review see Menke et al 1998). Whereas −KTS variants are usually much more potent transcriptional regulators in co-transfection studies, +KTS isoforms seem to be able to bind to RNA. Moreover, the nuclear localization of WT1 seems to change depending on the presence or absence of the three amino acids KTS. Isoforms lacking the KTS sequence show a more diffuse staining whereas +KTS variants localize in speckles, a pattern reminiscent of splicing factors (Larsson et al 1995, Englert et al 1995). Finally, recent biochemical results suggest that +KTS products are associated with splicing complexes (Davies et al 1998, Ladomery 1997).

WT1 mutations and urogenital abnormalities

WT1 has been identified as a gene mutated in Wilms' tumour, an embryonic kidney tumour affecting 1 in 10 000 children (Haber et al 1990, Gessler et al 1990). Soon after cloning it became clear that in addition to being a tumour suppressor, *WT1* fulfils additional functions during development. Firstly, patients with heterozygous

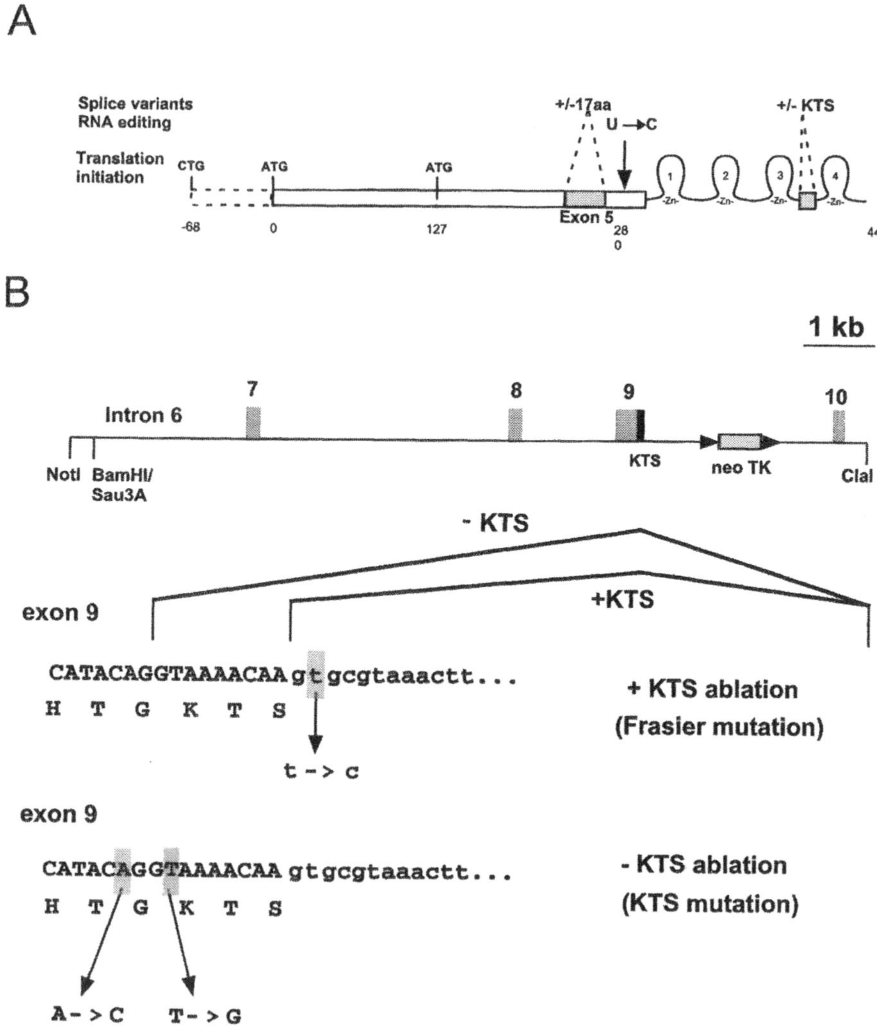

FIG. 1. Structure of WT1 and its various isoforms. (A) Through a combination of alternative splicing (exon 5 and exon 9) RNA editing (exon 6) and three alternative translation start sites, as many as 24 different isoforms of WT1 can be produced. (B) Schematic representation of the two targeting constructs designed to interfere with the alternative splice donor sites at the end of exon 9. Frasier mice mimic a mutation in Frasier patients and produce only WT1−KTS variants. Mutations in KTS mice abolish the first splice donor site and result in +KTS products only.

deletions of WT1 showed mild abnormalities in gonadal development, such as hypospadias and cryptorchidism. Secondly, dominant point mutations in *WT1* have been associated with Denys–Drash syndrome (DDS) (Pelletier et al 1991) and Frasier syndrome (Klamt et al 1998, Barbaux et al 1997), which are characterized by urogenital abnormalities ranging from hypospadias or sex reversal to gonadal dysgenesis. Mutations in Frasier patients are intronic and affect the alternative splicing of *WT1* within the zinc finger region (Fig. 1). As a consequence no +KTS isoforms are produced from the mutated allele. Interestingly, Frasier mutations are dominant and both + and −KTS variants are still expressed from the wild-type allele. We can therefore conclude that the ratio between +KTS/−KTS is important for normal development in human. The essential function for WT1 in gonad formation and survival was finally demonstrated using the knockout approach (Kreidberg et al 1993). Homozygotes showed gonadal dysgenesis due to massive apoptosis in the gonadal primordium.

Splice-specific knockouts demonstrate distinct functions *in vivo*

We have seen overwhelming evidence *in vitro* that + and −KTS products have distinct biochemical and cellular properties. To address whether the two alternatively spliced isoforms also serve distinct functions *in vivo*, we have generated mouse strains lacking either + or −KTS variants (Hammes et al 2001). For easier distinction of the two models we have named the mouse strain with the mutation mimicking the mutation found in human Frasier patients as Frasier mice and animals lacking −KTS products as KTS mice (Fig. 1B). In both models the observed phenotype of homozygous animals was less severe than that observed in complete knockout mice and the induction of kidney development occurred normally. The two splice variants must therefore be able to complement for each other at least to some extent. At later stages however there were clear-cut differences in particular during gonad formation. KTS mice showed a dramatic increase in apoptosis at E11.5 of the developing gonad suggesting that this isoform has an important function for cell survival. Interestingly, a recent publication by Richard et al (2001) describes −KTS products as an important factor for cell survival together with the prostate apoptosis response factor Par4. Frasier homozygotes (lacking +KTS products) did not show an increase in apoptosis and XX gonads developed normally. Frasier XY gonads, however, never formed sex cords and developed along the female pathway. This male-to-female sex reversal was also demonstrated on the molecular level. Expression of *Sox9* and *Amh* (*Mis*) was completely absent from Frasier XY gonads and *Dax1* showed the female specific expression pattern.

What is the function of the WT1+KTS protein during sex determination? Kim et al (1999) have shown that WT1−KTS isoforms can activate the *Dax1* promoter at least *in vitro*. They speculated that a reduction of +KTS isoforms may lead to an increase of −KTS variants and consequently an up-regulation of *Dax1*. Overexpression of *Dax1* could indeed interfere with male development, as has been demonstrated in transgenic studies (Swain et al 1998). Using a real-time PCR approach we did not detect any significant increase of *Dax1* expression suggesting a distinct mechanism for the observed sex reversal in Frasier mice.

Another proposed target gene for WT1 is the sex-determining gene *Sry* (Hossain & Saunders 2001). Again the transcriptionally active form in their experiments was the −KTS variant, whereas +KTS proteins had no stimulating effect on *Sry* transcription. Interestingly, when we tested Frasier homozygous animals we found a dramatic decrease of *Sry* expression indicating that WT1+KTS is the more important isoform for *Sry* regulation *in vivo*. At present we do not know whether +KTS variants are involved in transcriptional activation of the *Sry* gene or whether they may act through a different mechanism. Given the evidence from *in vivo* studies, which indicate a role for +KTS in RNA binding (Kennedy et al 1996, Caricasole et al 1996), it is tempting to speculate that it may be involved in stabilising the *Sry* mRNA by binding to it. Future experiments will be aimed to address this question.

***Sox9* is sufficient to induce testis formation in XX mice**

We have seen that WT1+KTS is required for the expression of high levels of the sex determining gene *Sry*. Shortly after the induction of *Sry* expression, *Sox9* becomes activated in the male gonad (Kent et al 1996, Morais da Silva et al 1996). Several studies both *in vitro* and *in vivo* document the importance of this gene for male development. Firstly, human patients with mutations in *SOX9* suffer from Campomelic Dysplasia, a condition often associated with male-to-female sex reversal (Foster et al 1994, Wagner et al 1994). Secondly, SOX9 is able to bind and activate the anti-Müllerian hormone (*AMH*; also known as Müllerian inhibiting substance, *MIS*) promoter both *in vitro* (De Santa Barbara et al 1998) and *in vivo* (Arango et al 1999). The sex reversal found in human patients suggested that *SOX9* might also serve other functions besides the activation of *AMH* during sex determination, since *Amh* knockout mice show pseudohermaphroditism rather than a complete sex reversal. To answer this question we brought the mouse *Sox9* gene under control of an ectopic promoter expressed in both male and female gonads (Fig. 2; Vidal et al 2001). As *Wt1* is expressed in XX and XY animals from the earliest stages of urogenital development (E9.5), we decided to introduce the *Sox9* gene into a yeast artificial chromosome (YAC) construct containing the mouse *Wt1* locus (Scholz et al 1997). We expected that such a YAC knock-in approach would result in the expression of *Sox9* in a *Wt1* specific pattern. XY transgenic animals generated with this

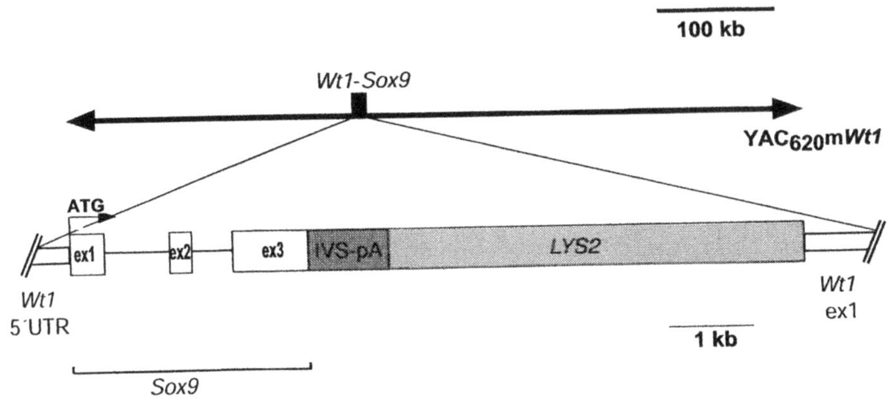

FIG. 2. YAC knock-in approach to address *Sox9* function *in vivo*. The *Sox9* genomic locus was homologously recombined into a mouse *Wt1* YAC and transgenic mice were generated with this construct. Regulation of *Sox9* occurred through WT1 regulatory regions encoded on the YAC and consequently expression in both XX and XY gonads was detected.

construct developed normally and were fertile. In contrast XX mice transgenic for *Wt1-Sox9* developed testes, with apparently normal Sertoli and Leydig cells. Germ cells were almost entirely absent, due to the presence of the two X chromosomes (Hunt et al 1998). Taken together these data suggest that *Sox9* can substitute for *Sry* and induce testis formation.

Conclusions

Wt1 and *Sox9* are key players during embryonic development. Here we have shown yet another facet of the variety of actions these genes can fulfil in gonad formation and sex determination. Taken together our data suggest a new model for the involvement of *Wt1* and *Sox9* in gonad formation (Fig. 3). Proliferation of the coelomic epithelium leads to the development of the undifferentiated gonad. −KTS isoforms are required for the survival of the gonadal primordium and KTS mice show increased apoptosis. In male gonads the sex determination process is initiated by the expression of *Sry*. +KTS variants are required for high levels of *Sry* expression and consequently the activation of other male specific genes such as *Sox9* and *Amh*. It seems that *Sry* is only required for a very short time, possibly for the activation of *Sox9*. Once activated *Sox9* on its own or through interaction with other proteins regulates genes such as *Amh*, but also other genes important during sex determination. Future research will focus on the identification of these downstream targets and how they initiate Sertoli cell differentiation.

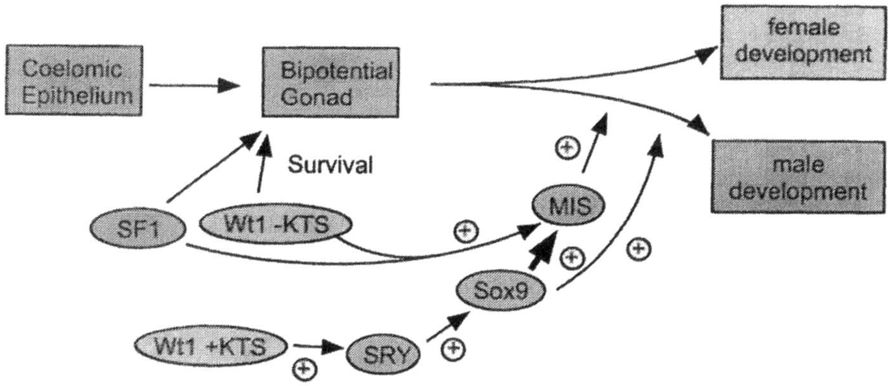

FIG. 3. Model for the role of *Wt1* and *Sox9* in gonad formation and sex determination. WT1−KTS variants possibly together with SF1 are required for the survival of the gonadal mesenchyme. During male sex determination WT1+KTS isoforms are required for the activation or stability of *Sry*, which subsequently leads to the activation of *Sox9*. *Sox9*, possibly with the help of SF1 and WT1−KTS, activates the *Mis* (*Amh*) promoter, which in turn leads to degeneration of the Müllerian duct. Moreover, *Sox9* can initiate testis differentiation and must therefore have additional target genes, which regulate Sertoli cell development. The absence of *Sry* in XX mice leads to the development of ovaries. Similarly, Frasier mutations interfere with the activation of *Sry* and, hence, block male development.

References

Arango NA, Lovell-Badge R, Behringer RR 1999 Targeted mutagenesis of the endogenous mouse Mis gene promoter: in vivo definition of genetic pathways of vertebrate sexual development. Cell 99:409–419

Barbaux S, Niaudet P, Gubler MC et al 1997 Donor splice-site mutations in WT1 are responsible for Frasier syndrome. Nat Genet 17:467–470

Birk OS, Casiano DE, Wassif CA et al 2000 The LIM homeobox gene Lhx9 is essential for mouse gonad formation. Nature 403:909–913

Caricasole A, Duarte A, Larsson SH et al 1996 RNA binding by the Wilms tumor suppressor zinc finger proteins. Proc Natl Acad Sci USA 93:7562–7566

Davies RC, Calvio C, Bratt E, Larsson SH, Lamond AI, Hastie ND 1998 WT1 interacts with the splicing factor U2AF65 in an isoform-dependent manner and can be incorporated into spliceosomes. Genes Dev 12:3217–3225

De Santa Barbara P, Bonneaud N, Boizet B et al 1998 Direct interaction of SRY-related protein SOX9 and steroidogenic factor 1 regulates transcription of the human anti-Mullerian hormone gene. Mol Cell Biol 18:6653–6665

Englert C, Vidal M, Maheswaran S et al 1995 Truncated WT1 mutants alter the subnuclear localization of the wild-type protein. Proc Natl Acad Sci USA 92:11960–11964

Foster JW, Dominguez-Steglich MA, Guioli S et al 1994 Campomelic dysplasia and autosomal sex reversal caused by mutations in an SRY-related gene. Nature 372:525–530

Gessler M, Poustka A, Cavenee W, Neve RL, Orkin SH, Bruns GA 1990 Homozygous deletion in Wilms tumours of a zinc-finger gene identified by chromosome jumping. Nature 343:774–778

Haber DA, Buckler AJ, Glaser T et al 1990 An internal deletion within an 11p13 zinc finger gene contributes to the development of Wilms' tumor. Cell 61:1257–1269
Hammes A, Guo JK, Lutsch G et al 2001 Two splice variants of the Wilms' tumor 1 gene have distinct functions during sex determination and nephron formation. Cell 106:319–329
Hossain A, Saunders GF 2001 The human sex-determining gene *SRY* is a direct target of *WT1*. J Biol Chem 276:16817–16823
Hunt PA, Worthman C, Levinson H et al 1998 Germ cell loss in the XXY male mouse: altered X-chromosome dosage affects prenatal development. Mol Reprod Dev 49:101–111
Kennedy D, Ramsdale T, Mattick J, Little M 1996 An RNA recognition motif in Wilms' tumour protein (WT1) revealed by structural modelling. Nat Genet 12:329–331
Kent J, Wheatley SC, Andrews JE, Sinclair AH, Koopman P 1996 A male-specific role for SOX9 in vertebrate sex determination. Development 122:2813–2822
Kim J, Prawitt D, Bardeesy N et al 1999 The Wilms' tumor suppressor gene (*wt1*) product regulates Dax-1 gene expression during gonadal differentiation. Mol Cell Biol 19: 2289–2299
Klamt B, Koziell A, Poulat F et al 1998 Frasier syndrome is caused by defective alternative splicing of WT1 leading to an altered ratio of WT1 +/−KTS splice isoforms. Hum Mol Genet 7:709–714
Kreidberg JA, Sariola H, Loring JM et al 1993 WT-1 is required for early kidney development. Cell 74:679–691
Ladomery M 1997 Multifunctional proteins suggest connections between transcriptional and post-transcriptional processes. BioEssays 19:903–909
Laity JH, Dyson HJ, Wright PE 2000 Molecular basis for modulation of biological function by alternate splicing of the Wilms' tumor suppressor protein. Proc Natl Acad Sci USA 97:11932–11935
Larsson SH, Charlieu JP, Miyagawa K et al 1995 Subnuclear localization of WT1 in splicing or transcription factor domains is regulated by alternative splicing. Cell 81:391–401
Luo X, Ikeda Y, Parker KL 1994 A cell-specific nuclear receptor is essential for adrenal and gonadal development and sexual differentiation. Cell 77:481–490
Menke AL, van der Eb AJ, Jochemsen AG 1998 The Wilms' tumor 1 gene: oncogene or tumor suppressor gene? Int Rev Cytol 181:151–212
Miyamoto N, Yoshida M, Kuratani S, Matsuo I, Aizawa S 1997 Defects of urogenital development in mice lacking Emx2. Development 124:1653–1664
Morais da Silva S, Hacker A, Harley V, Goodfellow P, Swain A, Lovell-Badge R 1996 Sox9 expression during gonadal development implies a conserved role for the gene in testis differentiation in mammals and birds. Nat Genet 14:62–68
Pelletier J, Bruening W, Kashtan CE et al 1991 Germline mutations in the Wilms' tumor suppressor gene are associated with abnormal urogenital development in Denys-Drash syndrome. Cell 67:437–447
Richard DJ, Schumacher V, Royer-Pokora B, Roberts SG 2001 Par4 is a coactivator for a splice isoform-specific transcriptional activation domain in WT1. Genes Dev 15:328–339
Scholz H, Bossone SA, Cohen HT, Akella U, Strauss WM, Sukhatme VP 1997 A far upstream cis-element is required for Wilms' tumor-1 (*WT1*) gene expression in renal cell culture. J Biol Chem 272:32836–32846
Shawlot W, Behringer RR 1995 Requirement for Lim1 in head-organizer function. Nature 374:425–430
Swain A, Narvaez V, Burgoyne P, Camerino G, Lovell-Badge R 1998 Dax1 antagonizes Sry action in mammalian sex determination. Nature 391:761–767
Tsang TE, Shawlot W, Kinder SJ et al 2000 *Lim1* activity is required for intermediate mesoderm differentiation in the mouse embryo. Dev Biol 223:77–90

Vidal V, Chaboissier MC, de Rooij D, Schedl A 2001 Sox9 induces testis development in XX transgenic mice. Nat Genet 28:216–217
Wagner T, Wirth J, Meyer J et al 1994 Autosomal sex reversal and campomelic dysplasia are caused by mutations in and around the SRY-related gene SOX9. Cell 79:1111–1120

DISCUSSION

Short: I know we are not supposed to be discussing the kidney, but does *Wt1* do anything to the development of the mesonephric kidney? One might imagine that lesions in the mesonephric kidney would seriously interfere with genital ridge formation.

Schedl: That is a good point. The *Wt1* knockout mice have fewer mesonephric tubules. Other than this, I don't think much work has been done on the mesonephros.

Short: Does the genital ridge form normally in these mice?

Schedl: They have a genital ridge, but this undergoes apoptosis at about day 11.5. This is very similar to what we see in the −KTS knockout.

Koopman: What is the effect of ectopic expression of *Sox9* in the kidney?

Schedl: There is no effect. The mice seem to be completely normal. In the kidney *Sox9* is expressed at the ureteric tip, whereas *Wt1* is expressed in the metanephric mesenchyme. I think *Sox9* has to work in the epithelial component, at least in the kidney.

Lovell-Badge: In these experiments, can you distinguish the transgene expression from that of the endogenous gene, and do you see activation of the endogenous gene?

Schedl: We started to do this experiment, but the first trial failed. I can't comment on this. In principal, we should be able to distinguish between the two.

Renfree: In your transgenic sex-reversal mice, are the testes smaller? You said that the number of germ cells is reduced: are they completely abolished or do they disappear in the long-term?

Schedl: They are a lot smaller. The size is pretty much the same during embryonic development, but then when proliferation of the germ cells occurs in wild-type mice, germ cells in knockout mice undergo apoptosis. The reduction of size is almost certainly due to the fact that there is a second X chromosome. This is also seen in the *Sry* sex-reversal mice.

McLaren: Do you know whether the functional difference between the two isoforms is due to the presence or absence of KTS amino acids, or is it a spacing phenomenon?

Schedl: Nick Hastie's lab has done some experiments that address this question (Davies et al 2000). It looks as if it is a spacing effect. Puffer fish also has *Wt1*, with one of the amino acids replaced.

Behringer: Do you think that any SOX protein expressed with the *Wt1* promoter would cause sex reversal?

Schedl: I don't think that just any SOX protein would, but I do think that SOX8 and SOX10 would.

Carmerino: What happens to the adrenals when *Sox9* is ectopically expressed?

Schedl: Ectopic expression of *Sox9* in the adrenal doesn't seem to have any effect. We have tried to express *Sox9* under an adrenal-specific promoter, which didn't work because no decent promoters are available. Interestingly, the *Wt1* gene is not expressed in the adrenal glands once they are distinguishable from the gonads. Since *Wt1* is important for adrenal formation, I assume it must be expressed very early on in the adrenal/genital primordium.

Capel: If you think the +KTS isoform normally binds RNA and you believe that this is the isoform that is important for testis determination, has anyone looked to see whether *Wt1* binds the *Sry* RNA, for example?

Schedl: We are doing this at the moment.

Capel: Robin Lovell-Badge's lab made a construct that he used to express the human *SRY* gene, and Peter Koopman did a similar experiment and got a different result. What is different about these two constructs? Could it have something to do with the RNA?

Greenfield: Robin, with your mouse *Sry* transgenes, have you any evidence that the CAG-rich domain might be required for transcriptional regulation?

Lovell-Badge: We have not tried to address this yet. We know that if you delete the 3′ end then you don't get expression. If you delete the 5′ end you can still get expression. It is clear that there is something about the 3′ end that is important.

Koopman: Some of the transgenic mouse experiments that we have done have implicated the 3′ UTR in the regulation of the function of *Sry*.

Wilkins: When a molecule shows multiple functions, there has usually been a sequence of acquisitions of these capacities. Has anyone looked at *Wt1* functions in other vertebrates, in particular in fish?

Schedl: Nick Hastie's lab has done an evolutionary study on *Wt1*. I have told you about two alternative splices. The first one is exon 5, and this seems to be very mammalian specific. It doesn't occur in alligator and fish. The KTS sequence is present wherever *Wt1* is found, so this seems to be a hallmark of *Wt1*.

Harley: Has anyone done RNA splicing assays with KTS?

Schedl: Nick Hastie's lab is trying to do this. What they find is that it co-purifies with the active splicing component, but there is no functional evidence that it actually does anything (LaDomery et al 1999). If it is working on a specific molecule or RNA, then this will be very difficult to see: you would first have to identify the target to know what to put into the splicing assay.

Behringer: Do I understand correctly that overexpression of *Sox9* in the male gonad is not detrimental?

Schedl: We can't really conclude this because we don't know whether there is an autoregulatory loop that switches off the endogenous *Sox9* gene. I don't think we have huge amounts of *Sox9* expression in our transgenic animal.

Swain: In your transgenic experiment how do you know that SOX9 is acting like the endogenous SOX9 protein and not just acting through its HMG box domain? It has been suggested that the only active part of SRY is the HMG box domain, and because your levels are low, it could be argued that SOX9 is just working as SRY, by providing an HMG box domain.

Schedl: So you are suggesting that SOX9 can substitute for SRY function.

Swain: You need to repeat your experiment with a SOX9 protein that lacks the transactivation domain in order to argue that SOX9 is actually working as the endogenous SOX9.

Goodfellow: You need to look at the timing of the expression. If you are right, it should be expressed very early.

Swain: Right, and you wouldn't need that much. This is also consistent with your results.

Renfree: Andreas Schedl, what do you think *Wt1* is doing in the Sertoli cells? Is it produced all the way through to adult life, or just at certain stages?

Schedl: *Wt1* is produced throughout adult life in the Sertoli cells, but I have no idea what it does. It seems to be involved in fertility. Our own unpublished data suggest that mice with low levels of *Wt1* expression in the adult testis are initially fertile and then become quite rapidly infertile. Sertoli cells are supporting cells, and they are important for getting the germ cells to develop into sperm.

Lovell-Badge: What is the role of *Wt1* in apoptosis?

Schedl: Again, we don't really know much about this. About 50 target genes have been identified *in vitro* using transfection assays of cells. But I don't believe most of them. *Bcl2* is one of the target genes identified that way. Unfortunately, *Wt1* will affect almost any promoter, because there is a binding site on almost any promoter: it is a CG-rich binding site and most genes have CPG islands. If you put large amounts of DNA and protein into a cell, they will bind to each other.

Goodfellow: I think the experiment that Andreas Schedl has done is a crucial one. We are all circling around the same point, which is that we need to understand whether this is a box effect. This is one of the big puzzles about SRY: it just looks like a box. We need to understand the biochemistry of what happens when you deliver a box. Presumably, if its action is due to a box effect, it must be soaking up some limiting factor at a crucial point in time. The experiments performed by Eva Eicher imply that the box is not a specific component, although it would have been more compelling if she had used a more diverse box than the SOX9 box. If I have understood properly, you could set up this experiment to get the answer. The prediction is that you will see early expression of the transgene that will switch on the normal expression of the endogenous *Sox9* gene, and this is what is causing sex reversal. If it is not doing this, then the interpretation must be that *Sox9* itself is responsible for the sex reversal.

References

Davies RC, Bratt E, Hastie ND 2000 Did nucleotides or amino acids drive evolutionary conservation of the WT1 +/−KTS alternative splice? Hum Mol Genet 9:1177–1183

LaDomery MR, Slight J, Mc Ghee S, Hastie ND 1999 Presence of WT1, the Wilm's tumor suppressor gene product, in nuclear poly(A)(+) ribonucleoprotein. J Biol Chem 274: 36520–36526

General discussion I

The mechanism of action of SRY

Koopman: There seem to be two schools of thought on the function of SRY protein. One is that SRY is just an HMG box with some dangly bits, and the other is that SRY is an HMG box in combiantion with another important part at the C-terminus. We seem to be going off on the 'SRY as just a box' tangent without thinking this through carefully enough. My understanding of the biochemistry suggests that all SOX proteins (and SRY would have to be included) achieve target specificity and complex protein–protein interactions that allow them to function as individual proteins with diverse roles in development by having an HMG box and other important protein domains. To assume that any of the SOX proteins, including SRY, acts just by binding and bending, doesn't explain how SRY does what it does — how it binds to certain target genes and not others.

Goodfellow: We don't know that it binds to any target genes. What is the evidence that it does?

Koopman: There's a lot of evidence that the HMG box binds to specific DNA targets.

Goodfellow: There is evidence that the HMG box can bind to DNA, but there is no evidence that it binds to specific targets.

Koopman: It binds to specific target sequences in DNA, not necessarily specific DNA targets.

Lovell-Badge: If mutations occur which affect all the properties that you are talking about, then sex reversal occurs.

Koopman: We know that certain classes of mutations in humans that affect DNA binding cause sex reversal.

Goodfellow: But you don't know what else they affect. You don't know that they don't affect protein–protein interactions.

Harley: It is interesting that SRY mutations cause complete gonadal dysgenesis, despite the fact that point mutations can have quite variable biochemical activities, from wild-type-like activity (in terms of DNA binding) to complete abolition of that activity. Surely this suggests the existence of other activities that as yet are unknown.

Goodfellow: In the experiments in which you showed that SRY lost the ability to bind and bend to DNA, could you rule out stability of the protein structure as a contributory factor?

Harley: No.

Goodfellow: If that is the case, the argument that Peter Koopman is making doesn't rule out it being an effect on protein–protein interaction.

Harley: We have data showing that a couple of sex-reversing campomelic dysplasia (CD) mutations do affect stability, and this in turn affects its ability to be recognized by its nuclear receptor importin β and be translocated into the nucleus. Regardless of DNA binding activity, if it is not getting into the nucleus it can't do its job.

Lovell-Badge: Is anyone aware of mutations outside the HMG box that are really having an effect on its potential DNA binding activity?

Harley: I have measured two sex-reversing mutations in SRY, one in the N-terminal region and the other in the C-terminal region, and neither have effects on DNA binding or bending. One is a familial mutation with wild-type-like DNA binding activities.

Poulat: We say that SRY is only a box. We can exchange this box with other boxes; we find patients with CD who have a truncation of the SOX9 C-terminal domain. Basically we have a truncated SOX9 protein, which is also more-or-less only a box: nevertheless, in this case we have sex reversal. It would be interesting to see whether this kind of truncated protein has any effect when expressed under the SRY promoter. Perhaps we are focusing too much on the box. I can't believe that proteins such as SRY could be maintained like that in evolution. When you look at mammals, the N-terminal and C-terminal domains are still there. If these regions were of no interest, they may well have been lost.

Goodfellow: That is exactly what has happened.

Poulat: Something has been retained; at least at the C-terminus.

Goodfellow: There's very little homology at the C-terminus.

Poulat: We are attempting to explain the function of proteins just by comparing the sequence. Perhaps the important molecular structures are what have been retained.

Lovell-Badge: The experiments I talked about in which I replaced the human ORF in the mouse regulatory region show that the human protein can work in mice. You could imagine a scheme where there are interacting partners present that interact with the C-terminal domain, which just happen to be there in the mouse but aren't normally used with the mouse protein, but it is a little weird.

Poulat: For sure, but there is a sex reversal that has been described where the C-terminal part of SRY is cut by a stop codon.

Lovell-Badge: There has to be protein stability.

Poulat: But can we explain everything by stability?

Lovell-Badge: We know that SRY has to be expressed above a particular threshold. It is not normally expressed much above that threshold. Even just a small reduction in activity is sufficient to prevent it working properly. A lot of the effects that we are looking at may be very weak ones. We know from transgenic experiments that just expressing 50% of the normal level is frequently insufficient to give you sex reversal.

Goodfellow: It's clear that we can't rule out a specific sequence to which it binds and so on, but the dosage argument suggests that we are dealing with some other limiting component. The arguments that Robin Lovell-Badge has mentioned all suggest the model where SRY is titrating a component in competition which would have bound to something else. This is the theory that is most consistent with the lack of conservation of the sequence outside the box. It is also consistent with the suggestion from the current data that any box will do. What we should be looking for are the components that form part of a complex with the box. It is even possible that you need DNA or RNA in order to form the complex, but the specific sequence at which SRY binds could be irrelevant, because what it is doing is absorbing some limiting factor.

Swain: It may also be that transcription at the *Sox9* locus is particularly sensitive to chromatin changes. The binding of an HMG box to the locus could make a big difference, which might be dependent on levels of protein. You could argue that the phenotype seen in Colin Bishop's transgenic experiment (Bishop et al 2000) was due to a change in chromatin that occurred distally to the gonad-specific promoter elements, which are much closer to the start of transcription.

Greenfield: Robin, are you going to use your Myc epitope-tagged SRY to do chromatin immunoprecipitation? This would be a good experiment to identify the regulatory regions where it presumably binds.

Lovell-Badge: We are working up to do this.

Vilain: I know of two human sex-reversal cases with duplications of relatively large chromosomal regions. One is a duplication of 22q that leads to XX sex reversal (one case of XX hermaphrodite and one case of XX male). In the other, there is duplication of 17p leading to an XX male. What is interesting about these two cases is that these regions both contain *SOX* genes: 22q has *SOX10* and 17p has *SOX20*. One could argue that any region that is duplicated in humans that contains a *SOX* gene that happens to be expressed in the gonad may be able to replace *SRY* just by dosage effects. If you just double the dose you would be able to replace *SRY* artificially. One way to test this would be in transgenic animals.

Lovell-Badge: Does anyone know what happens in the mole vole, which lacks *SRY*? Could this be a duplication?

Graves: I don't think anyone knows yet. Perhaps there is another *SOX* gene.

Greenfield: There seems to be a picture emerging in which there are two temporally distinct steps. Step one would be the appearance in the genital ridge in the appropriate cell type of a stable HMG box protein. This may or may not bind certain targets specifically; it may or may not interact with other molecules.

Step two is the subsequent appearance of another HMG box protein that is transactivation dependent. Is this correct?

Koopman: All I am arguing against is the all importance of the HMG box for SRY function. I am not saying necessarily that it needs a transactivation domain. I'm suggesting that there is some sort of protein interaction or target specificity required.

Greenfield: How do you explain Robin's successful sex reversal in the absence of a protein encoding that domain?

Koopman: Because, as I said before, Robin's data are not incompatible with the idea that you need an HMG box of one type or another and a C-terminal part of the protein that separate from the HMG box.

Greenfield: So you are not specifying the kind of C-terminal protein.

Lovell-Badge: We would like to do similar experiments putting another SRY protein in the context of the mouse regulatory sequences, such as the marsupial SRY protein. If that works, then you would have to argue that all possible C-terminal domains could work.

Koopman: But they are not just any old C-terminal domains: they are C-terminal domains of SRY. These may not need to look similar at the sequence level to fulfil the same function.

Harley: Berta cloned a C-terminal interacting factor, which was a PDZ-like protein. They have the property of binding almost any C-termini and then being involved in intracellular signalling. Perhaps if you have a C-terminus it doesn't have to have sequence conservation, but some kind of recognition motif.

Behringer: Why can't the HMG domain bind co-factors?

Koopman: It could. All I am saying is that I think there are other things going on in addition to whatever functions the HMG box might mediate.

Goodfellow: Basically we can't distinguish between any of the hypotheses. SRY could be binding in a sequence-specific manner and acting as an activator or a repressor, or by blocking something else that would have bound to the site that SRY binds to. Then you can work the same trick all the way through. The specificity of the binding may actually be such that in the absence of other information in the molecule, all HMG boxes will bind to a related set of sequences, which are actually modified by the co-factors. Then you go for the hypothesis that instead of blocking something that is binding, you are removing a cofactor which would be used by another factor.

Behringer: What would be the definitive experiments that we could all try?

Goodfellow: I think we need to find out which SOX boxes actually work and which ones don't. The problem with the current experiments is that we are in a SOX9/SRY loop.

Behringer: So we should do the assay that Eva Eicher did, but just try more SOX proteins.

Goodfellow: I think this would at least give us a clue about where to look. Clearly, if the chromatin precipitation experiments could actually get to target sequences, this would be a remarkable step forward. I don't like the experiments that involve testing promoters for binding of SRY. Unless you know this happens *in vivo*, you just get trapped in a loop that doesn't take you forward.

Harley: Returning to the HMG box, most activities that I have looked at don't vary between HMG boxes. However, when you look at its intrinsic ability to recognize specific sequences in DNA from random pools, there is a conserved six bases and then a wobble at each end. This wobble seems to be SOX specific. SOX9 prefers an AG at one end and a GG at the other, whereas SRY prefers an AT and a TT. This may be symptomatic of something in chromatin, but is the only difference that I have seen between those HMG boxes. This could also explain why three–fivefold mouse SOX3 or SOX9 HMG box expression is required to replace SRY HMG box in Eva's experiments.

Graves: My attitude to SRY has hardened somewhat by realizing what a very recent gene it is. It has only been around for about 130 million years. It is disappearing fast and has already gone in voles. It is probably is just a HMG box that happened to be in the right place at the right time. And almost anything has been attached to it in different species. In one marsupial a new intron has been introduced probably only 14 million years ago and there is a new C-terminal region that is completely unrelated to the regions in other species. If this sort of thing can happen, I can't see that evolutionarily it is terribly important what other functions are added. Perhaps they are rather marginal functions. Perhaps it is important to have a transactivation function in mouse, but not in other species. Perhaps rodents have invented a completely different way to use SRY by adding other functions on.

Short: How do we explain facultative sex reversal in fish? Is it possible to explain how, on a social whim, fish can change from testis to ovary and back to testis again?

Fernald: The problem is that we don't know the mechanisms at the genetic level in those animals. Many species have both gonads present in primitive forms and one or the other gets turned on depending on the social situation. There isn't genetic understanding of the process.

Goodfellow: A tangential thought. If we look at nuclear hormone receptors, the coactivators and corepressors that bind to the nuclear hormones are very ligand dependent. One can start to define the biological activity of the nuclear hormone receptor in terms of which drug is binding to the receptor, which coactivators are binding, and then whether you are getting the full biological response. Our knowledge of nuclear hormone receptors is an order of magnitude further advanced than that of the SOX family of transcriptional regulators.

Capel: So you are suggesting a series of binary inputs into the activity of SRY that could interact with a number of other things: each can be on or off.

Goodfellow: I guess what I am saying is that we have ignored the cofactor molecules in transcription factors for too long. This is why I was emphasizing the possibility that we may be looking at soaking up a cofactor that is needed for expression of another gene. I agree that there is no more evidence for this than anything else.

Poulat: Concerning the level of expression of SRY, if SRY is just a box, it would mean that in terms of finding the targets, it would be extremely non-specific. We know that SRY is binding to 'AACAAT' that can be found in every gene. If you want to have some specific target in the nucleus for this box, you have to fill your nucleus with tons of box. The problem is that the expression level of SRY is extremely low.

Goodfellow: Specificity could change if there is a cofactor.

Poulat: You have to restrict the specificity of the protein.

Goodfellow: All our experiments are done in a test tube with milligram levels of SRY, or micrograms at least.

Harley: No, SRY binds at 10 nM levels, which is a respectable binding. SOX9 does as well. But these are to optimized binding sites. In the case of SOX9 binding to its Col2 sites, it binds about fourfold less than at optimal sites.

Capel: I think it is a mistake to overlook the 3′ UTR in *Sry*. When we were first working on the mouse transcript we identified the circular form that deletes the 3′ UTR. That transcript is not translated. For all we know, the 3′ UTR of *Sry* controls its translation in a specific region of the cell where its concentration is very high. We have not been looking at that level of regulation, and we should consider it more carefully.

Vilain: We know that 3′ UTR deletion in humans results in XY sex reversal.

Capel: And it also seems to be required in the mouse for efficient sex reversal. I'm convinced that this is an important element of the gene. For all we know, the RNA is localized through the 3′ UTR, and where it is localized it is translated. Perhaps this creates a localized high concentration. I'm just suggesting that there could be many elements of regulation here that we are not in touch with.

Swain: And WT1 might be involved in this process.

Goodfellow: Has any more work been done in those species where there has been *Sry* amplification, looking for potentially co-amplified genes? The rat has 20 copies of *Sry*, although it is not clear how many of these are active. Clearly, if one gene is amplified, another related gene might also be amplified.

Schedl: Is there any other *Sox* gene apart from *Sox9* expressed early on in the gonad? There are so many of them now.

Lovell-Badge: *Sox3* is expressed early on, albeit at a low level.

Schedl: I'm just asking because of this argument about boxes. If there is a gene expressed at high levels you can argue that it might be a different HMG box that doesn't do the same job as SRY. If you find one that has quite high conservation in terms of the HMG box then perhaps the regions outside this are important.

Graves: I continue to find *Sox3* very interesting, because it is expressed early on in the gonad and it is also pretty clear that it is the gene from which *Sry* evolved. The two do have overlapping expression, even in mouse where the window is so narrow. I wonder whether *Sox3* is also involved in the interactions between *Sry* and *Sox9*.

Lovell-Badge: We still don't know about the function of *Sox3*. We finally have a conditional mutation through the germline, so we can now address this.

Graves: The whole issue of dosage comes in here. All these interactions seem to be dose dependent, at least in humans. It is such a recurring theme in sex determination, for other genes as well as the *Sox* genes.

Goodfellow: Peter Koopman, have you looked at the human genome sequence to count the total number of *SOX* box genes?

Koopman: There are currently about 20 human *SOX* genes known.

Behringer: If I understand correctly, around the *Sox9* gene the chromosome is very gene vacant. This is interesting.

No one has really mentioned the double repressor model, which was pushed by the *odsex* paper (Bishop et al 2000). Robin Lovell-Badge's *Sox9* regulation data argue against this.

Capel: In what way?

Behringer: In the *odsex* paper there was a suggestion that the deletion, which is way upstream of where Robin is working, had taken out a *cis* element that would be required for this repression mechanism. Robin's can switch without that *cis* element.

Koopman: To me, the combined data suggest that there is a female-specific repressor of *Sox9* that is a megabase upstream, and a male-specific activator element within 70 kb of the transcription start site.

Behringer: Robin's female turns down *lacZ* expression.

Lovell-Badge: My view is that the sequences found a megabase upstream are actually all to do with chromatin domains. The critical element is where you may have an anti-testis gene product binding, and we proposed at one point that DAX1 could be involved here. Thus the role of SRY could still be to prevent that repressor from binding.

Behringer: Is anything known about gene-vacant areas in chromatin? Is *Sox9* unusual?

Lovell-Badge: How about *Wt1*? This has a long regulatory region that is gene vacant.

Schedl: But we don't know much about *Wt1* regulation, so it is difficult to draw any conclusions. I think *Pax6*, which is just next to it, also has very few genes nearby. *Wt1* and *Pax6* are about 700 kb apart. After *Pax6* follows a very gene-poor region and there is a regulatory element about 300 kb away, which seems to have some kind of tissue-specific element.

Scherer: As far as we have analysed the human *SOX9* region, there is a 2 Mb intergenic distance 5′ and 500 kb 3′. From the cytogenetic data and from the sequence it is a G band/R band transition zone, so the 5′ region is lower in GC content than the 3′ region. GC regions are known to be generally gene poor.

Goodfellow: So no other genes have been found in this region by sequence analysis.

Scherer: There are just two pseudogenes and a few non-coding transcripts with no open reading frames.

Reference

Bishop CE, Whitworth DJ, Qin Y et al 2000 A transgenic insertion upstream of *Sox9* is associated with dominant XX sex reversal in the mouse. Nat Genet 26:490–494

Anomalies of human sexual development: clinical aspects and genetic analysis

Eric Vilain

UCLA Department of Human Genetics, CA, USA

Abstract. Disorders of human sex determination result in malformations of the external and internal genitalia. These malformations may vary from sexual ambiguity to complete sex reversal (XY female, XX male). Most of the knowledge of the molecular mechanisms involved in the mammalian sex determination pathway has been derived from the genetic analysis of intersex patients. Clinical management of these conditions critically depends on a precise understanding of their pathophysiology. Until recently, only transcription factors such as SRY, SOX9, DAX1, WT1 and SF1 were known to be responsible for abnormal gonadal development and sexual ambiguity. Gonadal dysgenesis may be isolated, as in the case of SRY mutations, or associated with abnormal development of other organs, such as bone or adrenals, consistent with the spatial expression profile of the disrupted genes (*SOX9* or *SF1*). WNT4 is a new sex-determining signalling molecule. Deletions of *Wnt4* were shown to be responsible for the masculinization of XX mouse pups while its duplication and overexpression in humans leads to XY sex reversal. Similarly, duplications of loci containing *DAX1* or *SOX9* have also been shown to cause sex reversal. These results support the emerging concept that mammalian sex determination is dosage sensitive at multiple steps of its pathway.

2002 The genetics and biology of sex determination. Wiley, Chichester (Novartis Foundation Symposium 244) p 43–56

Sexual development is the process by which external and internal genitalia are formed. Its disruption results in various degrees of sexual anomalies. Sexual development may be viewed as being composed of two processes: sex determination and sexual differentiation. This distinction, although somewhat artificial, has nevertheless proved important to the understanding of the medical classification and management of abnormalities of human sexual development. Sex determination is the developmental decision that directs the orientation of the undifferentiated embryo into a sexually dimorphic individual. In mammals, this occurs during the development of the gonads. If mammalian embryos are castrated early in development and reimplanted into the uterus, they all develop into females, regardless of their genetic sex. This led the physiologist Jost to conclude that sex determination is synonymous with testis determination. In essence, once the testes are formed in males, sex is determined. Following this sex determination decision, the process of sexual differentiation begins and the testes start producing the male hormones testosterone and anti-Müllerian hormone (AMH, also known as Müllerian inhibiting substance, MIS), which are responsible for male sexual characteristics. This concept, verified in almost all mammalian species, led to the search for a sex-determining gene that was a testis-determining factor (TDF). When the karyotype of patients with Klinefelter syndrome who are male (47, XXY) and Turner syndrome who are female (45, X) were discovered, it became clear that the Y chromosome was sex-determining and that TDF had to be located on the Y chromosome.

Human pathologies of sexual development

Malformations of genitalia occur with an estimated frequency of 1% and are extremely varied in their presentation. In most cases, they are simple, isolated variations of the 'normal' anatomy of external genitalia, such as an enlarged clitoris, a small penis, an abnormal position of the urethral opening (known as hypospadias) or undescended testes (referred to as cryptorchidism). In more rare instances, these variations are so far from the normal anatomic standards that they are referred to as ambiguous genitalia or intersex conditions.

In the last few years, there has been considerable debate over the clinical management of these cases. In particular, the necessity of early sex assignment by surgical methods has been highly controversial, as outcome data on large cohorts of patients are still missing. In this context, understanding the mechanisms of sexual development and the pathophysiology of intersex conditions has become increasingly important. Endocrine and genetic advances in the biology of sexual development are becoming an integral part of the decision-making process in intersex cases.

Pathologies of sexual differentiation are the most frequent and best understood. The gonads develop normally, but the subsequent development of internal or external genitalia fails. For instance, in an XY individual with a disorder of sexual differentiation, testes develop normally but testosterone fails to act normally, either because of a defect of its biosynthesis, or because of a defect in its receptor. As a consequence, external genitalia are feminized. In XX individuals with disorders of sexual differentiation, ovaries are normal but the external genitalia are masculinized because of an excessive impregnation by exogenous androgens, or more commonly, of adrenal origin (congenital adrenal hyperplasia).

Pathologies of sex determination are characterized by an abnormal development of the gonads (gonadal dysgenesis). They are poorly understood and are intensely investigated. They are caused by the defective action of genes involved in sex determination. Most of them have been identified in humans with disorders of sex determination, also known as sex reversal, by a positional cloning approach. These individuals have a discordance between their phenotypic and their genotypic sex. They are XX males, XX true hermaphrodites, or XY females with gonadal dysgenesis. XX males typically have normal male genitalia, small azoospermic testes and no Müllerian structures (uterus, Fallopian tubes, upper part of vagina), but may also present at birth with severe hypospadias or sexual ambiguity. XX true hermaphrodites present with ambiguous genitalia, persistence of some Müllerian structures, and are defined pathologically by the presence of both ovarian and testicular tissue in their gonads. XY females with pure gonadal dysgenesis have normal female genitalia, including a normal uterus due to lack of AMH production, and fibrous streak gonads in place of the ovaries. When the gonadal dysgenesis is partial, these patients may present with sexual ambiguity. These disorders are difficult to diagnose, as little is known about their pathogenesis. These pathologies, occurring with a frequency of approximately 1 in 20 000 have allowed the mapping of sex determining genes, and TDF in particular. However, a large majority (about 75%) of sex-reversed patients cannot yet be explained at the molecular level, suggesting the existence of a number of unknown sex determining genes (Vilain & McCabe 1998). Female gonadal development remains mostly mysterious at the molecular level. Long considered as a 'default pathway', it now appears to be an active process, as 'anti-testis' genes that may also be 'pro-ovary' are being identified. We will review several sex-determining genes and the pathologies they induce in humans when their action is disrupted. They are summarized in Table 1. A number of genes responsible, when deleted, for abnormal gonadal development in mice, have not been shown to be involved in human pathologies of sex determination as yet. They include *Lim1*, *M33* and *Fgf9*.

Genes involved in early gonadal development

A number of genes encode transcription factors required for early morphogenesis of the gonads. They may also play a role throughout testis and ovary development. SF1 (steroidogenic factor 1), an orphan member of the nuclear receptor superfamily initially identified as a regulator of cytochrome P450 hydroxylases (Honda et al 1993, Lala et al 1992), is able to activate various gonadal and adrenal steroid hydroxylases (Morohashi et al 1993). Further studies demonstrated that *SF1* was expressed in the developing hypothalamus, pituitary, adrenals and gonads as early as E9 in genital ridges, and its homozygous deletion resulted in the absence of development of the gonads and the adrenals, as well as abnormal gonadotropic function (Ingraham et al 1994). SF1 therefore acts at multiple levels of the reproductive axis, including during the early stages of gonadal and

TABLE 1 Genes involved at the initial stages of sexual development: chromosomal localization, gene family and presumed functions

Gene	*Localization*	*Gene family*	*Putative function*	*Phenotype of mutations*
SF1	9q33	Nuclear receptor	Transcription factor	Gonadal dysgenesis and adrenal insufficiency
WT1	11p13	Zinc finger protein	Transcription factor	Denys–Drash and Frasier syndromes
SRY	Yp11	HMG protein	Transcription factor	XY gonadal dysgenesis
DAX1	Xp21.3	Nuclear receptor	Transcription factor	Duplication: XY gonadal dysgenesis Mutation: adrenal hypoplasia congenita
SOX9	17q24	HMG protein	Transcription factor	Duplication: XX sex reversal Mutation: campomelic dysplasia with XY gonadal dysgenesis
DMRT1	9p24	DM domain protein	Transcription factor	XY gonadal dysgenesis
AMH	19q13	Transforming growth factor (TGF)β	Growth factor	Persistent Müllerian duct syndrome
WNT4	1p35	Wnt	Growth factor	XX sex masculinization in mouse

(Adapted from Vilain & McCabe 1998)

adrenal development. In humans, a mutation in *SF1* was identified in a patient with adrenal insufficiency and XY sex reversal, confirming the role of SF1 in human gonadal and adrenal development (Achermann et al 1999).

WT1 is a transcription factor expressed as early as E9.5 in the intermediate mesoderm of both males and females (Pelletier et al 1991). Knockout mice homozygous for a *Wt1* null mutation have kidney and gonadal agenesis (Kreidberg et al 1993). In humans, mutations in *WT1* were identified in patients with Denys–Drash syndrome and Frasier syndrome, who present with severe renal failure caused by mesangial sclerosis and XY gonadal dysgenesis (Pelletier et al 1991, Barbaux et al 1997). This suggests a crucial role for WT1 in human kidney and gonad development.

Sex-determining genes

Genes directly responsible for the decision to form either a testis or an ovary have primarily been identified by the genetic mapping of patients with sex reversal. They are summarized in Table 1, along with the phenotypes they induce when mutated. While some genes such as *SRY* and *SOX9* have been shown to influence sex determination towards maleness, others such as *DAX1* and *WNT4* have been shown to prevent it, or even to possibly influence ovarian formation.

SRY

By positional cloning, a small fragment of the Y chromosome (35 kb), translocated on the X chromosome of XX males and true hermaphrodites, was found to contain TDF. *SRY* was identified as a conserved sequence within these 35 kb (Sinclair et al 1990). It encodes a 204 amino acid protein with the ability to bind and bend DNA through an HMG (High Mobility Group) conserved motif (Harley et al 1992). Several convergent arguments proved that *SRY* was TDF. SRY protein has the biochemical properties of a transcription factor (Harley et al 1992); it is localized in the expected portion of the Y chromosome (Sinclair et al 1990); and its temporal profile of expression is appropriate, since murine *Sry* is expressed between E10.5 and E12.5, just prior to the appearance of seminiferous tubules (Koopman et al 1990). More importantly, an XX mouse transgenic for 14 kb of a genomic Y chromosome fragment containing *Sry* developed as a male (Koopman et al 1991). Finally, we and others provided multiple genetic evidence that *SRY* was indeed the testis-determining factor in humans. Point mutations in *SRY* were shown to divert the fate of the bipotential gonad of an XY fetus from testicular to ovarian tissue (review in Vilain & McCabe 1998). These mutations were found in XY females with pure gonadal dysgenesis.

SRY analysis is inadequate to explain the phenotype of all the patients with pathologies of sex determination. For instance, we have shown that a completely normal male phenotype could occur in an XX patient without any Y chromosome sequences including *SRY* (Vilain et al 1994). Genetic studies have also shown that while *SRY* is present in 90% of XX males without ambiguities, it is detected in only 10% of XX true hermaphrodites and in only 10% of XX ambiguous males (McElreavey et al 1995). Conversely, *SRY* mutations are found in only 25% of XY females with gonadal dysgenesis (McElreavey et al 1995, Vilain & McCabe 1998). This suggests that genes other than *SRY* are needed for normal male development.

SOX9

Like *SRY*, *SOX9* is a male-determining gene. Chromosomal rearrangements of chromosome 17 were observed in patients with campomelic dysplasia (Tommerup et al 1993), a severe skeletal dysplasia in which a majority of XY patients are phenotypic females. This allowed the cloning of *SOX9*, a member the *SOX* gene family of transcription factors related by the presence of an HMG box (Foster et al 1994, Wagner et al 1994). Point mutations in *SOX9* associated with campomelic XY females showed that it was a sex-determining gene (Foster et al 1994, Wagner et al 1994). It also binds to the same DNA targets as SRY *invitro*. Although the physiological target of SOX9 remains unknown, there is some evidence that it can regulate the transcription of *AMH* in association with SF1 (de Santa Barbara et al 1998). However, several indirect arguments challenge the hypothesis that SOX9 regulates AMH expression directly. The fact that XY patients with mutations in *SOX9* are sex reversed while XY patients with mutations in *AMH* are male (Behringer et al 1994) suggests the existence of a number of genetic intermediates between these two genes. In addition, it was shown in chickens that *AMH* is expressed prior to *SOX9* (Oreal et al 1998), suggesting alternative regulation in this species. Recently, an XX male patient was shown to carry a large duplication of chromosome 17 including *SOX9* (Huang et al 1999). This is the first example of XX sex reversal not caused by *SRY* in humans. It suggests that *SOX9*, like *SRY*, has the capability to induce male development in an XX individual.

DAX1: an 'anti-testis' gene

Duplications of a region of the short arm of the X chromosome (Xp21.3) were found in several XY females with gonadal dysgenesis (Bardoni et al 1994). The shortest duplicated region of the X responsible for sex reversal was found to be 160 kb, and was named DSS (dosage-sensitive sex reversal) (Bardoni et al 1994). *DAX1*, a gene in which mutations also lead to adrenal hypoplasia congenita, was cloned within DSS (Zanaria et al 1994). *DAX1* encodes an unusual member of the nuclear hormone receptor superfamily, with a typical ligand-binding domain but a novel putative DNA-binding domain containing 3.5 repeats of 65–67 amino acids that may represent zinc finger structures (Zanaria et al 1994). Although its physiological target is still unknown, DAX1 was shown to bind to single-strand hairpin DNA motifs and to act as a repressor of transcription. It was also shown recently that DAX1 could act as an RNA-binding protein (Lalli et al 2000). *Dax1* knockout resulted in a defect of spermatogenesis (Yu et al 1998). No sex reversal was observed, but neither was any overt adrenal phenotype, suggesting that this milder-than-expected phenotype was caused by a hypomorphic allele of *Dax1*. The murine pattern of expression of *Dax1* is consistent with its role in sex determination. It is expressed at E11.5 in the gonads of both sexes (Swain et al 1996). In males, this corresponds to the peak of expression of *Sry* and to the period immediately prior to the first signs of testis differentiation. At E12.5, *Dax1* is turned off in the testis, but remains on in the ovary (Swain et al 1996). This suggests a possible role for *Dax1* in ovarian formation. In addition, transgenic XY mice carrying additional copies of *Dax1* develop as females, suggesting that *Dax1* antagonizes the action of *Sry* and can be considered an 'anti-testis' gene, and possibly a 'pro-ovary gene' (Swain et al 1998, Goodfellow & Camerino 1999).

Other sex-determining genes

Several other sex-determining loci are known, based on sex reversed patients with chromosomal abnormalities. They include 9p24, a region deleted in some XY females (Bennett et al 1993) that contains the transcription factors DMRT1 and DMRT2 (Raymond et al 1998, Ottolenghi et al 2000). They also include 10q, a region deleted in several XY females (Wilkie et al 1993), and 22q, a region duplicated in an XX true hermaphrodite (Aleck et al 1999) and an XX male (Seeherunvong et al 2000).

A genetic model for mammalian sex determination

Based on the pattern of inheritance of XX sex-reversal in humans, we proposed a new model for sex determination in mammals (Vilain et al 1993, McElreavey et al 1993). In order to explain the mechanisms of the recessive mode of inheritance of XX males without *SRY*, we proposed that *SRY* might antagonize a gene, termed *Z*, which would in turn inhibit male-specific genes. The observation of XY females with a duplication of DSS (Bardoni et al 1994) and the antagonistic effects of *Sry* and *Dax1* in mice (Swain et al 1998) support this hypothesis, and suggest that *Z* is, in fact, *DAX1*. Our working model is that *SRY* would inhibit the action of *DAX1*, which would in turn prevent testis formation. When *DAX1* is duplicated, the doses of *SRY* would not be high enough to antagonize the increased *DAX1* activity. *DAX1* is therefore still active and continues to prevent testis formation. This results in the development of an XY female. This model became more complex as more sex-determining genes were discovered. In fact, *DAX1* is part of a complex network of interaction between a number of sex-determining genes. *SF1* up-regulates the expression of *DAX1* in an adrenocortical carcinoma cell line (Vilain et al 1997), probably by binding to an SF1-response element in the *DAX1* promoter. DAX1 and SF1 also interact at the protein level as part of a multi-protein complex. It was demonstrated that SF1 acts synergistically with WT1 to up-regulate *AMH* expression, and that this activation could be blocked by DAX1 (Nachtigal et al 1998).

Signalling sex determination

Until recently, all known sex-determining genes were transcription factors. WNT4, a member of the WNT family of locally acting cell signals was shown to

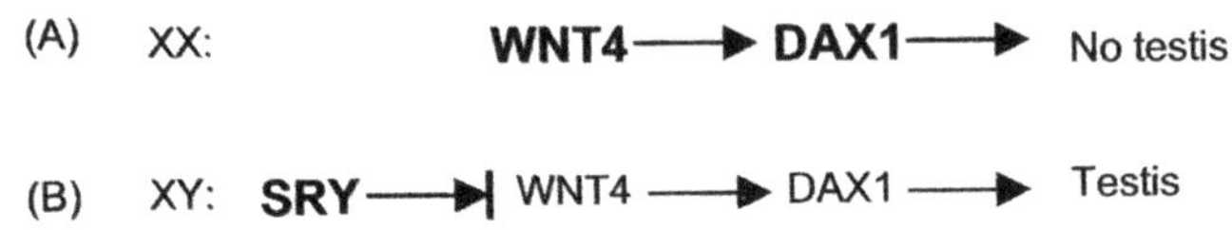

FIG. 1. Hypothetical schematic diagram of the mammalian sex determination pathway. (A) In XX individuals, *WNT4* expression up-regulates the expression of *DAX1*. Then, *DAX1* expression prevents the formation of testes and allows the normal formation of ovaries. (B) In XY individuals, expression of *SRY* inhibits the action of WNT4 and, consequently, of DAX1. Low levels of DAX1 cannot fully inhibit the formation of the testes, leading to normal male development. Genes in bold are 'on'. Genes not in bold are 'off'.

be a new signalling molecule involved in sex determination in mice. *Wnt4* is expressed as early as E9.5 in the mesonephros and in the coelomic epithelium of the presumptive gonad (Vainio et al 1999). *Wnt4* expression is then down-regulated in the developing male gonad, but persists in the developing ovary. Targeted deletion of *Wnt4* results in the masculinization of XX mice. We have recently shown that overexpression of *WNT4* in humans results in XY sex reversal, as observed with overexpression of *DAX1* (Jordan et al 2001). These results suggest that *WNT4* could act as an 'anti-testis' gene like *DAX1*. WNT4 is part of the family of cysteine-rich glycosylated secreted ligands involved in cell proliferation and differentiation of a variety of organisms, from *Caenorhabditis elegans* and *Drosophila* to mammals. In their canonical pathway, WNT molecules bind to Frizzled receptors, which activate a signalling cascade that includes dishevelled (DSH), glycogen synthase kinase 3 (GSK3), and β-catenin/TCF (T cell factor), which binds to a TCF response element. Interestingly, the TCF response element (AACAAAG) is known to bind members of the TCF/LEF family, which contain an HMG box. In an alternate pathway, the transcriptional activation is thought to occur as a result of G protein-mediated modulation of internal Ca^{2+} concentrations. We have shown that in a mouse Sertoli cell line, WNT4 can up-regulate *Dax1* expression. One hypothetical model is that WNT4 acts as a molecular link between SRY and DAX1. SRY would inhibit the action of DAX1 via WNT4 (Fig. 1).

Conclusion

Sex-determining genes direct the fate of the bipotential gonad into either testis or ovary. They can be categorized into (1) transcription factors involved throughout gonadal morphogenesis (e.g. *SF1*, *WT1*), (2) inducers of testicular development (*SRY*, *SOX9*), and (3) 'anti-testis' genes and potential promoters of ovarian development (*DAX1* and *WNT4*). All these genes are expressed in the developing genital ridges, and their products interact with each other as part of a complex genetic pathway leading to gonadal differentiation into one sex or another.

Duplication of chromosomal regions containing sex-determining genes lead to XY sex reversal (*DAX1* and *WNT4*) or XX sex reversal (*SOX9*). A new concept is emerging, as modification of the copy number of key sex-determining genes changes the fate of the gonadal sex: mammalian sex determination appears to be sensitive to gene dosage at important steps of its pathway. However, the precise molecular mechanisms of gene dosage in sex determination are not known. Disruption or overexpression of sex-determining genes results in sex reversal in humans (XX males and XY females), but a majority of patients with abnormal gonad development remain unexplained genetically.

Identifying new sex-determining genes will not only enhance the understanding of gonadal development, but will also provide molecular tools to help diagnose and manage patients affected with disorders of sexual development.

References

Achermann JC, Ito M, Ito M, Hindmarsh PC, Jameson JL 1999 A mutation in the gene encoding steroidogenic factor-1 causes XY sex reversal and adrenal failure in humans. Nat Genet 22:125–126

Aleck KA, Argueso L, Stone J, Hackel JG, Erickson RP 1999 True hermaphroditism with partial duplication of chromosome 22 and without SRY. Am J Med Genet 85:2–4

Barbaux S, Niaudet P, Gubler MC et al 1997 Donor splice-site mutations in WT1 are responsible for Frasier syndrome. Nat Genet 17:467–470

Bardoni B, Zanaria E, Guioli S et al 1994 A dosage sensitive locus at chromosome Xp21 is involved in male to female sex reversal. Nat Genet 7: 497–501

Behringer RR, Finegold MJ, Cate RL 1994 Müllerian-inhibiting substance function during mammalian sexual development. Cell 79:415–425

Bennett CP, Docherty Z, Robb SA, Ramani P, Hawkins JR, Grant D 1993 Deletions 9p and sex reversal. J Med Genet 30:518–520

de Santa Barbara P, Bonneaud N, Boizet B et al 1998 Direct interaction of SRY-related protein SOX9 and steroidogenic factor 1 regulates transcription of the human anti-Müllerian hormone gene. Mol Cell Biol 18:6653–6665

Foster JW, Dominguez-Steglich MA, Guioli S et al 1994 Campomelic dysplasia and autosomal sex reversal caused by mutations in an SRY-related gene. Nature 372:525–530

Goodfellow PN, Camerino G 1999 DAX-1, an 'antitestis' gene. Cell Mol Life Sci 55:857–63

Harley VR, Jackson DI, Hextall PJ et al 1992 DNA binding activity of recombinant SRY from normal males and XY females. Science 255:453–455

Honda S, Morohashi K, Nomura M, Takeya H, Kitajima M, Omura T 1993 Ad4BP regulating steroidogenic P-450 gene is a member of steroid hormone receptor superfamily. J Biol Chem 268:7494–7502

Huang B, Wang S, Ning Y, Lamb AN, Bartley J 1999 Autosomal XX sex reversal caused by duplication of SOX9. Am J Med Genet 87:349–353

Ingraham HA, Lala DS, Ikeda Y et al 1994 The nuclear receptor steroidogenic factor 1 acts at multiple levels of the reproductive axis. Genes Dev 8:2302–2312

Jordan BK, Mohammed M, Ching ST et al 2001 Up-Regulation of WNT-4 signaling and dosage-sensitive sex reversal in humans. Am J Hum Genet 68:1102–1109

Koopman P, Munserberg A, Capel B, Vivian N, Lovell-Badge R 1990 Expression of a candidate sex-determining gene during mouse testis differentiation. Nature 348:450–452

Koopman P, Gubbay J, Vivian N, Goodfellow PN, Lovell-Badge R 1991 Male development of chromosomally female mice transgenic for Sry. Nature 351:117–121

Kreidberg JA, Sariola H, Loring JM et al 1993 WT-1 is required for early kidney development. Cell 74:679–691

Lala DS, Rice DA, Parker KL 1992 Steroidogenic factor1, a key regulator of steroidogenic enzyme expression, is the mouse homolog of fushi tarazu factor 1. Mol Endocrinol 6: 1249–1258

Lalli E, Ohe K, Hindelang C, Sassone-Corsi P 2000 Orphan receptor DAX-1 is a shuttling RNA binding protein associated with polyribosomes via mRNA. Mol Cell Biol 20:4910–4921

McElreavey K, Vilain E, Abbas N, Herskowitz I, Fellous M 1993 A regulatory cascade hypothesis for mammalian sex determination: SRY represses a negative regulator of male development. Proc Natl Acad Sci USA 90:3368–3372

McElreavey K, Barbaux S, Ion A, Fellous M 1995 The genetic basis of murine and human sex determination: a review. Heredity 75:599–611

Morohashi K, Zanger UM, Honda S, Hara M, Waterman MR, Omura T 1993 Activation of CYP11A and CYP11B gene promoters by the steroidogenic cell-specific transcription factor, Ad4BP. Mol Endocrinol 7:1196–1204

Nachtigal MW, Hirokawa Y, Enyeart-VanHouten DL, Flanagan JN, Hammer GD, Ingraham HA 1998 Wilms' tumor 1 and Dax-1 modulate the orphan nuclear receptor SF-1 in sex-specific gene expression. Cell 93:445–454

Oreal E, Pieau C, Mattei MG et al 1998 Early expression of AMH in chicken embryonic gonads precedes testicular SOX9 expression. Dev Dyn 212:522–532

Ottolenghi C, Veitia R, Quintana-Murci L et al 2000 The region on 9p associated with 46, XY sex reversal contains several transcripts expressed in the urogenital system and a novel doublesex-related domain. Genomics 64:170–178

Pelletier J, Bruening W, Li FP, Glaster T, Haber DA, Housman D 1991 WT1 mutations contribute to abnormal genital system development and hereditary Wilms' tumour. Nature 353:431–434

Raymond CS, Shamu CE, Shen MM et al 1998 evidence for evolutionary conservation of sex-determining genes. Nature 391:691–695

Seeherunvong T, Perera E, Benke P, Benigno A, Donahue R, Berkovitz G 2000 46, XX Sex Reversal with Duplication of Chromosome 22q. In: Proceedings of the Endocrine Society's 82nd Annual Meeting. June 2000, Toronto

Sinclair AH, Berta P, Palmer MS et al 1990 A gene from the human sex-determining region encodes a protein with homology to a conserved DNA-binding motif. Nature 346:240–244

Swain A, Zanaria E, Hacker A, Lovell-Badge R, Camerino G 1996 Mouse Dax1 expression is consistent with a role in sex determination as well as in adrenal and hypothalamus function. Nat Genet 12:404–409

Swain A, Narvaez V, Burgoyne P, Camerino G, Lovell-Badge R 1998 Dax1 antagonizes Sry action in mammalian sex determination. Nature 391:761–767

Tommerup N, Schempp W, Meinecke P et al 1993 Assignment of an autosomal sex reversal locus (SRA1) and campomelic dysplasia (CMPD1) to 17q24.3-q25.1. Nat Genet 4:170–174

Vainio S, Heikkilä M, Kispert A, Chin N, McMahon AP 1999 Female development in mammals is regulated by Wnt-4 signalling. Nature 397:405–409

Vilain E, Guo W, Zhang YH, McCabe ER 1997 DAX1 gene expression upregulated by steroidogenic factor 1 in an adrenocortical carcinoma cell line. Biochem Mol Med 61:1–8

Vilain E, McElreavey K, Herskowitz I, Fellous M 1993 La determination du sexe: faits et nouveaux concepts. Médecine/Sciences 8:I–VII

Vilain E, Lefiblec B, Morichon-Delvallez N et al 1994 SRY-negative XX fetus with complete male phenotype. Lancet 343:240–241
Vilain E, McCabe ERB 1998 Mammalian sex determination: from gonads to brain. Mol Genet Metab 65:74–84
Wagner T, Wirth J, Meyer J et al 1994 Autosomal sex reversal and campomelic dysplasia are caused by mutations in and around the SRY-related gene SOX9. Cell 79:1111–1120
Wilkie AO, Campbell FM, Daubeney P et al 1993 Complete and partial XY sex reversal associated with terminal deletion of 10q: report of 2 cases and literature review. Am J Med Genet 46:597–600
Yu RN, Ito M, Saunders TL, Camper SA, Jameson JL 1998 Role of Ahch in gonadal development and gametogenesis. Nat Genet 20:353–357
Zanaria E, Muscatelli F, Bardoni B et al 1994 An unusual member of the nuclear hormone receptor superfamily responsible for X-linked adrenal hypoplasia congenita. Nature 372:635–641

DISCUSSION

Wilkins: You described various human babies who show sexual ambiguity. Presumably these all come from parents who are sexually normal enough to be fertile. So are they newly arising mutants?

Vilain: Yes, they are almost always newly arising mutants. The only case that is suspicious is the familial case I described earlier, in which there is an extra dose of *SRY* in the father of the two XX masculinized children. The father is a normal, fertile male who has *SRY* on his Y chromosome and also on his X chromosome. His mother must have had *SRY* on one of her X chromosomes, but we were unable to access her DNA to show this. One could argue that she may have been a true hermaphrodite who was fertile, and was only very mildly masculinized.

Zarkower: I have seen at least one other report of an XY female duplication on distal 1p. Have you looked at this?

Vilain: Yes, there are four reports I know of showing duplication of 1p. Two of them are sex reversed, the other two have only cryptorchidism, which is a non-specific sign of any chromosomal abnormality. Those two clearly do not include WNT4. There is one other case from a German laboratory in Magdeburg, and I will soon have access to these cells to study.

Short: Could you tell us more about the clinical management of patients born with micropenis? I have seen it stated recently that if the phallus is less than a certain length, a gender reassignment is carried out almost routinely (Dreger 1998). This sounds horrific. What is the general practice in the USA if a boy is born with basically male external genitalia but a micropenis?

Vilain: If there is no ambiguity, the clinical management is to let the patients stay male. There was a famous example that has been in the media spotlight of a botched circumcision in one of two male twins, Bruce Reimer. The doctors decided to make this boy a girl, and Bruce became Brenda. The hypothesis was that nurture would always overcome nature. This was an immense failure, and Brenda grew up to be a very depressed woman, who eventually changed back to his original gender and took the name David. Medical practice is now very cautious and does not use surgery without at least a specialized team of people from various disciplines, including endocrinologists, urologists, psychologists and geneticists, to make the most precise diagnosis and adequate gender assignment. The problem is, we know almost nothing about brain gender 'imprinting' during fetal life. We don't know the effect of the presence or absence of a Y chromosome, or testosterone levels, on brain sexual differentiation. This is a virgin field to be explored.

Short: John Colapinto's recent book describing John Money's alleged mismanagement of that famous case is an amazing account of how one can be sucked into a major clinical error by prejudice (Colapinto 2000). The statement that I was referring to was in a book called *Hermaphroditism and the medical invention of sex* (Dreger 1998). The statement was made that in the USA it was better for an XY child born at term with a phallus length of less than 2.5 cm (when stretched) to be made into a girl!

Vilain: This is no longer routine in major centres in the USA. The problem is isolated surgeons making decisions on their own without understanding what is going on in the field.

Sinclair: What about the option of not intervening at all and allowing children to decide for themselves when they reach sexual maturity?

Vilain: This is becoming an option that can be proposed to the parents. However, I'm not entirely convinced that letting the child choose for him of herself until he or she is 18 years of age is always going to be the best choice in terms of quality of life. With our highly sexually dimorphic culture, there are practical issues such as where they would go to the toilet at school. Unfortunately, we are not ready to deal with this in our society. In theory it is a great solution, because we just wait until the child tells us.

Sinclair: This is what the Intersex Society is actually suggesting that patients should do.

Vilain: I understand this. However, there are no outcome studies that tell us whether the children are actually happier when no gender assignment decision has been made. This is a big problem in this field. We only listen to the angry patients 20–30 years after mistakes have been made. We would have to do retrospective studies with lots of ethical issues, going back to ask patients how they feel now. Many of my patients would not like their families to know, and some patients aren't even aware themselves. What is left are prospective studies, and we hope these will provide the answer in a number of years.

Josso: I think the idea of letting people choose their gender at age 18 is crazy: it is totally impossible.

Vilain: There is an intermediate position. At five or six years of age children have the ability to tell whether they feel like boys or girls.

Josso: Another important factor in the decision is how things will go at puberty. Physicians themselves have a hard enough time making these decisions, so how can a child of five or six decide? We find that many parents have strong opinions themselves, and this can influence clinical decisions.

McLaren: How can non-gender assignment work in practice? Do the parents refer to the child as 'he' or 'she'?

Vilain: I am not proposing that this should be the case, but it does happen. I don't know how the family refers to the child. There are ambiguous first names that can be used.

Goodfellow: The dialogue that occurs between the medical profession and patient groups is something that the medical profession has to listen to. Not just with respect to this very difficult area, but generally. Treatment can reflect the social prejudices of the treaters. When a particular treatment is chosen because of the prejudices of the people who are performing that treatment, there has to be a social dialogue. The responsibility for the treatment of patients in the UK has changed in my lifetime. Thirty years ago you could not see your medical records because they belonged to the doctor, not you. This has changed. Clearly, there is no easy solution to this problem, because unless social attitudes change dramatically we are dealing with individuals who fall outside societal norms. Each case must also be treated individually on the basis of the medical needs: some of these intersex individuals have medical problems independent of the gender issue. It is difficult to come to a general conclusion, but we would be wrong not to engage in dialogue with those to be treated.

Camerino: In Italy we don't have many prospective studies. This means that we don't know how patients feel after they have become independent of their families. I don't think that the problem has been studied very scientifically.

Short: The more people who get involved in the discussion the better. To read that Colapinto account of how John Money handled the Reimer case makes your hair stand on end.

Goodfellow: It was not just him: people who I respect greatly, suddenly found their whole professional basis—how they treated and supported patients—suddenly thrown into question. In some ways, I thought that book detracted from the issue because it painted a very black and white picture.

Short: I heard Milton Diamond, one of John Money's critics, voicing some of those concerns many years ago.

Fernald: You gave a figure of 1% for sex determination problems in humans. This struck me as very high: which population does this refer to? Is this fraction the same in Asia, for example?

Vilain: The figure of 1% is probably true if you take into account all the minor disorders such as undescended testis or hypospadias. It is a figure quoted in literature reviews and is essentially in western populations.

Behringer: I question whether *WNT4* really is a sex-determining gene. My bias is that it is not. At least with the mouse, a *Wnt4*-deficient mouse would be classified as a female pseudohermaphrodite. Does a Leydig cell make a testis? Are they really Leydig cells?

Short: How would you diagnose a Leydig cell in the absence of a seminiferous tubule adjacent to it?

Vilain: The main question in this *Wnt4* knockout is why there aren't any Sertoli cells. If you consider that the Sertoli cells are the absolute requirement to determine a testis, it is not a sex-determining gene. Andy McMahon questions the possibility of Leydig cells being at the same level as Sertoli cells in terms of precursors of the testis. The question is, is it a dogma that the Sertoli cells are the first cells, or can Leydig cells also be the first apparition of testicular cells. There is masculinization in these mice; there is no question about this.

Behringer: But the loss of the Müllerian ducts is because of the lack of *Wnt4* expression in these Müllerian ducts. Then the stabilization of the Wolfian ducts is because of the hormones being produced.

Vilain: There are male hormones.

Behringer: It seems more like a differentiation than a primary sex determination. It depends whether those are Leydig cells.

Capel: Along the lines of whether *Wnt4* might be a good candidate for *Tda1*, have you looked to see whether there is any difference in *Wnt4* between B6 and DBA2 mice?

Vilain: We have looked at the presence of polymorphism in the coding sequence. The answer is that there is no difference. We are now looking at levels of expression in the gonad at 11.5 and 12.5 days, to see whether there are variations of levels and timing between B6 and DBA for *Sry* and *Wnt4* expression. There was a recent paper by Nagamine et al (1999) showing that there are variations in expression of *Sry* in the developing gonads between various B6 strains of mice. One hypothesis is that *Tda1* would counteract this difference in dosage.

References

Colapinto J 2000 As nature made him: the boy who was raised as a girl. HarperCollins, New York

Dreger A 1998 Hermaphrodites and the medical invention of sex. Harvard University Press, Cambridge, MA

Nagamine CM, Morohashi K, Carlisle C, Chang DK 1999 Sex reversal caused by *Mus musculus domesticus* Y chromosomes linked to variant expression of the testis-determining gene Sry. Dev Biol 216:182–194

The molecular action of testis-determining factors SRY and SOX9

Vincent R. Harley

Prince Henry's Institute of Medical Research, Monash Medical Centre, 246 Clayton Road, Melbourne, VIC 3168, Australia

Abstract. Despite 10 years of work since the discovery of SRY, little is known about its biochemical function. The HMG domain, a DNA-binding and DNA-bending motif, plays a central role, being the only region conserved between species and the site of almost all clinical mutations causing XY gonadal dysgenesis. By contrast, SOX9 harbours a number of highly conserved regions, including two domains required for maximal transactivation. The heat shock protein HSP70 recognizes a specific region of SOX9 hitherto of unknown function which may facilitate the assembly of multi-protein complexes at promoter/enhancer regions. The SRY and SOX9 HMG domains carry two nuclear localization signals (NLSs), one at each end which function independently and by distinct mechanisms. The N-terminal NLS is bound by calmodulin while the C-terminal NLS is bound by importin β. Four XY gonadal dysgenesis patients with mutations in SRY NLS regions showed reduced nuclear import accompanied in some cases by reduced importin β recognition. A campomelic dysplasia patient with SOX9 mutation outside the NLS regions also showed defective SOX9 nuclear import implying that nuclear import defects could be a common explanation for SRY and SOX9 HMG domain mutations.

2002 The genetics and biology of sex determination. Wiley, Chichester (Novartis Foundation Symposium 244) p 57–67

Despite 10 years of work since the discovery of SRY, little is known about its biochemical function. The HMG domain plays a central role because it is the only protein domain conserved between species and almost all clinical mutations causing XY gonadal dysgenesis reside in this domain. The HMG box is capable of binding and bending DNA *in vitro* and so SRY has been proposed to act as an architectural transcription factor which elicits its effect by remodelling chromatin, but this remains speculative in the absence of a downstream target. Similarly, while the clinical consequence of SOX9 mutations causing campomelic dysplasia/autosomal sex reversal is clear, the biochemical action of SOX9 during sex determination is less so. Here, I describe recent information on the normal role of SRY and SOX9 in sex determination as gleaned from the identification of some of the protein partners of SRY and SOX9 and studies investigating the biochemical consequences of human mutations.

Human SRY

SRY is expressed in the gonadal ridges of humans prior to overt differentiation of the gonad at about 7 weeks' gestation (Hanley et al 2000). SRY protein appears to be localized in the nucleus of pre-Sertoli cells consistent with a transcriptional function. XX mice transgenic for the human SRY open reading frame cause sex reversal (R. Lovell-Badge, personal communication) which suggests that the HMG domain might be the only part of the protein required for sex-determining activity since this is the only domain conserved between humans and mice. That the N- and C-terminal flanking regions are required for sex reversal remains unclear. In support of a role, missense SRY mutations in XY gonadal dysgenesis have been observed in both regions. Yet with respect to DNA binding and bending properties, the entire open reading frame of *SRY* appears to show the same activity as its HMG domain alone. Furthermore, clinical mutations outside the HMG domain (e.g. S18N) do not affect DNA binding or DNA bending (C. Mitchell & V.R. Harley, unpublished results). *In vitro* analysis of recombinant SRY protein suggests that its HMG domain has the ability to recognize specific sequences in DNA (Harley et al 1994) with affinity enhanced through phosphorylation (Desclozeaux et al 1998). This appears to be a property of the entire class of SOX (SRY-type HMG BOX) proteins, which bind the consensus sequence AACAAT.

SRY/SOX proteins differ in their intrinsic DNA sequence specificity. For example SRY shows a preference *in vitro* for A/T A/T AACAATAG while SOX9 prefers AGAACAATGG (Mertin et al 1999). This might reflect a difference in DNA binding specificity *in vivo*, and while SRY and SOX9 are capable of binding the same target sequences *in vitro*, their affinities are different. *In vivo* support for this observation came from work by Eicher and colleagues which showed that only when overexpressed could SOX9 or SOX3 HMG domains substitute for that of SRY to cause sex reversal in XX transgenic mice (Bergstrom et al 2000). Swapping the HMG domain of SOX9 with that of SOX1 greatly affects SOX9's transactivation potential (Kamichi et al 1999) confirming that the SOX9 HMG domain appears to possess some DNA target specificity not present in other SOX proteins.

Patients with XY gonadal dysgenesis carry point mutations in their *SRY* open reading frame. A number of mutations have been characterized and in some cases DNA-binding or DNA-bending activity are reduced (Harley et al 1992, Pontiggia et al 1994). A selection of mutants is shown in Table 1. Surprisingly a number of XY females carry SRY mutations which do not affect their DNA binding or DNA

TABLE 1 DNA binding and bending activities from selected XY gonadal dysgenesis patients

SRY variant	*Inheritance*	*%DNA binding*	*%DNA bending*	*Reference*
Wild-type		100%=32 nM	100%=55	Pontiggia et al 1994
S18N	Familial	~WT	~WT	Domenice et al 1998, Mitchell & Harley 2001
F67V	Mosaic father	~WT	nd	Tho et al 1998
F109S	Familial	~WT	100%	Jager et al 1992, Schmitt-Ney et al 1995
R76P	n/a	~WT	~WT	Mitchell & Harley 2001
M64T	n/a	~WT	nd	Tho et al 1998
M78T	n/a	~WT	~WT	Mitchell & Harley 2001
R133W	n/a	80%	~WT	Harley et al 2001
P125L	n/a	~WT	nd	Tho et al 1998
A113T	*de novo*	~WT	~WT	Zeng et al 1993
190M	Familial	>50%	~WT	Pontiggia et al 1994

nd, not determined; WT, wild-type.

bending activities *in vitro* which suggest that other essential activities of SRY must exist. Also, the results confirm our earlier suggestion that SRY functions at a biochemical threshold and that familial mutations are close to this threshold level and manifest in certain genetic backgrounds (Harley et al 1992).

Evidence that human SRY is a transcription factor is slow to arrive and reporter gene studies in transfected cell cultures include reports of mild activation (Cohen et al 1994). The main limitations of these studies have been the lack of a relevant DNA target and/or inappropriate cell lines. Some support for SRY being a repressor of a repressor comes from the Odsex mouse (an XX male) where it has been proposed that an SRY binding element is deleted in the *Sox9* regulatory region and therefore in the normal mouse XY gonads, SRY normally disrupts the binding of the repressor (Bishop et al 2000). In contrast, studies on the mouse SRY show that it has a strong ability to activate transcription in GAL4 fusions of its C-terminal glutamine-rich region (Dubin & Ostrer 1994) which is completely absent in SRY of other species. Thus SRY is likely to activate transcription in non-mouse species but through a different mechanism. For example, in humans transcriptional effects might be somehow exerted through the N or C terminal domains—indeed, one point mutation in XY females has been reported in each domain and the C-terminal domain interacts with a Pdz protein *in vitro* pointing to functional roles for these domains (Poulat et al 1997). However it is hard to envisage a different mechanism operating in every species through these regions, given the poor conservation outside the HMG domain among mammals arising from the increased rate of evolution of the Y chromosome. More likely the SRY HMG domain itself carries the necessary information for transcription through structure and sequence specific DNA recognition together with co-activator proteins to establish the correct architecture in chromatin, analogous to LEF1 (Grosschedl et al 1994).

Human SOX9 and its protein partners

In stark contrast to SRY protein sequence, the extraordinary conservation of SOX9 protein sequence throughout vertebrates may have been maintained through interactions of its functional domains with a suite of cellular proteins. In addition to DNA sequence-specificity, interactions between SOX9 and other proteins are likely to be involved in regulation by SOX9. That SOX9 cannot mediate transcriptional activation on its own can be seen from the fact that type II collagen is not transcribed in the testis and conversely, anti-Müllerian hormone (AMH; also known as Müllerian inhibitory substance, MIS) is not transcribed by chondrocytes. Few interactions have been described; SOX9 interacts with SF1 during MIS/AMH regulation (de Santa Barbara et al 1998) and phosphorylation of SOX9 involves interaction with protein kinase A (Huang et al 2000). Recently we have confirmed that the interaction between SRY and the Ca^{2+}-binding protein calmodulin (CaM) (Harley et al 1996) is conserved for SOX9 and that antiCaM drugs block nuclear import (Argentaro et al 2001). We have also demonstrated that the HMG boxes of SOX9 and SRY interact with importin β (Preiss et al 2001). Intriguingly all three of these protein–protein interactions occur via the HMG domain of SOX9; however, much of this highly conserved protein still has no function assigned to it.

The SOX9 C-terminal region, rich in proline, glutamines and serines (the so-called PQS domain) is the major transcriptional activation domain (Sudbeck et al 1996). The adjacent region rich in prolines, glutamines and alanines (the PQA domain) is also required for maximal activation (McDowall et al 1999). This domain is only conserved in mammals and therefore may relate in some way to organisms with an SRY sex-determining mechanism. The mechanism and factors through which SOX9 activates the pre-initiation transcription complex via recognition of PQS and/or PQA regions remain unknown.

Through *in vitro* and *in vivo* studies, we have identified the heat shock protein HSP70 as an interacting partner for SOX9 in testicular and chondrocyte cell lines (Marshall & Harley 2001). HSP70 forms a ternary complex with DNA-bound SOX9. The interaction between HSP70 and SOX9 is ATP-independent, in contrast to the ATP-dependent interaction of the substrate-binding domain of

HSP70 with denatured proteins. The interaction involves the C-terminal of HSP70 with a 100 amino acid region of SOX9 between the HMG box and the PQA domain, hitherto of unknown function but highly conserved among Group E SOX proteins. The regulation of the *AMH* gene is controlled not only by SOX9 and SF1 (de Santa Barbera et al 1998, Arango et al 1999) but also by WT1 (Nachtigal et al 1998). While binding sites to both SOX9 and SF1 are conserved within the *AMH* promoter, WT1 has no conserved binding site and it interacts only very weakly with SF1. Considering that WT1 strongly interacts with HSP70 *in vivo* (Maheswaran et al 1998) it is possible to speculate that WT1 binding at the *AMH* promoter is stabilized by the formation of a SOX9–HSP70–WT1 protein complex. The fact that SOX9 and SF1 also interact suggests that the four proteins may form a tightly associated complex at the promoter.

Nuclear import of SRY and SOX9

In the developing embryonic gonad in humans and mice, the subcellular location of SOX9 protein in pre-Sertoli cells is initially cytoplasmic until at the sex determining period, co-incident with SRY expression in males, SOX9 is localized to the nucleus (de Santa Barbara et al 2000, Morais da Silva et al 1996). Sexually dimorphic subcellular expression of SOX9 suggests mechanisms by which SOX9 activity is regulated. The signals for nuclear import reside in the HMG domain and are highly conserved among SRY/SOX family members. We present evidence for two possible mechanisms of nuclear import of SRY/SOX9 (Fig. 1).

The SRY/SOX HMG domain carries two highly conserved nuclear localization signals (NLSs), one at each end of the HMG domain (Sudbeck & Scherer 1997). We observed that each NLS of SRY, when fused to fluorescence-labelled β galactosidase and incubated with mechanically perforated cells, is independently capable of rapidly (i.e. a few minutes) localizing this large carrier protein into the nucleus (Harley et al 2001). Conventional NLS-containing proteins are imported following recognition by importins and RanGTP, then the complex is docked at the nuclear pore complex (NPC) and translocated. Once in the nucleus, GTP is converted, the complex dissociates and is recycled. We used an ELISA-based assay to determine which importins recognize SRY NLSs. Using peptides we found that the SRY C-terminal NLS bound importin β almost as well as the intact SRY HMG domain, whereas the N-terminal NLS peptide bound neither importin α nor importin β. We had shown previously that the N-terminal NLS binds to CaM *in vitro* (Harley et al 1996) so we undertook some experiments with SOX9 which also has this NLS and for which we have more sophisticated assays. SOX9 could bind CaM *in vitro* on native gels and binding could be blocked with CaM antagonists such as calmidazolium (CDZ). In our reporter gene assay, SOX9

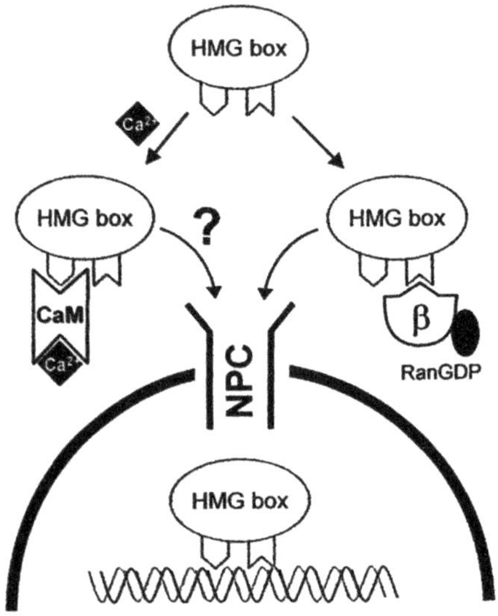

FIG. 1. The HMG domain of SRY and SOX9 carries two nuclear localization signals (NLSs). The C-terminal NLS is recognized by importin *β* and translocated into the nucleus through interactions of importin *β* with Ran-GTP and with components of the nuclear pore complex (NPC). Once inside the nucleus DNA recognition may facilitate release of importin *β* in addition to conversion of GTP to GDP. Some XY gonadal dysgenesis patients with nuclear import defects show reduced binding to importin *β*. The N-terminal NLS is recognized by CaM in the presence of Ca^{2+}. By some undefined mechanism not involving importins, the complex is translocated into the nucleus — a process that can be blocked with CaM antagonists.

failed to activate transcription in cultured cells in the presence of CDZ arising from a reduction in nuclear accumulation (Argentaro et al 2001). This suggests CaM plays a role in nuclear import and transcriptional activity of SRY and SOX9.

We studied four SRY clinical mutations with amino acid substitutions in their NLS regions to test the possibility that SRY might not transport properly to the nucleus prior to DNA binding. *SRY* was transfected into COS cells that were then stained for SRY protein by immunohistochemistry. From confocal microscopy images we measured the amount of SRY protein that accumulated in the nucleus relative to the cytoplasm. We observed that all four SRY mutant proteins showed reduced accumulation in the nucleus when compared to SRY from normal males (Harley et al 2001). This suggested that both NLS signals are required during sex determination for optimal transport and provide a cellular basis for XY sex reversal in these cases.

Since SRY binds importin *β* via its C-terminal NLS, we tested binding of importin *β* to SRY from an XY female with a C-terminal NLS mutation. As expected this mutant showed significantly reduced binding as a direct consequence of the mutation. Thus failed importin *β* recognition is the likely biochemical defect in this SRY mutant whose DNA binding and DNA bending activity were near wild-type. We tested SRY from three XY females with mutations in their N-terminal NLS regions and to our surprise, one of these also showed reduced importin β binding. One explanation for this is that the N-terminal NLS, although unable to bind importin β, is in close proximity to the C-terminal NLS and the mutations sterically hinder the ability of the C-NLS to be recognized by importin β.

Compound effects in SOX9 protein from campomelic dysplasia patients

We recently reported the identification of the novel amino acid substitution mutations, F154L and A158T, in the SOX9 HMG domain of two patients with campomelic dysplasia, the former an XX female and the latter a sex reversed XY female (Preiss et al 2001). On the basis of our molecular model of the SOX9 HMG domain (McDowall et al 1999), we postulated that F154 and A158 form part of a hydrophobic core region and would play a role in stabilizing the 3D alignment of the three helices of the HMG domain. However, tryptophan fluorescence studies did not demonstrate significant changes in tertiary structure of either mutant. The mutations would appear not to perturb the environment or tertiary structure (relative orientation of the helices), which could suggest functional redundancy in the amino acids forming the hydrophobic core. In contrast, our circular dichroism results indicated that both F154L and A158T mutations caused a loss of secondary structure, mainly in helix 3. The NLS located at the end of helix 3 is a conventional basic amphipathic helix which, in SRY, mediates nuclear import via direct interaction with importin β (see above). In SOX9, the A158T mutant showed decreased nuclear accumulation. This mutation is more proximal to the helix 3 NLS than F154L (whose nuclear import was normal) and might disrupt the function of this NLS. However A158T bound with wild-type affinity to importin β, suggesting that while this recognition step is normal, other components of the importin β-mediated nuclear import pathway could be affected. Further studies are required to elucidate the component of nuclear import that presumably fails to efficiently recognize the A158T mutant. The demonstration that a mutation outside the NLS regions affects nuclear localization raises the possibility that a large number of *SRY*, *SOX9* and *SOX10* clinical mutations could affect nuclear import in addition to, or distinct from, DNA binding and bending.

This study also presented data for the first time on the effect of point mutations in campomelic dysplasia on transactivation activity in cultured cells and allows us to correlate this with DNA binding activity *in vitro*. In the A158T mutant, a sixfold loss of DNA binding activity together with a twofold loss of nuclear import led to only a 30% loss of transcriptional activation. Similarly, in the F154L mutant, a 20-fold loss of DNA binding activity led to only a 66% loss of transcriptional activation activity. Our binary *in vitro* system is simplistic given that SOX9 acts in the context of a multiprotein complex *in vivo*. Our data is consistent with that for yeast ROX1, the only other HMG domain protein for which *in vitro*/*in vivo* correlations have been reported. In ROX1, substitutions causing a large reduction in DNA binding activity *in vitro* produce a small effect upon *ANB1* repressor activity *in vivo* (Deckert et al 1999). For example, the analogous change to SOX9 A158T in ROX1 affects DNA binding 10-fold and repression *in vivo* fourfold. In ROX1 F154 is W and a substitution to L affects DNA binding 1000-fold but repression only 50-fold. Thus a reduction in DNA binding *in vitro* produces only a small effect *in vivo* but this is presumably sufficient to account for the phenotypic effects. On this basis, small changes in DNA binding activity of SOX9 mutants may show undetectable changes in transactivation and lead to wild-type phenotype. However, small changes in SOX9 DNA-binding activity *in vitro* do not seem to correlate with milder symptoms or with whether campomelic dysplasia is accompanied by XY sex reversal; in these cases it could be that alterations of non-DNA binding functions of the HMG domain underlie the defect. Our data shows that A154T mutant has 60% of wild-type activation function in cultured cells. Given that this observation reflects the situation *in vivo* in campomelic dysplasia/SRA1 where one allele is mutant for SOX9 and the other is wild-type, our study raises the possibility that a high level of SOX9 transactivation activity is normally required for proper testis and bone formation. It is likely that interactions with transcriptional co-activators or components of the basal transcriptional machinery may mediate the effect of mutation.

Acknowledgement

I thank Michael Clarkson for critical reading of this manuscript.

References

Arango NA, Lovell-Badge R, Behringer RR 1999 Targeted mutagenesis of the endogenous mouse Mis gene promoter: in vivo definition of genetic pathways of vertebrate sexual development. Cell 99:409–419

Argentaro A, Mertin S, Chai A, Jans D, Harley VR 2001 SOX9 interacts with calmodulin during nuclear import. Submitted

Bergstrom DE, Young M, Albrecht KH, Eicher E 2000 Related function of mouse SOX3, SOX9 and SRY HMG domains assayed by male sex determination. Genesis 28:111–124

Bishop CE, Whitworth DJ, Qin Y et al 2000 A transgenic insertion upstream of sox9 is associated with dominant XX sex reversal in the mouse. Nat Genet 26:490–494

Cohen DR, Sinclair AH, McGovern JD 1994 SRY protein enhances transcription of fos-related antigen 1 promoter constructs. Proc Natl Acad Sci 91:4372–4376

Deckert J, Khalaf RA, Hwang SM, Zitomer RS 1999 Characterisation of the DNA binding and bending HMG domain of the yeast hypoxic repressor Rox1. Nucleic Acids Res 27:3518–3526
de Santa Barbara P, Bonneaud N, Boizet B et al 1998 Direct interaction of SRY-related protein SOX9 and steroidogenic factor 1 regulates transcription of the human anti-Mullerian hormone gene. Mol Cell Biol 18:6653–6665
de Santa Barbara P, Moniot B, Poulat F, Berta P 2000 Expression and subcellular localization of SF-1, SOX9, WT1, and AMH proteins during early human testicular development. Dev Dyn 217:293–298
Desclozeaux M, Poulat F, de Santa Barbara P et al 1998 Phosphorylation of an N-terminal motif enhances DNA-binding activity of the human SRY protein. J Biol Chem 273:7988–7995
Domenice S, Yumie Nishi M, Correia Billerbeck AE et al 1998 A novel missense mutation (S18N) in the 5′ non-HMG box region of the SRY gene in a patient with partial gonadal dysgenesis and his normal male relatives. Hum Genet 102:213–215
Dubin RA, Ostrer H 1994 Sry is a transcriptional activator. Mol Endocrinol 8:1182–1192
Grosschedl R, Giese K, Pagel J 1994 HMG domain proteins: architectural elements in the assembly of nucleoprotein structures. Trends Genet 10:94–100
Hanley NA, Hagan DM, Clement-Jones M et al 2000 SRY, SOX9 and DAX1 expression patterns during human sex determination and gonadal development. Mech Dev 91:403–477
Harley VR, Jackson DI, Hextall PJ et al 1992 DNA binding activity of recombinant SRY from normal males and XY females. Science 255:453–456
Harley VR, Lovell-Badge R, Goodfellow PN 1994 Definition of a consensus DNA binding site for SRY. Nucleic Acids Res 22:1500–1501
Harley VR, Lovell-Badge R, Goodfellow PN, Hextall PJ 1996 The HMG box of SRY is a calmodulin binding domain. FEBS Lett 391:24–28
Harley VR, Layfield S, McDowall S, Jans D 2001 STY nuclear import defects cause XY females. Submitted
Huang W, Zhou X, Lefebvre V, de Crombrugghe B 2000 Phosphorylation of SOX9 by cyclic AMP-dependent protein kinase A enhances SOX9's ability to transactivate a *Col2a1* chondrocyte-specific enhancer. Mol Cell Biol 20:4149–4158
Jager R, Harley V, Pfeiffer RA, Goodfellow PN, Scherer G 1992 A familial mutation in the testis-determining gene SRY shared by both sexes. Hum Genet 90:350–355
Kamachi Y, Cheah KS, Kondoh H 1999 Mechanism of regulatory target selection by the SOX high-mobility-group domain proteins as revealed by comparison of SOX1/2/3 and SOX9. Mol Cell Biol 19:107–120
Marshall OJ, Harley VR 2001 Identification of an interaction between SOX9 and HSP70. FEBS Letts 496:75–80
Maheswaran S, Englert C, Zheng G et al 1998 Inhibition of cellular proliferation by the Wilms tumor suppressor WT1 requires association with the inducible chaperone HSP70. Genes Dev 12:1108–1120
McDowall S, Argentaro A, Ranganathan S et al 1999 Functional and structural studies of wild type SOX9 and mutations causing campomelic dysplasia. J Biol Chem 274:24023–24030
Mertin S, McDowall SG, Harley VR 1999 The DNA-binding specificity of SOX9 and other SOX proteins. Nucl Acids Res 27:1359–1364
Morais da Silva S, Hacker A, Harley V, Goodfellow P, Swain A, Lovell-Badge R 1996 Sox9 expression during gonadal development implies a conserved role for the gene in testis differentiation in mammals and birds. Nat Genet 14:62–68
Nachtigal MW, Hirokawa Y, Enyeart-VanHouten DL, Flanagan JN, Hammer GD, Ingraham HA 1998 Wilms' tumor 1 and Dax-1 modulate the orphan nuclear receptor SF-1 in sex-specific gene expression. Cell 93:445–454
Pontiggia A, Rimini R, Harley VR, Goodfellow PN, Lovell-Badge R, Bianchi ME 1994 Sex-reversing mutations affect the architecture of SRY-DNA complexes. EMBO J 13:6115–6124

Poulat F, Barbara PS, Desclozeaux M et al 1997 The human testis-determining factor SRY binds a nuclear factor containing PDZ protein interaction domains. J Biol Chem 14:7167–7172
Preiss S, Argentaro A, Clayton A et al 2001 Compound effect of point mutations causing campomelic dysplasia/autosomal sex reversal upon SOX9 structure, nuclear transport, DNA binding and transcriptional activation. J Biol Chem 276:27864–27872
Schmitt-Ney M, Thiele H, Kaltwasser P, Bardoni B, Cisternino M, Scherer G 1995 Two novel SRY missense mutations reducing DNA binding identified in XY females and their mosaic fathers. Am J Hum Genet 56:862–869
Sudbeck P, Scherer G 1997 Two independent nuclear localization signals are present in the DNA-binding high-mobility group domains of SRY and SOX9. J Biol Chem 272:27848–27852
Sudbeck P, Schmitz ML, Baeuerle PA, Scherer G 1996 Sex reversal by loss of the C-terminal transactivation domain of human SOX9. Nat Genet 13:230–232
Sweitzer TD, Hanover JA 1996 Calmodulin activates nuclear protein import: a link between signal transduction and nuclear transport. Proc Natl Acad Sci USA 93:14574–14579
Tho SP, Zhang YY, Hines RS, Hansen KA, Khan I, McDonough PG 1998 Analysis of DNA binding activity of recombinant SRY protein from XY sex reversed female mutant for SRY (n=4). J Soc Gynec Invest Suppl 5:138A
Zeng YT, Ren ZR, Zhang ML, Huang Y, Zeng FY, Huang SZ 1993 A new de novo mutation (A113T) in the HMG box of the SRY gene leads to XY gonadal dysgenesis. J Med Genet 30:655–657

DISCUSSION

Behringer: Are there any post-translational modifications of SRY?

Harley: We haven't looked at this.

Schedl: Is tissue specificity important? I am asking because of this finding that SOX9 is initially cytoplasmic but then in the male it gets translocated to the nucleus. I wonder whether there would be any male-specific importins that could do that job?

Harley: Yes, there are about six isoforms of importin β, and one of them is a testicular isoform, so there are testis-specific importins.

Lovell-Badge: It doesn't have to be sex specific; it could just be a timing mechanism.

Harley: Or there could be a retention factor in the cytoplasm.

Koopman: Can CaM binding modulate other properties of the SRY protein, such as DNA-binding affinity?

Harley: It certainly competes for DNA binding. It induces an incredible conformational change in SRY upon binding, and consequently it could recruit other proteins to SRY.

Lovell-Badge: Are any of these interacting factors involved in degradation of proteins? It looks like SRY protein is fairly short lived within a cell, suggesting rapid turn over.

Harley: I don't know.

Scherer: I have a question about the HSP70 interaction that you mentioned. Would you argue that this is functionally relevant? If so, would you expect some amino acid substitutions to be found in that interacting region?

Harley: I don't know of any — but you might have some!

Scherer: I know of only a single one in SOX9, which we have found but which hasn't yet been published. It is outside the HMG domain, in the N-terminal region that is possibly a dimerization domain. This is the only one I know of.

Harley: HSP70 is quite promiscuous, so it may have quite a bit of tolerance. We need to do more work along these lines to see whether there is such a multiprotein complex that exists on the *AMH* promoter.

Scherer: Have you tested SOX8 and SOX10?

Harley: We almost did this, but we haven't quite completed it yet. Most of the HSP70 binding region is conserved.

Short: To go back to Peter Koopman's question in the general discussion about the box and the dangly bit of SRY, how do you now see this panning out in terms of important sites for action?

Harley: Apart from the PDZ *in vitro* binding domain, I don't know what the function of the C-terminus is. When we compare the DNA binding and bending activities, they don't seem to be altered by the presence of the C-terminal region. It doesn't seem to be any more or less stable as a pure protein. It could relate to RNA stability.

Koopman: What about the part N-terminal to the HMG box. Is there a mutation associated with XY gonadal dysgenesis?

Harley: Yes, this was S18N. However, we haven't been able to detect any DNA binding change.

Perhaps I could ask generally: have people done screens using this N-terminal region?

Poulat: A long time ago we screened with the N-terminal region of SRY, but we didn't see anything.

Concerted regulation of gonad differentiation by transcription factors and growth factors

Taiga Suzuki*†, Hirofumi Mizusaki*, Ken Kawabe*‡, Megumi Kasahara§, Hidefumi Yoshioka§|| and Ken-ichirou Morohashi*||[1]

**Department of Developmental Biology, National Institute for Basic Biology, Myodaiji 38, Okazaki 444-8585, †Institute for Virus Research, Kyoto University, Kawahara-cho, Shogoin, Sakyo-ku, Kyoto 606-8397, ‡Third Department of Internal Medicine, Faculty of Medicine, Kyushu University, Higashi-ku, Fukuoka 812-8582, §Department of Natural Sciences, Hyogo University of Teacher Education, Yashiro-cho, Katoh-gun 673-1494 and ||Core Research for Evolutional Science and Technology (CREST), Japan Science and Technology, Japan*

Abstract. It is well known that signals from growth factors regulate gene transcription thus initiating certain steps of cellular and tissue differentiation during development. In gonad differentiation several transcription factors have been identified as the genes underlying human diseases displaying gonadal defects and as the genes necessary for gonad differentiation as demonstrated by gene disruption studies. In addition, one of the growth factors, WNT4, is known to be involved in gonadal differentiation. However, it remains unclear which gene is directly downstream of the WNT4 signal. We have recently demonstrated that *Dax1* (*NR0B1*) gene transcription is significantly up-regulated by the presence of SF1 (NR5A1). Functional analysis showed that DAX1 acts as a repressor against SF1 through direct interaction between the repeated sequences at the N-terminus of DAX1 and a ligand-binding domain of SF1. Considering that the expressions of these factors during gonad differentiation show a sexually dimorphic pattern, it is likely that the *Dax1* gene transcription is up-regulated by WNT4 signal and thereafter DAX1 suppresses the genes downstream of SF1 such as *Amh* and steroidogenic genes in female gonads.

2002 The genetics and biology of sex determination. Wiley, Chichester (Novartis Foundation Symposium 244) p 68–78

Several transcription factors are involved in the process of gonadal differentiation. Some of these factors, such as *SRY* (Gubbay et al 1990), *WT1* (Call et al 1990,

[1]The chapter was presented at the symposium by Ken-ichirou Morohashi, to whom correspondence should be addressed.

Gessler et al 1990), *DAX1* (*NR0B1*) (Nuclear receptor nomenclature committee 1999, Zanaria et al 1994, Muscatelli et al 1994) and *SOX9* (Wanger et al 1994, Foster et al 1994) have been identified as the genes responsible for various human diseases that display structural and functional defects in tissues including the gonads. The essential functions of other transcription factors such as SF1 (also known as Ad4BP and NR5A1) (Luo et al 1994, Shinoda et al 1995, Sadovsky et al 1995), EMX2 (Miyamoto et al 1997), M33 (Katoh-Fukui et al 1998) and LHX9 (Birk et al 2000) were identified by the phenotypes of mice disrupted for these genes (Morohashi 1997, Swain & Lovell-Badge 1999). In addition, the expression profiles with respect to their distribution and sexual dimorphism strongly suggest that they have functional significance at an early stage of gonadal differentiation. However, it remains to be clarified how the above transcription factors regulate their target genes and how the genes encoding the transcription factors are regulated. When considering a gene regulatory cascade that supports differentiation of the gonadal tissues, approaches taking into account both of these aspects are important (Morohashi & Omura 1996). Consequently, we investigated the functions of SF1 and DAX1, and the regulation of the genes encoding these factors.

Functional correlation between SF1 and DAX1

SF1 and DAX1 are both classified as members of the nuclear receptor superfamily since they contain a ligand-binding domain (LBD) at the C-terminus (Fig. 1A). However, interestingly, DAX1 has unusual repeated sequences at the N-terminus instead of a zinc finger DNA binding domain, which, in addition to the LBD, is a structure common to all members of the nuclear receptor superfamily. With respect to their functions, it has been reported that SF1 acts as an activator for transcription of the steroidogenic genes and anti-Müllerian hormone (AMH; also known as Müllerian inhibiting substance, MIS), whereas DAX1 acts as a suppressor against SF1-mediated transcription (Crawford et al 1998). Although their transcriptional activities have been identified by reporter gene assays, the molecular mechanism by which DAX1 suppresses the transcription activity of SF1 remains to be elucidated.

To address the issue, we initially examined the function of the unusual repeated sequences at the N-terminus of DAX1. Expression constructs encoding N-terminal or C-terminal halves were constructed to determine the part of DAX1 implicated in the suppressive function (Fig. 1C). As indicated in Fig. 1B, transcription of the steroidogenic *Cyp11A* gene activated by the function of SF1 decreased following the addition of an expression vector for the whole molecule of DAX1. Interestingly, similar suppression was observed when the N-terminal repeated region but not the C-terminal LBD was expressed, indicating that a

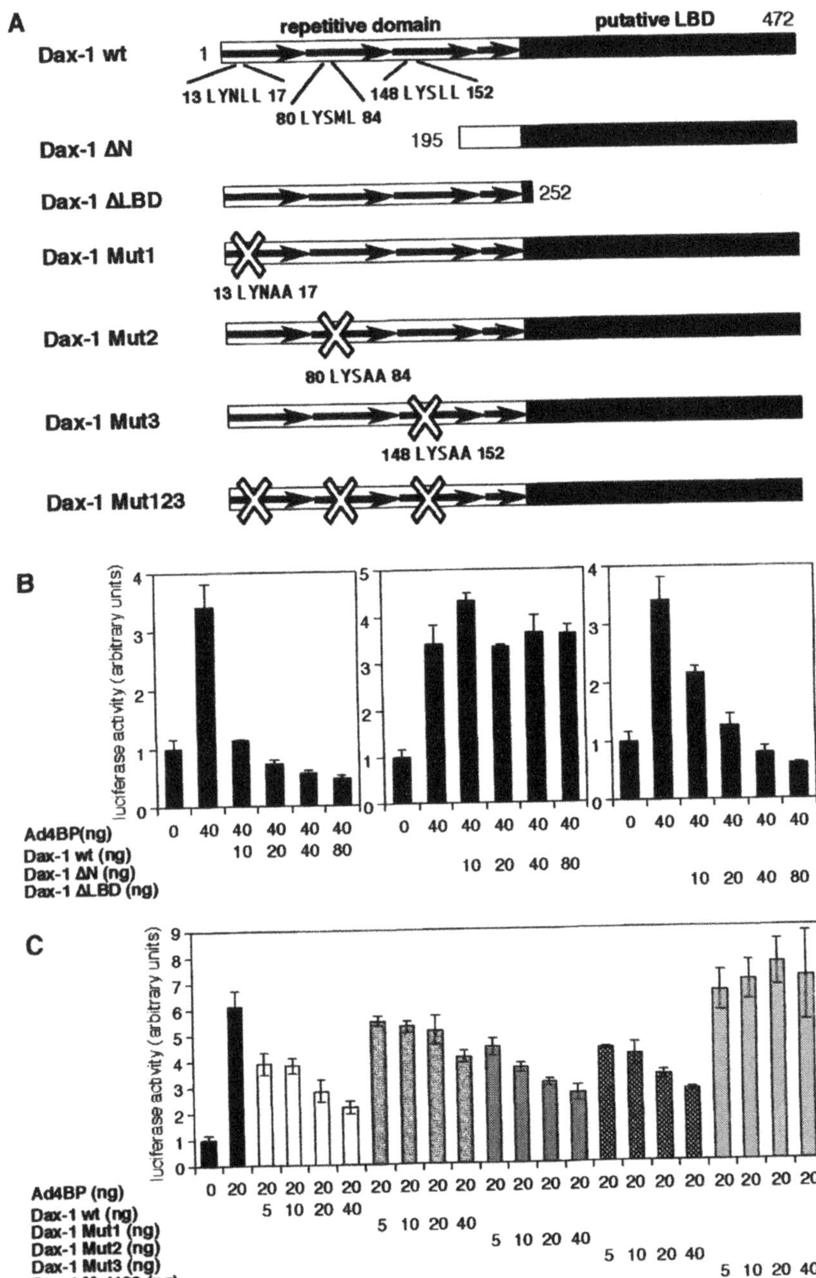
A
repetitive domain
putative LBD
472
Dax-1 wt
1
13 LYNLL 17
80 LYSML 84
148 LYSLL 152
Dax-1 ΔN
195
Dax-1 ΔLBD
252
Dax-1 Mut1
13 LYNAA 17
Dax-1 Mut2
80 LYSAA 84
Dax-1 Mut3
148 LYSAA 152
Dax-1 Mut123
B
luciferase activity (arbitrary units)
Ad4BP(ng) 0 40 40 40 40 40 0 40 40 40 40 40 0 40 40 40 40 40
Dax-1 wt (ng) 10 20 40 80
Dax-1 ΔN (ng) 10 20 40 80
Dax-1 ΔLBD (ng) 10 20 40 80
C
luciferase activity (arbitrary units)
Ad4BP (ng) 0 20 20 20 20 20 20 20 20 20 20 20 20 20 20 20 20 20 20 20 20 20
Dax-1 wt (ng) 5 10 20 40
Dax-1 Mut1 (ng) 5 10 20 40
Dax-1 Mut2 (ng) 5 10 20 40
Dax-1 Mut3 (ng) 5 10 20 40
Dax-1 Mut123 (ng) 5 10 20 40

sequence responsible for the suppressive function resides in the repeated region. Since amino acids responsible for crucial roles are generally conserved among animal species, the primary structures corresponding to the repeated region were compared among humans, mice, rats and pigs. Comparison revealed that a few stretches of amino acids are conserved in the repeated regions. Among them, we noted that one of the conserved sequences contains an LxxLL motif, because this motif was originally identified as a sequence in co-activators of the p160 protein family responsible for interaction with nuclear receptors (Torchia et al 1997, Heery et al 1997, Voegel et al 1998, Xu et al 1999). To clarify the function of this motif, we examined whether DAX1 containing amino acid substitutions in this motif continued to act as the transcriptional suppressor. As expected, the mutation impaired the suppressor activity when it was introduced in the three motifs simultaneously (Fig. 1C). Since DAX1 carrying a single motif mutation retained a significant (but not full) suppressive activity, the three motifs seem to be complementary to the function. Binding through the LxxLL motifs with SF1 was investigated with yeast and mammalian two-hybrid assays, and *in vitro* assay. Taken together, the results clearly indicated that all three LxxLL motifs interact with the C-terminal half of SF1, and thereby DAX1 acts as the suppressor.

We further characterized the binding specificity of the DAX1 LxxLL motifs by comparing it with those of co-activators. Examination of their preference of interaction revealed that the motif in DAX1 is quite distinct from those in co-activators. Therefore, in the next step, we determined the part of the motif that is responsible for such a distinct preference for interaction. Since mutually distinct amino acids are located between the leucines and those surrounding the motif, we interchanged these amino acids and analysed the interactions. Binding preference was predominantly affected by the substitution of amino acids between the leucines, while alteration of the surrounding amino acids yielded modest effects.

The present study revealed that direct interaction through the LxxLL motif of DAX1 results in the suppression of SF1 mediated transcription. It was reported recently that DAX1 interacts directly with the oestrogen receptor (ER) to inhibit

---

FIG. 1. Identification of the suppressor domain of DAX1. Structures of mouse DAX1 (DAX1 wt) and its mutated forms are schematically presented (A). Arrows at the N-terminal indicate repetitive sequences, and amino acids corresponding to the three LxxLL motifs are shown. The suppressor functions of these forms of DAX1 were evaluated from reduction of SF1 (Ad4BP)-mediated transcription of the steroidogenic cytochrome P450 gene (B and C). As indicated in B, the whole molecule (DAX1 wt) and N-terminal repeated region of DAX1 (DAX1 ΔLBD) act as the suppressor, whereas the activity was not detectable in the C-terminal LBD region (DAX1 ΔN). The suppressive activity was slightly decreased with mutation of one of the three LxxL motifs (DAX1 Mut1, DAX1 Mut2, and DAX1 Mut3), while DAX1 with mutations at all the motifs (DAX1 Mut123) did not act as a suppressor (C).

the ligand-dependent transcription (Zhang et al 2000). Similarly to the interaction with SF1, the LxxLL motif is responsible for the interaction with the AF2 C-terminal of ER. This is consistent with the reported observations that AF2 and LxxLL motif are involved in the interaction between nuclear receptors and co-activators. Taken together, it is likely that the transcription activities driven by a certain class of nuclear receptors are modulated by competitive interactions between DAX1 and co-activators. However, as described above, it should be noted that the amino acids located between the leucines determine in part the preference of the interaction. In addition, the transcription by SF1 is regulated under stimulation through PKA (Morohashi et al 1993), PKC (Leers-Sucheta et al 1997), and MAPK (Hammer et al 1999) activation, probably without reduction of the amount of DAX1. Therefore, the fine transcriptional regulation by SF1 could be explained as a concerted mechanism through multiple components of transcriptional regulators.

Regulation of *Dax1* gene transcription by SF1 and WNT

As reported previously, SF1 is an indispensable component for *Dax1* gene transcription (Yu et al 1998, Kawabe et al 1999). In fact, multiple binding sites recognized by SF1 in the upstream region of the *Dax1* gene are necessary for transcriptional activation. The *in vitro* observation using reporter gene assays was confirmed subsequently by an *in vivo* study using *SF1* gene disrupted mice, which lacked DAX1 expression in the developing genital ridge. Although these results strongly indicated that *SF1* gene is genetically located upstream from the *Dax1* gene, their expression profiles in terms of distribution and sexual dimorphism do not necessarily agree with our findings (Ikeda et al 2001). In this regard, a recent gene disruption study implicated *Wnt4* in gonadal sex differentiation (Vainio et al 1999). Normally, the steroidogenic *3β-HSD* and *Amh* genes are expressed in the developing fetal gonads of males but not females. Interestingly, however, the expression was detected in the fetal ovary of the gene-disrupted mice, suggesting that the WNT4 represses *3β-HSD* and *Amh* gene transcription in the fetal ovaries of the wild-type mouse. Considering that some of the WNT signals activate downstream gene transcription through stabilization of *β* catenin (Kühl et al 2000), it is unlikely that the signal represses the *3β-HSD* and *Amh* gene transcription.

To explain this, we hypothesized that WNT4 expressed in the developing gonad up-regulates a suppressor molecule and thereby down-regulates *3β-HSD* and *Amh* gene transcription. Since transcription of both genes is regulated in a positive fashion by SF1 (Leers-Sucheta et al 1997, Santa Barbara et al 1998), it was reasonable to assume that DAX1 plays a role as the suppressor. To confirm this assumption, we examined whether *β* catenin activates *Dax1* gene transcription.

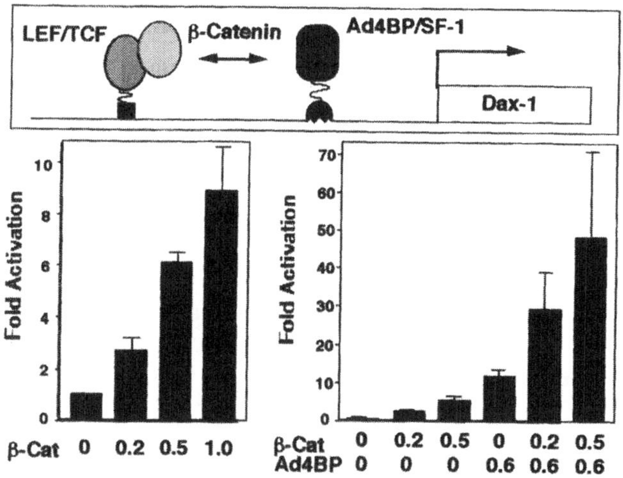

FIG. 2. Regulation of *Dax1* gene transcription by SF1 (Ad4BP) and β catenin. The *Dax1* promoter activity was analysed by luciferase reporter gene expression in cultured cells. Following the transfection of the expression vector for a stabilized form of β catenin (β-Cat) with the *Dax1* reporter gene construct, the cell extracts were prepared and subsequently subjected to the analysis of luciferase activity. The *Dax1* promoter was activated by the addition of the β catenin expression vector in a dose-dependent manner (0 to 1 μg) (left panel). Synergistic activation of *Dax1* gene transcription was observed when the expression vectors for β catenin and SF1 (Ad4BP) were transfected simultaneously (right panel). Data are mean ± SEM of three experiments.

As indicated in Fig. 2, *Dax1* gene transcription was activated in the presence of β catenin in a dose-dependent manner. Interestingly, the action of β catenin is further up-regulated in the presence of SF1, indicating that the two factors, β catenin and SF1, synergistically activate the *Dax1* gene transcription. We searched the binding sequence of the HMG box containing the transcription factor LEF/TCF, which heterodimerizes with β catenin, in the upstream region. In fact, several candidate sequences were identified and found capable of binding LEF/TCF. Therefore, we then investigated whether the LEF/TCF binding sites on the *Dax1* gene promoter are functional by disrupting the sequences. Interestingly, the activation by β catenin did not disappear completely, implicating another factor in the transcription activation by the WNT signal. In contrast, mutation at the sequences recognized by SF1 completely abolished the synergistic effects as well as activation by SF1 alone. Taken together, these results strongly suggest that β catenin interacts with SF1 as well as LEF/TCF, which in part leads to synergistic activation of *Dax1* gene transcription.

In parallel studies, we used yeast two-hybrid screening to isolate molecules that interact with transcription factors expressed in the developing gonads. When a

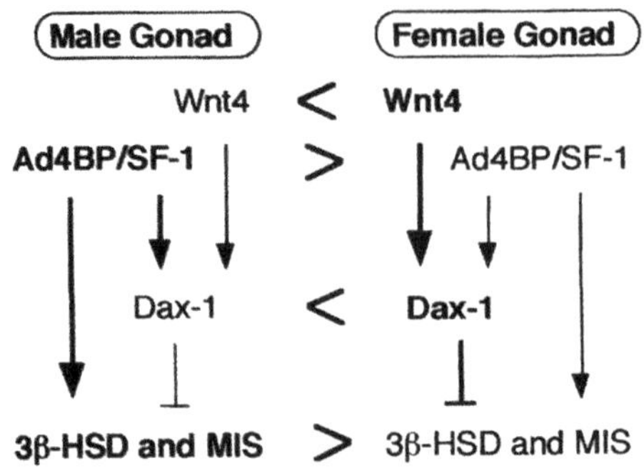

FIG. 3. Transcriptional regulation of *Dax1* gene at early stages of gonadal differentiation. In the sexually differentiating gonad and mesonephros, WNT4 is expressed in females more than in the male, whereas the amount of SF1 (Ad4BP) in the male gonad is higher than the female. Since *Dax1* gene transcription is regulated synergistically by SF1 (Ad4BP) and β catenin, *Dax1* is more abundantly transcribed in the female gonad than in the male. Consequently, the transcription of downstream *3β-HSD* and *Amb/Mis* genes is largely activated in the male but only slightly in the female gonad.

hinge region between the zinc finger DBD and LBD of SF1 was used as a bait plasmid, clones encoding β catenin were isolated. Although fine mapping of the regions implicated into the interaction is under investigation, the interaction between the two molecules strongly supported the observation described above. Recent studies have so far reported that members of the nuclear receptor superfamily, retinoic acid receptor (RAR) (Easwaran et al 1999) and androgen receptor (AR) (Truica et al 2000), interact with β catenin directly and result in down-regulation and up-regulation, respectively, of the transcription driven by the β cateninLEF/TCF complex. Although endogenous downstream genes have not yet been identified in both cases, the present study apparently indicated that SF1 and WNT signals converge into *Dax1* gene transcription.

The mechanisms of *Dax1* gene regulation governing its sexually dimorphic characteristics are summarized in Fig. 3. As described previously, it is difficult to explain the whole regulatory mechanism of the *Dax1* gene transcription by SF1 alone. For instance, SF1 is expressed in the male developing gonads more abundantly than in the female. Nevertheless, the amount of Dax1 in the female developing gonad is higher than that in the male gonad. In the case of WNT4 expression in the developing gonads and mesonephros, *in situ* examination revealed that the amount expressed in the female tissues is higher than in the male. With respect to the distribution of *DAX1*, strong signals were detected in the gonadal regions facing the mesonephros although such an expression domain was not observed in the case of SF1. In such inconsistent distribution, it is interesting to note that the expression of WNT4 in the gonads was more abundant at the region proximal rather than that distal to the mesonephros. Therefore, to understand the mechanism underlying DAX1 expression, we propose that SF1 plays basal and fundamental roles and that the WNT4 signal modulates the transcription mediated by SF1. Although the regulation above is likely to function in the sexually differentiating gonads of both sexes, the mechanisms of other transcription factors such as SOX9 and EMX2 are not fully understood. In addition, it should be noted that other cell growth factors as well as other forms of WNT molecules are expressed in the developing gonads and mesonephros. Further studies of the functional relationship between growth factors and transcription factors should identify the finely tuned, sophisticated mechanisms underlying the sex differentiation of the gonads.

Acknowledgements

This work was supported by CREST of Japan Science and Technology, and by a Grant-in-Aid for Scientific Research from the Ministry of Education, Sports, and Culture of Japan.

References

Birk OS, Casiano DE, Wassif CA et al 2000 The LIM homeobox gene *Lhx9* is essential for mouse gonad formation. Nature 403:909–913

Call KM, Glaser T, Ito CY et al 1990 Isolation and characterization of a zinc finger polypeptide gene at the human chromosome 11 Wilms' tumor locus. Cell 60:509–520

Crawford PA, Dorn C, Sadovsky Y, Milbrandt J 1998 Nuclear receptor DAX-1 recruits nuclear receptor corepressor N-CoR to steroidogenic factor 1. Mol Cell Biol 18:2949–2956

De Santa Barbara P, Bonneaud N, Boizet B et al 1998 Direct interaction of SRY-related protein SOX9 and steroidogenic factor 1 regulates transcription of the human anti-Müllerian hormone gene. Mol Cell Biol 18:6653–6665

Easwaran V, Pishvaian M, Salimuddin, Byers S 1999 Cross-regulation of β-catenin-LEF/TCF and retinoid signaling pathways. Curr Biol 9:1415–1418

Foster JW, Dominguez-Steglich MA, Guidi S et al 1994 Campomelic dysplasia and autosomal sex reversal caused by mutations in an SRY-related gene. Nature 372:525–530

Gessler M, Poustka A, Cavenee W, Neve RL, Orkin SH, Bruns GA 1990 Homozygous deletion in Wilms tumors of a zinc-finger gene identified by chromosome jumping. Nature 343:774–778

Gubbay J, Collingnon J, Koopman P et al 1990 A gene mapping to the sex-determining region of the mouse Y chromosome is a member of a novel family of embryonically expressed genes. Nature 346:245–250

Hammer GD, Krylova I, Zhang Y et al 1999 Phosphorylation of the nuclear receptor SF-1 modulates cofactor recruitment: integration of hormone signaling in reproduction and stress. Mol Cell 3:521–526

Heery D, Kalkhoven E, Hoare S, Parker MG 1997 A signature motif in transcriptional co-activators mediates binding to nuclear receptors. Nature 387:733–736

Ikeda Y, Takeda Y, Shikayama T, Mukai T, Hisano S, Morohashi K 2001 Comparative localization of Dax-1 and Ad4BP/SF-1 during development of the hypothalamic-pituitary-gonadal axis implies their closely related and distinct functions. Dev Dyn 220:363–376

Katoh-Fukui Y, Tsuchiya R, Shiroishi T et al 1998 Male-to-female sex reversal in M33 mutant mice. Nature 393:688–692

Kawabe K, Shikayama T, Tsuboi H et al 1999 *Dax-1* as one of the target genes of Ad4BP/SF-1. Mol Endocrinol 13:1267–1284

Kühl M, Sheldahl LC, Park M, Miller JR, Moon RT 2000 The Wnt/Ca2+ pathway: a new vertebrate Wnt signaling pathway takes shape. Trends Genet 16:279–283

Leers-Sucheta S, Morohashi K, Mason JI, Melner MH 1997 Synergistic activation of the human type II 3beta-hydroxysteroid dehydrogenase/delta5-delta4 isomerase promoter by the transcription factor steroidogenic factor-1/adrenal 4-binding protein and phorbol ester. J Biol Chem 272:7960–7967

Luo X, Ikeda Y, Parker KL 1994 A cell-specific nuclear receptor is essential for adrenal and gonadal development and sexual differentiation. Cell 77:481–490

Miyamoto N, Yoshida M, Kuratani S et al 1997 Defects of urogenital development in mice lacking Emx2. Development 124:1653–1664

Morohashi K 1997 The ontogenesis of the steroidogenic tissues. Genes Cells 2:95–106

Morohashi K, Omura T 1996 Ad4BP/SF-1, a transcription factor essential for the transcription of steroidogenic cytochrome P450 genes and for the establishment of the reproductive function. FASEB J 10:1569–1577

Morohashi K, Zanger UM, Honda S, Hara M, Waterman MR, Omura T 1993 Activation of CYP11A and CYP11B gene promoters by the steroidogenic cell-specific transcription factor, Ad4BP. Mol Endocrinol 7:1196–1204

Muscatelli F, Strom TM, Walker AP et al 1994 Mutations in the DAX-1 gene give rise to both X-linked adrenal hypoplasia congenita and hypogonadotropic hypogonadism. Nature 372:672–676

Nuclear receptor nomenclature committee 1999 A unified nomenclature system for the nuclear receptor superfamily. Cell 97:161–163

Sadovsky Y, Crawford PA, Woodson KG et al 1995 Mice deficient in the orphan receptor steroidogenic factor 1 lack adrenal glands and gonads but express P450 side-chain-cleavage enzyme in the placenta and have normal embryonic serum levels of corticosteroids. Proc Natl Acad Sci USA 92:10939–10943

Shinoda K, Lei H, Yoshii H et al 1995 Developmental defects of the ventromedial hypothalamic nucleus and pituitary gonadotroph in the Ftz-F1 disrupted mice. Dev Dyn 204:22–29

Swain A, Lovell-Badge R 1999 Mammalian sex determination: a molecular drama. Genes Dev 13:755–767

Torchia J, Rose DW, Inosroza J et al 1997 The transcriptional co-activator p/CIP binds CBP and mediates nuclear-receptor function. Nature 387:677–684

Truica CI, Byers S, Gelmann EP 2000 *β*-Catenin affects androgen receptor transcription activity and ligand specificity. Cancer Res 60:4709–4813

Vainio S, Heikkila M, Kispert A, Chin N, McMahon AP 1999 Female development in mammals is regulated by Wnt-4 signalling. Nature 397:405–409

Voegel JJ, Heine MJS, Tini M, Vivat V, Chambon P, Gronemeyer H 1998 The coactivator TIF2 contains three nuclear receptor-binding motifs and mediates transactivation through CBP binding-dependent and-independent pathways. EMBO J 17:507–519

Wanger T, Wirth J, Meyer J et al 1994 Autosomal sex reversal and campomelic dysplasia are caused by mutations in and around the SRY-related gene SOX9. Cell 79:1111–1120

Xu L, Glass CK, Rosenfeld MG 1999 Coactivator and corepressor complexes in nuclear receptor function. Curr Opin Genet Dev 9:140–147

Yu R, Ito M, Jameson JL 1998 The murine Dax-1 promoter is stimulated by SF-1 (steroidogenic factor-1) and inhibited by COUP-TF (chicken ovalbumin upstream promoter-transcription factor) via a composite nuclear receptor-regulatory element. Mol Endocrinol 12:1010–1022

Zanaria E, Muscatelli F, Bardoni B et al 1994 An unusual member of the nuclear hormone receptor superfamily responsible for X-linked adrenal hypoplasia congenita. Nature 372:635–641

Zhang H, Thomsen JS, Johansson L, Gustafsson JA, Treuter E 2000 DAX-1functions as an LXXLL-containing corepressor for activated estrogen receptors. J Biol Chem 275: 39855–39859

DISCUSSION

Schedl: Have you found any *in vivo* interaction between β catenin and SF1? This yeast two-hybrid system is nice, but you need to confirm it.

Morohashi: We haven't tried doing that yet, but we would like to.

Zarkower: What cell types did you do your co-transfections in, and does the cell type matter?

Morohashi: We used 293 cells. We haven't tried any other cell lines.

Sinclair: The unusual repeat binding region at the N-terminus of DAX1 appears to be absent in chickens and alligators. How do you explain how it might function without this binding region?

Morohashi: Chicken DAX1 has a single repeat containing the LxxLL motif.

Vilain: Along the same lines, all missense mutations in human *DAX1* that result in adrenal hypoplasia congenita are in the putative ligand-binding domain. None of them are in the N-terminal domain and the LxxLL motif. Some have been studied *in vitro*, and they modify the inhibition of SF1 transactivation. If you introduce those mutations in the ligand-binding domain, it will disrupt the normal inhibition of SF1 by DAX1. Those mutations are in the C-terminal domain. This is the same effect that you observe in your experiments when you put in the LxxLL domain. Do you think we are looking at the same thing, or two separate molecular mechanisms?

Morohashi: We haven't yet analysed the function of the C-terminal ligand-binding domain, so I can't really answer the question. But it was reported that this domain is required for interaction with the corepressor NcoR and the binding leads to suppression of Ad4BP/SF1-mediated transcription, while our results indicate that N-terminal half acts as a suppressor by itself through its LxxLL motif. Then, we suppose that there are two pathways for the suppression. One is through the ligand-binding domain and NcoR, and the other is through the N-terminal repeated region, both of which require interaction through the LxxLL motif.

Goodfellow: Are you saying that DAX1 is ligand dependent?

Morohashi: There are no data to show that.

Vilain: My point was that in humans, the mutation that results in the abolition of the inhibition of SF1 by DAX1 are all in the ligand-binding domain. None of them are in the N-terminal domain. Ken Morohashi's experiments showed that the mutations resulting in the same effect are in this LxxLL motif. How do we reconcile the fact that we observe the same *in vitro* effect with mutations that are very far away from each other?

Wilkins: A mutation in the C-terminal part could interact with the LxxLL motif.

Vilain: It could, but there is no structural information to indicate that this might actually happen.

Lovell-Badge: You are less likely to find point mutations in a repeated motif. The fact that they aren't found in human patients doesn't mean they don't occur.

Poulat: You said that β catenin is able to activate DAX1. Have you tried to see whether there is nuclear localization of β catenin in the genital region specifically in the female? Have you seen some free β catenin in the nucleus or in the cytoplasm?

Morohashi: This is a key issue, but we haven't addressed it.

Capel: We have done this, and we see much more β catenin in the male than in the female. It is highly expressed in the coelomic epithelium from the earliest stages, and there is not much in the interior part of the gonad. The domain is extended from the coelomic surface further into the gonad in the male than it is in the female. Of course, it is the nuclear localization of β catenin that we need to worry about. The problem with antibodies is that this is hard to determine, especially if there is a lot of cytoplasmic β catenin around, which there seems to be. I don't know whether the reason for the presence of this cytoplasmic β catenin is because those cells are epithelializing and it has something to do with junctions. Alternatively it could have something to do with the cytoskeleton and establishing connections between the epithelial cells of the gonad, or possibly some signalling function. It is where SF1 is, so it is easy to imagine that it has an interaction with SF1. It is harder to imagine that it has a role interacting with WNT4, although I know that this is the classic pathway. The domain of expression doesn't seem to be the same. Perhaps it is another WNT.

Poulat: Have you tried to correlate this β catenin expression with cell proliferation?

Capel: It is in the cells that are proliferating in the male. However, it is also in the coelomic epithelium in the female, but at lower levels. There's a difference in levels of around sixfold.

Vilain: One thing that remains unknown is the identity of the pathway that WNT4 operates through. We think it should go through the β catenin/TCF pathway as it does in thymocytes, but this still has to be proven in the testicular cells (Sertoli or Leydig).

General discussion II

Koopman: We have had some discussion of SRY and how it might work, but we haven't really resolved this. In my opinion there are a couple of simple experiments that need to be done that would save a lot of theorizing and arguing.

There seem to be conflicting views of how SRY functions. One view says that it functions as an HMG box and that the rest of the protein doesn't matter that much. The other view is that SRY functions by virtue of having an HMG box that does a number of things, such as DNA binding, DNA bending, calmodulin binding, nuclear localization and phosphorylation, but that a C-terminal domain of some sort might be necessary for building up a transcriptional complex that allows SRY to function. This assumes, or course, that SRY is a DNA-binding transcription factor that is somehow triggering a transcriptional pathway. This second, more complex view seems to suggest SRY is acting by doing more than just getting in the way of something else.

Goodfellow: If you really want to start at the beginning, there are three key questions about SRY. Does it work by binding to DNA? Does it work by binding to RNA? Does it work by binding to other proteins? It could also be a mixture of these three. These are the formal possibilities. I don't know anyone who has looked seriously at RNA, for example. There are many examples where RNA and DNA binding may contribute to the activity of a molecule.

Koopman: There is certainly evidence that SRY may also bind to DNA *in vitro*. There is good evidence that some SOX proteins can bind to DNA *in vivo*. This hasn't been shown for SRY, but SRY is a SOX protein.

Goodfellow: Proteins which are known to have DNA-binding activity can bind to RNA. There's no evidence to date that SRY's function is through DNA binding.

Koopman: Are you making things 'unnecessarily' complicated, or are you being sensibly critical?

Goodfellow: I am being 'unnecessarily' complicated, because unfortunately simplicity isn't always the answer. We can go through the two hypotheses that you put up and they can both be interpreted with respect to those three components.

Harley: Can I just comment from the DAX1 example, where it recognizes RNA structures. The HMG box also has those properties. Given the propensity of RNA to form structures, I think Peter Goodfellow is being constructively critical.

Koopman: Is it realistic to propose that SRY doesn't bind to any nucleic acid at all?

Capel: It can interact with other proteins as well as nucleic acids. They are not mutually exclusive.

Goodfellow: The subhypothesis concerns whether the DNA binding is specific or not. I agree, it is likely that SRY has nucleic acid binding activity, otherwise the box-like structure wouldn't have been maintained. But this doesn't imply sequence-specific binding. It could bind to DNA and that binding could then cause the complex that reduces the concentration of another protein which will have bound somewhere else, giving you a specific activity.

Koopman: Once again, I would say that SRY is just another SOX protein. SOX proteins can bind *in vitro* to just about any variation of a consensus heptameric sequence.

Goodfellow: The big problem you have is that you are essentially trapped into the hypothesis that says that sequence specificity, if it exists for SRY, is the same for SRY and SOX9. You can switch the boxes and it still works. If sequence specificity exists, then other proteins are providing the sequence specificity. One experiment is to break out of the SOX9/SRY loop, because we are all a bit suspicious that these are related. We need to go for a SOX gene which we know is not likely to be involved in sex determination.

Wilkins: I think those results would still be interpretable in different ways. The problem here is that mammals are not *Drosophila*. If this was *Drosophila*, you could use a genetic approach, taking a weak allele and selecting for enhancers or suppressors, and then identify those gene products. It seems to me that there are some weak alleles in SRY that exist naturally. It is a pity Eva Eicher isn't here to discuss this, because perhaps something like that could be done to take this forward.

Goodfellow: Eva has been trying to do this for the last 20 years.

Wilkins: I'm aware of that, but the problem with the *in vitro* approach is that you can never be certain that what you find is applicable *in vivo*, no matter how good it looks. There is a real dilemma here.

Goodfellow: If you put the SOX27 box into SRY, and it still gave sex determination, you would really start to struggle with the sequence-specific component.

Koopman: I would argue that you could put in any old HMG box, but some other part of the SRY protein might be required for stability of the binding, for example.

Scherer: I would suggest the LEF1 HMG box.

Behringer: Isn't the central criticism of this field that for the last 10 years there has been no target identified for SRY? This would lead us forward.

Capel: This is because either there is no DNA target or we just haven't found it. Has anyone done an extensive screen for RNA targets or protein targets?

Behringer: I bet there are about a dozen people in this room who have done two-hybrid screens that haven't worked.

Capel: But there are plenty of examples where they don't work because you are missing a third component, either a nucleic acid or another protein.

Greenfield: Peter Koopman, I thought that one of the putative functions that you attributed earlier to the C-terminal domain was to promote protein stability.

Koopman: Not protein stability; stability of SRY binding to its target.

Greenfield: So it is definitely a proper function. It must be species-specific then.

Koopman: I don't think so. That is a sequence-oriented view; domains that differ in primary sequence between species could conceivably carry out similar stabilizing functions.

Scherer: Is it possible to do immunoprecipitation on genital ridges?

Lovell-Badge: We can try, yes. We are going to try doing immunoprecipitation of chromatin. We need quite a lot of genital ridges for this, but it is possible.

Poulat: Did you measure the nuclear localization of SRY in your Myc/Sry transgenic? Have you seen variation in the expression of the protein?

Lovell-Badge: It has always been nuclear.

Poulat: How long is the protein present for?

Lovell-Badge: In the developing genital ridge as a whole, it roughly correlates with the RNA expression overall, so we are talking about 36 h maximum. However, within an individual cell it is almost certainly much less than this. We see very few cells that are double-positive for SRY and SOX9. My feeling is that it is probably just a few hours per cell.

Behringer: Peter Koopman, in the transgenics that you and Jo Bowles made, you were getting ectopic expression of SRY. But the mice were normal. Does that suggest anything about the role of SRY? Are these other tissues expressing SRY doing anything?

Koopman: It is difficult to know. We started off making transgenic mice by taking the 14 kb large genomic fragment that we used in Robin Lovell-Badge's lab, and cutting it down in a nested 5′ deletion series to try to look for regulatory elements. The bottom line is that we didn't find any. In all cases there was expression in the genital ridge, but there was also expression in all other tissues we looked at. This didn't appear to have any detrimental effects on embryonic development, but the levels of expression in the non-gonadal tissues were about one fifth what they were in the genital ridge. In the genital ridge, as you know, they are already pretty low.

Swain: For what it is worth, we made transgenics with *Sry* driven by a ubiquitous promoter, and there was no effect.

Goodfellow: Did you get protein expression?

Swain: We didn't look at protein; we didn't have the antibodies.

Goodfellow: The problem with that experiment is that if it is really deleterious, then it would select for animals not able to make protein. Perhaps a worthwhile experiment is to overexpress *Sry* in a mouse.

Behringer: We are making a mouse which will conditionally express *Sry* upon Cre recombinase expression.

Scherer: I don't see where you are heading with the mouse *Sry*. The human *SRY* is expressed in many tissues anyway.

Capel: Does anyone know about translation in those tissues in humans?

Goodfellow: Basically the levels of expression are, 'let's push PCR until we start seeing bands'.

Poulat: Also the protein is cytoplasmic in most cases, so it will be inactive.

Wilkins: I think there's an interesting question about *Sry* levels: why are they low during the critical time? I would suggest that this probably means that higher levels will be deleterious in some way. It would be interesting to know whether if we raised *Sry* expression during the critical stage this would cause defects.

Burgoyne: *Sry* is grossly overexpressed in a number of transgenic lines and they are fine.

Goodfellow: Does it change the timing? You have a double whammy: concentration versus time. If you double the concentration, you may have changed the timing also.

Swain: I have only looked at RNA expression, and it is a lot higher than the endogenous level. I haven't looked at timing.

Goodfellow: I'd like to go back to the idea I had of trying to reverse-select *Sry*. Essentially, we are still hung up on exactly the point that is being made over and over again, that we can't identify the target. This is an experiment that Eva Eicher has been trying to do by using hypermorphic variants of *Sry*. Part of me feels that there should be a genetic solution: we should be able to set up a screen where we can get complementation in the recipient of *Sry*, whatever that is, to correct a defect in *Sry*.

Lovell-Badge: It is possible to do suppressor–enhancer screens in mice. We are doing one specifically to look for other genes involved in sex determination.

Wilkins: Doesn't that take a huge number of animals?

Lovell-Badge: It is not so bad. It is not the same as Eva's experiments with the QTL analysis. We reckon we'll need 600 cages of mice, maximum.

Goodfellow: In those screens you get problems with the fact that you need sex in order to reproduce the mice. I know there was some discussion in zebrafish. Has anyone gone back to some of those large screens and thought about sex?

Behringer: I think they were only looked at much earlier than the sex differentiation stage.

Goodfellow: Now some of those screens are being looked at much later. Last time we looked at this no one knew anything about sex determination in the zebrafish, so the experiment died.

Behringer: They have a *Sox9* gene.

Koopman: It is still a bit of a mess: no one knows what's going on in zebrafish sex determination.

Greenfield: The large-scale ENU mouse screen in Munich has systematically searched for XY sex reversal. The last time I spoke to them they had screened thousands of DNA samples and found nothing.

Harley: Robin, is your human *SRY* that is sex-reversing in mice a robust starting point from which to test weak alleles of human mutations in enhancer–suppressor screens?

Lovell-Badge: We have several lines. We have one which always sex reverses and one which only sex reverses occasionally. One could use this latter line.

Behringer: Peter Goodfellow, in the context of your screen I think *odsex* might be picked up, because it rescues *Sry* deficiency. There may be background in the system, like all screens.

Goodfellow: But the problem with the screen is that what you really want to do is make it conditional on the presence of *Sry*. This is because your screen may pick up things that are active downstream, which are independent of *Sry*. It requires quite a lot of thought to make sure that it is conditional on the presence of *Sry*.

McLaren: What is the difference between an *Sry* that works and an *Sry* from a different species (or subspecies) that doesn't?

Goodfellow: That's why I said that Eva Eicher's screen is similar to what we have been talking about: effectively that experiment is treating *Sry* like a QTL. It still comes down to taking an *Sry* that doesn't work quite as well as you would hope.

McLaren: What is known about the sequence of the *SRY* that doesn't work, both within the box and outside the box?

Lovell-Badge: In the case of *Mus domesticus poschiavinus* it is probably just a lower level of expression. It is a weak allele in terms of its expression.

McLaren: But why is it expressed weakly? Is there a regulatory difference?

Lovell-Badge: I assume that there is, but it hasn't been characterized.

Zarkower: It is known that there is no apparent correlation between coding sequences in sensitive and non-sensitive alleles of *SRY*, and thus the difference is thought to be regulatory (Albrecht & Eicher 1997).

Lovell-Badge: Yes, it is not within the coding region. We had some evidence a long time ago that there was something different 3′ to the gene in the *poschiavinus* Y compared to the 129 Y chromosome, but this was within the inverted repeat region and difficult to analyse.

Harley: I'd like to make a general point about QTLs. The large number of XY females that have intact SRY could just have very subtle polymorphisms in any number of the testis genes that we have been considering.

Goodfellow: Clearly, within the next two or three years, every ORF is going to be PCRable, and there are several institutions who are trying to set up those primer pairs for exactly this experiment—looking for the mutations associated with disease. This would be a clear case where that would be worthwhile.

Vilain: Except you may find many hundreds of small variations in many genes. It is going to be difficult to interpret.

Goodfellow: You are looking for *de novo* mutations, so you know there is certain background number of known mutations that are going to change the coding sequence. In this experiment you would be looking for one of these.

Short: Are we now in fact discussing the molecular basis for Haldane's law? He said that in interspecific hybrids, it is the heterogametic sex that is absent, rare or sterile. Are we now thinking that we can explain that on the basis of interspecific variations in *SRY*?

Wilkins: Surely Haldane's law applies to much broader groups than these that just use *SRY*?

Short: Yes, Haldane's law also applies to avian hybrids, but I'm just taking this narrow case: Anne McLaren raised the question of what happens when the *SRY* of one species is expressed in another. I was taking off from there. It is amazing that one can skew the primary sex ratio so much in interspecific hybrids. There should be an explanation for this.

Charlesworth: There's a great deal of work in the evolutionary genetics literature, mostly in *Drosophila*, which largely explains it in terms of recessivity of deleterious effects of the genes in hybrids. If you have a gene coming in on the X chromosome, it is fully expressed in the heterogametic. It can interact with heterozygous genes on the autosome and this gives a sterile hybrid male. There are some cases where Y-linked factors are implicated, but overall it is much broader than simply the effects of Y-linked genes.

Short: My understanding was that people thought it was an X-linked gene effect.

Goodfellow: Why not take *Sry* and overexpress it in *Drosophila*, zebrafish and chickens, to see whether anything happens?

Poulat: We have tried this. In *Drosophila* it doesn't do anything. It is interesting to compare this with the effect of overexpressing *Sox9*. Someone in the lab expressed *Sox9* in the eyes and this eliminated the eyes. He has tried the same with *Sry* and it has no effect of it, wherever it is expressed.

Wilkins: This could be because its potential partners were not expressed, so we can't conclude much from this.

Goodfellow: But if you subscribe to the view that *Sry* captured something from somewhere, then you might argue the case that is also capable of capturing it if you put it in the eye.

Poulat: We have also expressed a truncated form of *Sox9*, and in this case the phenotype was the same. It is likely to be an effect of the HMG domain.

Zarkower: We have also tried using *Dmrt1* in *C. elegans*. There was no obvious phenotype.

Goodfellow: The widely held hypothesis is that *Sry* captured sex determination. So why don't we go out and capture sex determination again? Is there any chance of putting it into duck-billed platypus?

Graves: That would be the obvious experiment!

McLaren: Just going back to Haldane's law, as far as mammals are concerned, I wonder in how many of the cases of the males being absent, anyone has looked to see whether it is actually sex reversal rather than sex-specific mortality?

Short: Professor 'Twink' Allen and I have looked at mules, where there is a marked deficiency of males. We blood sampled about 100 female mules and screened them for *SRY* expression, hoping that we would find some XY females. We didn't find any, but this wasn't a big enough sample for a significant test.

Graves: Did you karyotype them? If they are not expressing *SRY* you could miss it.

Short: No. It would be quite an easy experiment to do.

Perhaps this is a silly question, but does anyone foresee any therapeutic use of SRY?

Behringer: Possibly in ovarian cancer. These are mostly derived from the epithelial layer of the ovary. SRY has little toxicity, and if it could push the differentiation one way or the other the tumour cells might get confused and then undergo apoptosis. It's just a wild idea.

Reference

Albrecht KH, Eicher EM 1997 DNA sequence analysis of *Sry* alleles (subgenus *Mus*) implicates misregulation as the cause of C57BL/6J-Y(POS) sex reversal and defines the SRY functional unit. Genetics 147:1267–1277

Evolution of the testis-determining gene — the rise and fall of *SRY*

Jennifer A. Marshall Graves

Research School of Biological Sciences, The Australian National University, Canberra, ACT 2601, Australia

Abstract. The mammalian Y chromosome has been known for a long time to harbour a gene that triggers testis determination, and this testis-determining factor was identified as *SRY* in 1990. It has been supposed that *SRY* was the original mammalian sex-determining gene that initiated the differentiation of the Y from the X early in mammalian evolution, and this belief has been reinforced by an analysis of divergence times. However, I will argue here that *SRY* evolved quite recently in therian mammals and was not the original mammalian sex-determining gene that defined the X and Y. It arose as a degraded version of the X-borne *SOX* gene that is better qualified to be a brain-determining gene. It has no central role in sex determination, and can be replaced as a trigger and lost, as have many other Y-borne genes in recent evolutionary history. The mole vole has evidently accomplished this.

2002 The genetics and biology of sex determination. Wiley, Chichester (Novartis Foundation Symposium 244) p 86–101

SRY, the mammalian sex-determining gene

The testis-determining factor (TDF), known from deletion mapping to reside on the short arm of the human Y, was positionally cloned in 1990 (Sinclair et al 1990). The human *SRY* gene was shown unequivocally to control sex determination by mutation analysis, and its mouse homologue *Sry* was also shown to be testis determining by transgenesis (reviewed in Koopman 1995). Other species of eutherian mammals were found to have an equivalent gene on the Y chromosome.

SRY is a small, single exon gene encoding an 80 amino acid DNA-binding motif (HMG domain) similar to the HMG (high mobility group) proteins that are architectural factors. It defined the burgeoning *SOX* (for *SRY*-like HMG box containing) gene family that includes transcriptional activators and repressors. *SRY* is thought to act by the binding of the HMG box to a 6 bp DNA sequence, which bends DNA through a specific angle (Harley et al 1992). This may promote association of regulatory elements bound to far-flung regions of DNA, forming a complex that controls the activity of other genes.

Sry is expressed during a narrow window in the developing mouse gonadal ridge at 11.5 dpc (days *post coitum*), the time at which the first histological signs of testis differentiation are noted (reviewed in Koopman 1995). However, human *SRY* is transcribed in many embryonic tissues, albeit at a low level and is limited to the testis in adults. The significance of *SRY* transcripts in developing tissues other than testis is unclear. Does *SRY* have a function other than testis determination in humans?

Just which other genes are controlled by *SRY* is not yet clear, as the target of *SRY* has not been identified. However, it is likely that *SOX9*, an autosomal gene with a conserved function in testis determination (Foster et al 1996, Wagner et al 1995), is somehow controlled by *SRY*. Whether this control acts via activation or repression of other genes in the sex-determining pathway is not yet clear. The products of related *SOX* genes include both transcriptional activators and repressors (Uchikawa et al 1999). Suggestions include the direct or indirect activation of *SOX9* (Dubin & Ostrer 1994), or a double repression (McElreavey et al 1993), perhaps via repression of *SOX3* that in turn relieves the repression on *SOX9* (Graves 1998).

The rise of SRY

SRY has not always been the master switch that controls sex determination. Non-mammal vertebrates have no sex-specific *SRY* (Griffiths 1991). Birds and snakes have a ZZ male: ZW female chromosomal sex determination system which is quite unrelated to the mammal XX:XY system (Nanda et al 1999). Birds appear to rely on another gene, *DMRT1* (Raymond et al 1998, Smith et al 1999). This gene lies on human chromosome 9 (Raymond et al 1999) and acts downstream in the human sex-determining pathway (as shown by sex-reversed phenotype of mutants). This means that *SRY* must have evolved specifically in the lineage that led to eutherian mammals.

Origin of the mammalian SRY gene

We can discover where *SRY* came from and guess at how it acquired its sex-determining function by comparing its position, sequence and expression with those of related *SOX* genes.

SRY belongs to the intronless sub-family of *SOXB* genes. One of these, *SOX3*, was identified on the X chromosome in marsupials, and subsequently in all therian mammals, so must have been on the X in a common mammal ancestor. *SOX3* shows the highest sequence similarity to *SRY* within the HMG box (Bowles et al 2000), suggesting that *SRY* evolved from *SOX3* (Foster & Graves 1994). The SRY sequence outside the HMG box is poorly conserved between different species. This suggests that all the conserved sex-determining activity of SRY is in the HMG box, a conclusion reinforced by the finding that almost all of the known amino acid substitutions found in mutant SRY proteins from XY females lie within this region (Hawkins 1993). Thus *SRY* seems to be essentially a truncated *SOX3*.

In contrast, *SOX3* is highly conserved between species, both outside and within the HMG box. This suggests that it serves a critical function in mammalian development. In humans it is expressed in the developing brain, spinal cord, thymus and heart, as well as several adult tissues including testis. Two boys with *SOX3* deletions were severely mentally retarded, but showed testicular development, excluding *SOX3* from a necessary role in male sex determination (Stevanovic et al 1993). Mouse *Sox3* is expressed in the developing CNS, but there is some weak expression (comparable to that of *Sry*) in the indifferent genital ridge (Collignon et al 1996). A chicken homologue *cSOX3* is expressed only in the CNS, but an amphibian homologue, *xSOX3*, is expressed in the *Xenopus* ovary (Koyano et al 1997, Penzel et al 1997).

Thus the progenitor of *SRY* was more likely to have been involved in brain development than testis determination, although it appears to have a minor conserved role in differentiation of gonads as well as central nervous system. How could a brain-determining gene become a testis-determining gene?

SOX3 as a victim of Y chromosome degradation

The evolution of *SRY* from *SOX3* is readily understood in the context of Y chromosome mayhem. The mammalian Y is essentially a broken down X, and many or most genes on the Y are relics of genes on the X. This includes several genes whose X-borne copies are widely expressed but whose Y-borne equivalents are limited to testis and have putative roles in spermatogenesis. *SRY* proves to be no exception.

All the evidence supports the postulate (Ohno 1967) that sex chromosomes originated from a pair of autosomes when a gene took on a controlling function in sex determination. As genes with a sex-specific function accumulated, there was selection for repression of recombination to preserve a male-specific package (Charlesworth 1991). In turn, absence of recombination allowed mutations and deletions to persevere and led to rapid degradation of this region on the Y and loss of homology with the X. Progressive attrition was also offset by at least one major addition to the eutherian sex chromosomes (Graves 1995). Ohno's theory explains why many or most of the active genes and several pseudogenes on the Y have homologues on the X.

This inexorable degradation explains why there are so few genes left on the Y. Only about 30 have survived on the differentiated part of the Y out of the original

1400 represented on the X. These genes survived because they acquired a vital function in male determination or differentiation.

Acquisition of a sex-determining function by SOX3

Most genes on the human Y are somehow involved in sex determination and differentiation, largely spermatogenesis. This 'functional coherence' is quite unlike the multiplicity of functions of genes on any other chromosome, or even region (Lahn & Page 1997). However, this specialization is readily explained by selection acting on Y-borne genes.

To escape the inexorable degradation of genes on the Y there must be a strong selective force maintaining gene activity. Since only half the population possesses a Y chromosome, these genes cannot be vital for life. However, they can be selectively maintained if they are necessary for male (but not female) reproduction. In fact, they might even be disadvantageous for females. One example is *ZFY* (the original candidate for the testis determining factor), which is a ubiquitously expressed transcription factor in humans, but is testis-specific in mouse and likely to function in spermatogenesis. Other candidate spermatogenesis genes *RBMY*, *DFFRY* and *DBY* also appear to have diverged from ubiquitously expressed homologues on the X chromosome (Delbridge et al 1999, Lahn & Page 1999) and found a role in spermatogenesis.

So *SRY* is just like the other genes on the Y chromosome. It has been truncated—chopped off at the socks—to the point that there is nothing left except the HMG box. But has been retained because it found a male-specific function.

How did a brain-determining gene become a testis-determining gene?

SRY and *SOX3* expression profiles overlap at least to some degree. The minor expression of *SOX3* in the testis in mouse, and its expression in *Xenopus* ovary suggests that *SOX3* may have had at least a side-interest in sex for a long time. The minor expression of *Sry* in the mouse brain (Mayer et al 1998) suggests that *SRY* may retain some of its original brain-determining function. In fact its wide expression pattern in humans and marsupials makes us wonder if it has subsidiary functions in many tissues, although the absence of *SRY* has no phenotypic effect other than on sex determination. I suggest that the dual function of *SOX3* has been partitioned between brain (retained by *SOX3*) and testis (taken over by *SRY*).

Partitioning of function between the X- and Y-borne copies of a gene has occurred at least at one other locus. Mutations of the *ATRX* gene on the human X affects many systems, causing male-to-female sex reversal as well as mental retardation and alpha thallasaemia. This gene is expressed ubiquitously, in the gonad as well as many other tissues and organs. However, in marsupials, there are two copies of the gene, one on the X and one on the Y (Pask et al 2000). *ATRX* is expressed widely but not in the gonad, whereas *ATRY* is expressed only in testis.

A pathway by which the Y-borne allele of *SOX3* abandoned its brain-determining function and became essential instead in sex determination, is suggested by the hypothesis that *SOX3* acts as a negative regulator of *SOX9* in determining testis (Graves 1998). In females, in the absence of *SRY*, *SOX3* inhibits *SOX9* and no testis forms. In males, *SRY* inhibits *SOX3*, permitting *SOX9* to enact its testis-determining role. An intermediate in the process could have been a dosage-determined system based on *SOX3* in which homozygotes for wild-type *SOX3* were female, whereas heterozygotes for a null allele were male; the 2:1 dosage difference determined sex via a differential effect on *SOX9* activity. This system could readily evolve into a male-dominant system by the truncation of the null allele so that instead of merely being inactive, it actively inhibits *SOX3* and allows *SOX9* to function to produce a testis. In support of this idea is the observation that truncation of *SOX9* turns it from an activator into a repressor (Südbeck 1996).

Thus *SRY* has followed a common path travelled by several genes on the Y chromosome. From a widely expressed gene with functions in both sexes, it has become specialized for testis determination. It seems likely that its action may have changed with its truncation. Its expression has become limited in mouse, but is still wide in humans, although it appears to do the same job in both species, as a male-specific gene with a specific function in testis differentiation.

When did *SRY* evolve?

We can discover when this change occurred by comparing *SRY* in different mammals, with reference to the framework of relationships provided by fossil evidence and, increasingly, molecular phylogenies. There are three major groups of extant mammals. Two Infraclasses, Eutheria (placental mammals) and Metatheria (marsupials) diverged about 130 mya (million years ago). The Subclass Theria that contains them is generally thought to have diverged from Subclass Prototheria (the egg-laying monotremes) about 170 mya. Mammals evolved from a branch of reptiles (synapsids) that left no other descendants, and are equally distantly related to the other major branches of reptiles and birds, having diverged about 350 mya. Reptiles in turn diverged from amphibians, which evolved from a branch of the fish about 450 mya.

Since no non-mammalian reptile has an *Sry* gene, we must conclude that *SRY* evolved after the divergence of synapsid reptiles about 350 mya. And since humans and rodents, and all other orders of eutherian mammals tested have an *SRY* gene, we conclude that SRY evolved before the eutherian radiation about 80 mya. This leaves a very wide gap that can be filled by searching for *SRY* in marsupials and monotremes.

In marsupials, the Y chromosome is testis-determining, although it does not control all aspects of sexual differentiation (Sharman et al 1990) as it does, directly or indirectly, in eutherian mammals. It was significant, therefore, that a male-specific *SRY* sequence was discovered on the Y chromosome in marsupials (Foster et al 1992), particularly since the lack of a male-specific copy in marsupials had earlier sounded the death knell of the previous candidate *ZFY* (Sinclair et al 1988). This made a Y-borne *SRY* gene at least 130 million years old, and showed that it was part of the conserved ancient region of the mammalian Y (Waters et al 2001).

The presence of a Y-borne *SRY* gene in marsupials does not, however, prove that it has a male-determining function. The near-ubiquitous expression patterns of marsupial *SRY* do not necessarily point to a role for this gene in sex determination. In the tammar wallaby, *SRY* is transcribed in the embryo at every stage sampled, as well as in a wide range of adult tissues (Harry et al 1995). In the absence of mutation analysis and transgenesis, there is still no direct demonstration that *SRY* is male-determining in marsupials. Moreover, there is a competing candidate sex determining gene in marsupials in the testis-specific *ATRY* gene on the Y chromosome (Pask et al 2000) homologous to the X-specific sex-reversing *ATRX* gene on the human X. Is it possible that *ATRY* acts as the testis-determining switch in marsupials?

The presence of *SRY* on the Y chromosome in marsupials, even if it is not male-determining, dates *SRY* at more than 130 million years old, and suggests that it was a property of the Y chromosome of an ancestral therian mammal.

Can we trace *SRY* back any further? Lahn & Page (1999), on the basis of sequence differences between human *SRY* and *SOX3* compared to those between other XY shared genes, date the time of separation of the two alleles at 240–320 mya, long before the three groups of extant mammals diverged. By their calculations, *SRY* is a member of the most ancient stratum of the X chromosome, suggesting that the acquisition of a sex determining function by *SRY* was the defining event in the initiation of sex chromosome differentiation.

If SRY was the original sex-determining gene that defined the mammalian Y chromosome, it should also be present on the Y chromosome in the third group of mammals, the monotremes. Demonstration of a monotreme *SRY* would push back the date of *SRY* evolution to beyond 170 million years, the date at which monotremes diverged from the therian mammals (marsupials and eutherians).

Monotremes (the platypus and two echidna species) have an X chromosome present in two copies in females and a single copy in males (Murtagh 1977,

Wrigley & Graves 1988). Since monotreme sex chromosomes are involved in a translocation chain at meiosis with other unpaired chromosomes whose identities and relationships are unknown (Watson et al 1992), it is not certain which chromosome is the Y. The sex-determining mechanism in monotremes is quite unknown—in the absence of sex chromosome aneuploids, we cannot say whether the Y (whichever it is) or the dosage of the X is male-determining.

We have made many attempts to demonstrate and isolate an *SRY* gene in the platypus and echidna. Southern blotting of DNA from males and females of both species using human, mouse or marsupial *SRY* as probe detects several bands shared between males and females which are equivalent to the related *SOX* genes, but no male-specific band has been demonstrated. PCR with primers designed from several *SRY* and *SOX* sequences amplifies only fragments shared between male and female. Screening platypus genomic or cDNA libraries has isolated other *SOXB* genes that map to autosomes, but no SRY (P. Kirby, J. A. M. Graves, unpublished data). Screening Noah's Ark blots with these platypus *SOXB* genes also detects only bands shared between males and females. The simplest interpretation is that there is no *SRY* gene in the platypus, and sex is determined by another gene on either the X or the Y chromosome. Possible candidates are *ATRY* and *DMRT1*, which are presently being cloned and localized in the platypus.

Evolution of *SRY* function?

SRY shows major changes in structure and sequence between mouse and human. Do these reflect a change in *SRY* function that occurred in rodent or primate evolution?

Most eutherian *SRY* genes that have been analysed share one or more C-terminal protein-binding (PDZ) domains outside the HMG box. These domains bind PDZ proteins, and by analogy to other genes may act as an adaptor between *SRY* and other proteins in a transcription complex (Poulat et al 1997).

The mouse *Sry* gene is exceptional in its possession of a long (223 bp) 3′ domain containing a CAG repeat. Its product therefore contains a C terminal domain composed of 20 blocks of glutamine runs interspersed with spacers of polar amino acids (Bowles et al 1999). It is thought to have arisen by insertion of a core domain downstream of *Sry*, followed by amplification and mutation. The length and make-up of this domain varies among *Mus* species and strains leading to the suspicion that the CAG domain is non-functional—yet another nasty accident that occurred to the poor *Sry* gene. However, truncation mutations lacking this region were found to be unable to reverse sex of XX embryos. The glutamine-rich domain is therefore essential for sex determination in the mouse. Bowles et al (1999) suggest that the glutamine-rich domain of the mouse *Sry* forms a 'polar zipper' structure that mediates protein interactions, in lieu of the protein–protein interactions undergone by the PDZ domain of human *SRY*.

This suggests that the mouse *Sry* has acquired a new function since the divergence of rodents from primates.

The fall of *SRY*

The Y chromosome is evidently a dangerous place to be. Genes are subject to a barrage of mutation, and the lack of recombination means that the Y cannot be reconstituted. Most of the original 1500-odd genes have been irretrievably lost.

Some genes have disappeared from the Y in one lineage but not another. For instance, the *UBE1* gene coding for a ubiquitin activating enzyme has a Y-borne as well as an X-borne copy in mouse and marsupial, but not human. Evidently the copy disappeared from the human Y very recently, since its X-borne partner still escapes X chromosome inactivation. Similarly, the *RPS4* gene coding for a ribosomal subunit has copies on the X and Y chromosome in humans, but has lost its Y homologue in mouse.

Genes on the Y illustrate different stages of degradation and loss (reviewed in Graves 1995). Genes within the pseudoautosomal regions have homologues on both X and Y that pair and recombine at meiosis. Some genes like *SMCY* still maintain an active homologue on the Y: the double dosage in males is balanced by the escape of the X homologue from X inactivation in females. Other Y-borne genes, like *RPS4*, are active, but less so than their X homologues. Some originally ubiquitous genes, like mouse *Zfy*, have become testis-specific, and their X homologues recruited into the X inactivation system. Many genes, like STS, are represented only by pseudogenes on the Y. The overwhelming majority of genes have been completely deleted from the Y and their X homologues subject to X inactivation in females. Different stages of degradation and loss may be shown by the same gene in different species. For instance, *UBE1Y* is pseudoautosomal in monotremes, active but male-specific in mouse and marsupial, present only as pseudogene fragments in several primates, and has been completely lost from the human Y (Mitchell et al 1998).

The 26-odd genes that survive on the differentiated region of the human Y also show signs of attrition. Many are functionless pseudogenes, having suffered partial deletions (e.g. *STS*). Others, such as the candidate spermatogenesis genes *DAZ* and *RBMY*, have undergone mutation and exon amplification and these as well as others are amplified into gene families, only a few members of which are active.

SRY is no exception. Indeed, this gene appears to be a butchered copy of *SOX3*, evolving in the first place by truncation of sequences outside the HMG box. It shows a very high mutation rate, sparking initial speculation that variation at this locus drives speciation. However, careful analysis shows that the rate is typical of other genes on the Y, and mutation-causing base changes are no more frequent than expected (Pamilo & O'Neill 1997). Transgenesis with *Sry* from other mouse species fails to effect sex determination, and as we have seen, the CAG repeat has been internally amplified to different extents in different mouse strains and species. In some species of Old World mice, *Sry* has been amplified many times (Nagamine 1994). In one marsupial species, an intron has been created *de novo* in *SRY* (O'Neill et al 1998).

It would be hardly surprising to find that *SRY* has completely disappeared in some lineages. This seems to be exactly what has happened in the mole voles of Eastern Europe. *Ellobius lutescens* and *E. tancrei* undergo apparently normal sex determination. However, both species lack a Y chromosome, and animals of both sexes have an XX or XO sex chromosome constitution respectively. A third *Ellobius* species *E. fuscocapillus* has an intact Y that looks much like a mouse Y, and a perfectly normal *Sry* gene. However, no *SRY* gene homologue can be detected in either species lacking a Y (Just et al 1995), even using a probe amplified from the closely related *E. fuscocapillus*. Evidently some other gene has taken over the primary sex-determining function in triggering the male developmental pathway in these species. There are no outward signs of differentiation of another chromosome, and a search for sex-associated variants of *SOX9* and *AMH* (MIS) have proved negative (Baumstark et al 2001).

The future of mammalian sex determination

What sort of human sex-determining system would we find if we returned to earth in 100 million years or so? The continued degradation of the Y chromosome seems to be assured. At the rate it is going, the pseudoautosomal region is likely to be differentiated in a few million years, and the entire Y may not last much longer.

Complete differentiation of the X and Y evidently happened in marsupials, in which there is no detectable pseudoautosomal region. No cross-hybridization between the X and tiny Y can be seen with X or Y probes (Toder et al 2000), and no homologous pairing, synaptonemal complexes or chiasmata can be detected at male meiosis (Sharp 1982). However, loss of the pseudoautosomal region (PAR) may be opposed by strong selection for a pairing function, given that men and mice lacking a pseudoautosomal region of the Y are sterile. We may see, then, that the PAR takes on a life of its own, as appears to have happened in mouse, where the PAR is grimly hanging on. It has become GC-rich, probably to the detriment of *Sts*, the only gene it still harbours (Salido et al 1996).

The alternative is that the Y chromosome could be rescued once more by translocation of an autosomal region to the pseudoautosomal region. This will enlarge the pseudoautosomal region, provide new genes to be moulded into a sex-specific role, and delay the inevitable decay of the Y. Conceivably, many serial translocations could occur until the entire genome became part of the X chromosome. Ultimately, degradation would render humans, like bees and wasps, diploid in females and haploid in males.

Whichever scenario is followed, there is no guarantee that *SRY* will survive as the sex-determining gene for much longer. After all, *SRY* is not a very old gene, being absent in birds and reptiles and apparently in monotremes. It is not itself required for testis determination, as is shown by human XX males that lack *SRY*. It is simply a trigger, and could conceivably be replaced by any gene in the pathway. Evolution of a variant that short circuits *SRY* is not at all difficult to imagine, especially if it is true that *SRY* acts in a very roundabout way.

Mole voles demonstrate that loss of the entire Y is possible. Indeed, it is an almost inevitable outcome of continued Y degradation. Not only has *Sry* been lost but all the spermatogenesis genes on the mouse Y have disappeared with it. It is hard to imagine that these could all be lost simultaneously. Were the spermatogenesis genes picked off one by one as their functions were taken over by autosomal or X-linked genes? Was the mole vole Y in its death throes devoid of everything except *Sry*? Genetic archaeology of exotic species will be very rewarding, for it is likely that the mole vole will help us foretell the future of the human Y chromosome and the *SRY* gene.

References

Baumstark A, Akverdyan M, Schulze A et al 2001 Exclusion of SOX9 as the testis determining factor in *Ellobius lutescens*: evidence for another testis determining gene besides *SRY* and *SOX9*. Molec Genet Metab 72:61–66

Bowles J, Cooper L, Berkman J, Koopman P 1999 Sry requires a CAG repeat domain for male sex determination in Mus musculus. Nat Genet 22:405–408

Bowles J, Schepers G, Koopman P 2000 Phylogeny of the SOX family of developmental transcription factors based on sequence and structural indicators. Dev Biol 227:239–255

Charlesworth B 1991 The evolution of sex chromosomes. Science 251:1030–1033

Collignon J, Sockanathan S, Hacker A et al 1996 A comparison of the properties of *Sox3* with *SRY* and two related genes, *Sox1* and *Sox2*. Development 122:509–520

Delbridge ML, Lingenfelter PA, Disteche CM, Graves JAM 1999 The candidate spermatogenesis gene *RBMY* has a homologue on the human X chromosome. Nat Genet 22:223–224

Dubin RA, Ostrer H 1994 *SRY* is a transcriptional activator. Mol Endocrinol 8:1182–1192

Foster JW, Brennen FE, Hampikian GK et al 1992 Evolution of sex determination and the Y chromosome: *SRY*-related sequences in detects marsupials. Nature 359:531–533

Foster JW, Graves JAM 1994 An *SRY*-related sequence on the marsupial X chromosome: implications for the evolution of the mammalian testis-determining gene. Proc Natl Acad Sci USA 91:1927–1931

Foster JW, Dominguez-Steglich MA, Guioli S et al 1994 Campomelic dysplasia and autosomal sex reversal caused by mutations in an *SRY*-related gene. Nature 372:525–530

Graves JAM 1995 The origin and function of the mammalian Y chromosome and Y-borne genes—an evolving understanding. BioEssays 17:311–320

Graves JAM 1998 Interactions between *SRY* and *SOX* genes in mammalian sex determination. BioEssays 20:264–269

Griffiths R 1991 The isolation of conserved DNA sequences related to the human sex-determining region Y gene from the lesser black-backed gull (*Larus fuscus*). Proc R Soc Lond B Biol Sci 244:123–128

Harley VR, Jackson DI, Hextall PJ et al 1992 DNA binding activity of recombinant *SRY* from normal males and XY females. Science 255:453–456

Harry JL, Koopman P, Brennan FE, Graves JAM, Renfree MB 1995 Widespread expression of the testis determining gene *SRY* in a marsupial. Nat Genet 11:347–349

Hawkins JR 1993 Mutational analysis of *SRY* in XY females. Hum Mutat 2:347–350.

Just W, Rau W, Vogel W et al 1995 Absence of *Sry* in a species of the vole *Ellobius*. Nat Genet 11:117–118

Koopman P 1995 The molecular biology of *SRY* and its role in sex determination in mammals. Reprod Fertil Dev 7:713–722

Koyano S, Ito M, Takamatsu N, Takiguchi S, Shiba T 1997 The *Xenopus Sox3* gene expressed in oocytes of early stages. Gene 188:101–107

Lahn B, Page DC 1997 Functional coherence of the human Y chromosome. Science 278: 675–680

Lahn BT, Page DC 1999 Four evolutionary strata on the human X chromosome. Science 286:965–967

Mayer A, Lahr G, Swaab DF, Pilgrim C, Reisert I 1998 The Y-chromosomal genes SRY and ZFY are transcribed in adult human brain. Neurogenetics 1:281–288

McElreavey KE, Vilain E, Abbas N, Herskowitz I, Fellous M 1993 A regulatory cascade hypothesis for mammalian sex determination: *SRY* represses a negative regulator of male development. Proc Natl Acad Sci USA 90:3368–3372

Mitchell MJ, Wilcox SA, Watson JM et al 1998 The origin and loss of the ubiquitin activating enzyme gene on the mammalian Y chromosome. Hum Mol Genet 7:429–434

Murtagh CE 1977 A unique cytogenetic system in monotremes. Chromosoma 65:37–57

Nanda I, Shan Z, Schartl M et al 1999 300 million years of conserved synteny between chicken Z and human chromosome 9. Nat Genet 21:258–259

Nagamine CM 1994 The testis-determining gene, *SRY*, exists in multiple copies in Old World rodents. Genet Res 64:151–159

Ohno S 1967 Sex chromosomes and sex linked genes. Springer-Verlag, New York

O'Neill RJ, Brennan FE, Delbridge ML, Crozier RH, Graves JAM 1998 De novo insertion of an intron into the mammalian sex determining gene, *SRY*. Proc Natl Acad Sci USA 95:1653–1657

Pask A, Renfree MB, Graves JAM 2000 The human sex-reversing *ATRX* gene has a homologue on the marsupial Y chromosome. ATRY: implications for the evolution of mammalian sex determination. Proc Natl Acad Sci USA 97:13198–13202

Pamilo P, O'Neill RJW 1997 Evolution of the *Sry* genes. Mol Biol Evol 14:49–55

Penzel R, Oschwald R, Chen YL, Tacke L, Grunz H 1997 Characterisation and early embryonic expression of a neural specific transcription factor *xSOX3* in *Xenopus laevis*. Int J Dev Biol 41:667–677

Poulat F, Desantabarbara P, Desclozeaux M et al 1997 The human testis determining factor *SRY* binds a nuclear factor containing PDZ protein interaction domains. J Biol Chem 272: 7167–7172

Raymond CS, Shamu CE, Shen MM et al 1998 Evidence for evolutionary conservation of sex-determining genes. Nature 391:691–695

Raymond CS, Parker ED, Kettlewell J et al 1999 A region of human chromosome 9p required for testis development contains two genes related to known sexual regulators. Hum Mol Genet 8:989–996

Salido EC, Li XM, Yen PH, Martin N, Mohandas TK, Shapiro LJ 1996 Cloning and expression of the mouse pseudoautosomal steroid sulphatase gene (*Sts*). Nat Genet 13:83–86

Sharp P 1982 Sex chromosome pairing during male meiosis in marsupials. Chromosoma 86: 27–47

Sharman G, Hughes R, Cooper DW 1990 The chromosomal basis of sex differentiation in marsupials. Aust J Zool 37:451–466

Sinclair AH, Foster JW, Spencer JA et al 1988 Sequences homologous to *ZFY*, a candidate human sex-determining gene, are autosomal in marsupials. Nature 336:780–783

Sinclair AH, Berta P, Palmer MS et al 1990 A gene from the human sex-determining region encodes a protein with homology to a conserved DNA-binding motif. Nature 346:240–244

Smith CA, McClive PJ, Western PS, Reed KJ, Sinclair AH 1999 Conservation of a sex determining gene. Nature 402:601–602

Stevanovic M, Lovell-Badge R, Collignon J, Goodfellow PN 1993 *SOX3* is an X-linked gene related to *SRY*. Hum Mol Genet 2:2013–2018

Südbeck P, Schmitz ML, Baeuerle PA, Scherer G 1996 Sex reversal by loss of the C-terminal transactivation domain of human *SOX9*. Nat Genet 13:230–232

Toder R, Wakefield M, Graves JAM 2000 The minimal mammalian Y chromosome — the marsupial Y as a model system. Cytogenet Cell Genet 91:285–292

Uchikawa M, Kamachi Y, Kondoh H 1999 Two distinct subgroups of Group B Sox genes for transcriptional activators and repressors: their expression during embryonic organogenesis of the chicken. Mech Dev 84:103–120

Wagner T, Wirth J, Meyer J et al 1994 Autosomal sex reversal and campomelic dysplasia are caused by mutations in and around the *SRY*-related gene *SOX9*. Cell 79:1111–1120

Watson JM, Meyne J, Graves JAM 1992 Studies of the chromosomes of the echidna meiotic translocation chain. In: Augee M (ed) Platypus and echidnas. Royal Soc NSW, Australia, p 53–63

Waters P, Duffy B, Frost CJ, Delbridge ML, Graves JAM 2001 The human Y chromosome derives largely from a single autosomal region added 80–130 million years ago. Cytogenet Cell Genet 92:74–79

Wrigley JM, Graves JAM 1988 Karyotypic conservation in the mammalian Order Monotremata (subclass Prototheria). Chromosoma 96:231–247

DISCUSSION

Scherer: Did you say that *ATRY* is expressed in the gonads in platypus?

Graves: No, we haven't found *ATRX* or *Y* in platypus yet; this work is in marsupials where *ATRY* is expressed in gonads, and *ATRX* is not expressed in gonads but it is expressed everywhere else.

Mittwoch: When you say that *SOX3* may inhibit *SOX9*, by what sort of mechanism do you think this is happening? Do you think that *SOX9* may be accelerating cell proliferation, and that *SOX3* may retard it?

Graves: I would expect it to be much more direct than that. HMG box proteins seem to form quite large complexes together, and some *SOX* genes are inhibitors and others are activators. It may be that they are part of a much larger complex.

Burgoyne: I have evidence for an X-linked gene that potentiates *Sry* action, which is the opposite of what Jenny Graves was suggesting that *Sox3* might do. This came out of a project in which I created XOs with a paternal X and maternal X, and I had XXs in the same cross. I put in an incompletely penetrant *Sry* transgene. I expected them all to give the same proportions of males, females and hermaphrodites, but the results are absolutely clear cut. If you are an XO with an incompletely penetrant *Sry* transgene, you are much more often female than if you are an XX. XXs were getting about 50% males, whereas the XOs were female or hermaphrodite and almost never male. This tells us that there is something on the X chromosome, which apparently is not dosage compensated, which potentiates the action of this incompletely penetrant SRY transgene. In trying to find out what this gene would be, I though that if it is non-dosage compensated it might map to the pseudoautosomal region (PAR), because genes in the PAR are not dosage compensated. I therefore added a PAR to the XOs, but this made no difference whatsoever. As for the other X-linked genes that aren't dosage compensated, most of them have homologues on the Y chromosome. So I put back a Y chromosome lacking *Sry*, to see if this would make them go back to being male. The data so far suggest these XY *Sry*-negative mice carrying the incompletely penetrant *Sry* transgene are also developing as females. Of course, the only Y-linked gene that I haven't put back by adding this Y is *Sry*, of which the homologue is *Sox3*. This leads to the intriguing possibility that *Sox3* is potentiating *Sry* action in the XX *Sry* transgenics. Although *Sox3* is thought to be dosage compensated, Adam Hacker in Robin Lovell-Badge's lab showed that in early genital ridges, there are higher levels of *Sox3* in XX than in XY.

Graves: That would be even more interesting: maybe it used to work in the other way. Sharat Chandra proposed years ago that X inactivation was evolved as a sex-determining device, and in females it would kill off one copy of a gene such as *Sox3*, so you would have one active copy of *Sox3* in females and two in males (Chandra 1985). This really would be exciting.

Short: Jenny Graves, you said there is no PAR in marsupials, but is there an X–Y bivalent in meiosis?

Graves: Yes, there is. They don't pair and recombine; they sort of touch at the ends. It is called 'telomere attraction', whatever that means in molecular terms.

Zarkower: In your model, snakes and turtles are grouped together. Have you looked at where *Dmrt1* is located in snakes, and do you know anything about the molecular nature of the snake X and Y chromosomes?

Graves: There is a tiger snake in a bag on the television set that is making the kids and the dog very nervous in our collaborator's house, so we are about to do those experiments. We haven't been able to paint into snakes, so far, but we are trying. We were amazed that we were able to paint the chicken Z across into turtles, because they were supposed to be so distantly related. I no longer believe this: I think turtles are much more closely related to birds than has been appreciated in the past. I suspect the whole anapsid/diapsid dichotomy is junk.

Koopman: What is known about *Sox3* expression in the gonads in different species?

Graves: The unhappy fact is that it is not expressed in the gonads in marsupials (Pask et al 2000). This kills off the idea that it is an inhibitor in marsupials. However, we have no direct indication that *Sry* is sex-determining in marsupials. I don't know anything that has been done on *Sox3* expression in chickens.

Lovell-Badge: We have looked. *Sox3* is clearly expressed in germ cells, but not obviously in the somatic cells.

Graves: I think in *Xenopus* it is expressed in the ovary as well as the brain.

Lovell-Badge: I can't remember. *Xenopus* is difficult anyway, being a pseudo-tetraploid. There are extra copies of the genes.

Graves: From the data that we have looked at it is really not very clear to me that it is *Sox3*. If you don't have *Sox1* and *Sox2*, I don't think you can claim that you have got the right gene.

Lovell-Badge: There are definitely *Sox1*, *Sox2* and *Sox3* genes in birds.

Koopman: What is the situation with *Sox3* in mice? Is it expressed in fetal gonads or not?

Lovell-Badge: It is expressed in the gonad at a low level, but it isn't specific to the gonad. We did RNase protection assays to detect it. The *in situs* didn't look great.

Burgoyne: The expression wasn't germ cell dependent. Interestingly, as I said earlier, there was more in the female genital ridge than the male. It may escape X inactivation, at least in the genital ridge.

Wilkins: For model systems where there are genetic tests, we often isolate and identify particular genes, and assign them certain roles. We then tend to think, 'Ah, this gene must perform this function in a large number of organisms'. However, many genes are parts of gene families, and many different gene products — whether they are parts of the same family or are unrelated but can do the same thing — form groups of genes that are functionally related. What evolution seems to do is to play around with the members that have similar functional capacities to do certain things. In certain lineages, one member of that functional group will do one thing, and in other lineages, others will. We are terribly surprised when we get results such as Jenny Graves' demonstration that *Sry* is not the be-all and end-all of sex determination, when in fact this is probably a common theme in evolution. Often there is selection for maintenance of the function, but the players change.

Lovell-Badge: That has become clear with *Sox2* and *Sox3*. In birds, although both seem to be involved in neural induction, *Sox3* is expressed in the epiblast earlier than *Sox2*. In mammals, however, *Sox2* is expressed in the inner cell mass of blastocysts, much earlier than *Sox3* which only comes on later in the epiblast. With respect to germ cells, in the chick *Sox3* shows strong expression, whereas in the mouse it is *Sox2*.

Burgoyne: Why hasn't the emu W become wimpified?

Graves: That's an interesting question. The same thing happens in snakes. In different families there are different extents of differentiation of the W chromosome. It looks like it happens independently in different families.

Burgoyne: In the different groups of birds we are presumably talking about the same W.

Graves: Yes, but it has obviously been evolving quite independently for 80 million years since emus (ratites) diverged from chickens (carinates). I don't think we have the slightest idea about why it would go faster in some lineages than others. Can someone please explain why there are no sex chromosomes in frogs!

Charlesworth: The general idea for why crossing over is suppressed across the whole of the Y chromosome, rather than just around a sex-determining region, depends on the notion that there are genes which are advantageous in one sex and disadvantageous in another. You want to keep the ones that are good in males linked to the male allele at the sex-determining locus. It could simply be a matter of happenstance in different lineages as to whether these genes pop up as mutations and get recruited. I don't think in principle that it is terribly surprising that different lineages behave differently. Of course, there is no evidence one way or another as to exactly why it happens one way in one group and differently in another. Emus belong to a distinct branch, the ratites, which are a primitive type of birds that have been separated from the others for a long time. Things could have happened quite differently there.

Wilkins: A phylogenetic question. Your scheme assumes that monotremes are the most basal mammals. There was a report a while back showing molecular evidence that suggested that monotremes are a branch of the marsupials (Janke et al 1996).

Graves: This was based on whole mitochondrial sequencing. Clearly, the nodes are very close, and mitochondrial DNA brings out monotremes and marsupials as being sister taxa, both equally related to the eutherian mammals. Almost everything else says the opposite.

Renfree: The fossil record shows unequivocally that monotremes branched off the mammal-like reptile lineage very early on. Then, in the fossil record there are distinct therian mammals, a variety of small carnivorous-like mammals, that you cannot tell whether they are marsupial or eutherian. The most recent data show that only from 100 million years ago can you identify marsupials and eutherian animals. They appear simultaneously as distinct groups in the fossil record with therian mammals as their precursors. The monotremes were distinct from this lineage about 150 million years ago.

Graves: I'm looking for a unique genetic event that will distinguish those hypotheses. I think the location of genes such as Sox3, that are found on the X in human and marsupial but not monotremes, will provide this.

Capel: What is the information on ATR X in humans?

Vilain: It causes a classic X-linked mental retardation, associated with a thalassemia and dysmorphic features. In terms of sexual phenotype it is variable, from XY males who have cryptorchidism and hypospadias to severe ambiguity. I don't know whether pure gonadal dysgenesis has been reported.

Camerino: I think there was one case of this reported. The clinical data are incomplete. However, there was at least one case in which testicular dysgenesis was reported.

Vilain: I don't know whether complete sex reversal has been reported.

Josso: We have a patient with incomplete sex reversal, but who was raised as a girl. It seems to depend on the mutation.

Short: Jenny Graves, are you leaving it open as to whether temperature-dependent sex determination is the ancestral or a derived form?

Graves: I think switches in both directions probably occur. Obviously, you can get shifts from temperature dependence to genetic dependence, but I see no reason why you can't do it the other way round. After all, we know that there are species of fish and lizards that have closely related species with the opposite form of sex determination. There is an Australian lizard which in one population undergoes temperature-dependent sex determination (TSD) and another does genetic sex determination.

Zarkower: Jonathan Hodgkin (1983) has shown nicely that you can completely change the sex-determining system in *Caenorhabditis elegans* by single base mutations.

Short: Are there any examples of TSD in a snake?

Graves: I don't think so. There's always the suspicion that TSD may be lurking, even in birds, under the current sex-determining systems. I've heard that if the temperature of incubation is raised you get more males, although most of them are dead.

References

Chandra HS 1985 Is human X chromosome inactivation a sex-determining device? Proc Natl Acad Sci USA 82:6947–6949

Hodgkin J 1983 Two types of sex determination in a nematode. Nature 304:267–268

Janke A, Gemmell NJ, Feldmaier-Fuchs G, von Haeseler A, Paabo S 1996 The mitochondrial genome of a monotreme—the platypus (*Ornithorhynchus anatinus*). J Mol Evol 42:153–159

Pask AJ, Harry JL, Renfree MB, Marshall Graves JA 2000 Absence of SOX3 in the developing marsupial gonad is not consistent with a conserved role in mammalian sex determination. Genesis 27:145–152

A comparative analysis of vertebrate sex determination

Andrew Sinclair, Craig Smith, Patrick Western and Peter McClive

Department of Paediatrics, University of Melbourne and Murdoch Children's Research Institute, Royal Children's Hospital, Melbourne, Victoria 3052, Australia

Abstract. Sex determination in vertebrates is controlled by a variety of mechanisms. We compared the expression of *SF1*, *DAX1*, *DMRT1*, *SOX9* and *AMH* during gonadogenesis in the mouse, chicken and alligator embryo. In contrast to the expression profile of *Sf1* in mouse embryos, chicken and alligator embryos show higher levels of *Sf1* expression in the developing ovaries compared to testes. This may reflect the higher level of sex hormone synthesis in the ovary compared to the testis in chickens and alligators. The *DAX1* gene has a similar expression profile in all three vertebrate species but appears to have different gene structure. As in mouse, *DMRT1* was expressed at very high levels in the chicken and alligator male gonad. The male-specific up-regulation of *SOX9* expression appears to be a common feature in all three vertebrates. In the chicken and alligator *AMH* is expressed prior to *SOX9*, suggesting that in these species *SOX9* cannot initiate *AMH* expression as it does in mammals. *SOX9* acts at multiple points in the vertebrate testis pathway but it appears that only some of these functions have been conserved through evolution.

2002 The genetics and biology of sex determination. Wiley, Chichester (Novartis Foundation Symposium 244) p 102–114

Vertebrate sex-determining genes

In vertebrates, sex-determining genes must operate within the embryonic gonads, regulating ovarian versus testicular development. It has been postulated that the genetic pathway controlling gonadal sex differentiation is similar in all vertebrates, with only the initial sex-determining switch varying between groups (*SRY* in mammals, temperature in many reptiles, and an unknown genetic trigger in birds). Many of the genes now implicated in mammalian sex determination have orthologues that are also expressed in the embryonic gonads of birds (chickens) and reptiles (alligators). *SOX9*, for example, has a male-specific role in both mammals, birds (Kent et al 1996) and reptiles (Western et al 1999a). However, it is becoming apparent that the structure and/or expression patterns of these genes have not necessarily been conserved between the two groups. Our research has focused on the expression of *SF1*, *DAX1*, *DMRT1*, *SOX9* and *AMH* (MIS) in chicken and alligator embryos. These genes show some interesting similarities and differences to their mammalian counterparts that broaden our understanding of vertebrate sex determination.

Sex determination in birds is chromosomally based. The male carries two Z sex chromosomes, while the female carries one Z and one W sex chromosome. The sex chromosomes of birds and mammals are not homologous, having evolved from different autosomal pairs (Graves 1995). No *SRY* gene has been identified in birds and the basic mechanism of sex determination in these vertebrates remains unknown. Recent evidence suggests that Z-linked genes show dosage compensation, indicative of Z inactivation (McQueen et al 2001). Sex may be controlled by Z chromosome dosage (escape of Z inactivation) or it may depend upon a dominant ovarian determinant carried on the W chromosome. We have used the chicken embryo as a model to examine the expression of known (mammalian) genes with a role in sex determination.

In many reptiles the primary sex-determining trigger is regulated by egg incubation temperature. Temperature-dependent sex determination (TSD) occurs in all crocodilians and marine turtles examined to date and is common in terrestrial turtles and viviparous lizards (Wibbels et al 1998). We have focused on known (mammalian) genes with a role in sex determination and analysed their expression in the American alligator embryo. This species has a female:male:female pattern of TSD. Eggs incubated at 30 °C or 34.5 °C result in 100% or 95% female hatchlings, respectively, while incubation at 33 °C results in 100% male hatchlings (Lang & Andrews 1994). Temperature acts to determine the sex of the embryo during the middle third of development (stages 21–24 at 33 °C or stages 20–23 at 30 °C). This temperature-sensitive period (TSP) of gonadogenesis is the time during which the indifferent gonad is irreversibly committed to either testis or ovarian development (Lang & Andrews 1994).

SF1

Steroidogenic factor 1 (SF1) belongs to the large family of orphan nuclear hormone receptors, for which ligands have not been identified. In mammals, *SF1* is initially expressed in the undifferentiated gonads of both sexes, and null mutations in mice show that the gene is essential for the formation of the gonadal and adrenal primordia (Luo et al 1994). In mouse embryos, *Sf1* expression is maintained during testicular differentiation, but is down-regulated during ovarian differentiation. Furthermore, several lines of evidence indicate that SF1 (together with SOX9 and WT1) regulates *AMH* expression in the developing male gonad (Nachtigal et al 1998). These data have led to the proposal that SF1 has a dual role during gonadogenesis in the formation of the undifferentiated gonad and later male-specific differentiation. *SF1* has an important role in endocrine function. In addition to controlling *AMH* expression, SF1 also regulates genes encoding steroidogenic enzymes, including aromatase.

One important difference between the differentiation of the gonads in the mammals and other vertebrates is the effect of the oestrogenic enzyme, aromatase. In the non-mammalian vertebrates, aromatase activity and oestrogen synthesis are required for normal ovarian differentiation whereas in mammals ovary differentiation appears to be largely independent of oestrogen activity. The steroidogenic level of the developing ovary in the non-mammalian vertebrates is relatively high when compared to the testis. Since aromatase is critical to gonadal development in birds and reptiles, we have examined the expression profile of chicken and alligator *SF1* during embryogenesis (Smith et al 1999c, Western et al 2000).

As in mammals, *SF1* transcripts are detectable in embryonic chicken and alligator urogenital tissue from an early-undifferentiated stage. As development proceeds, expression becomes localized to the developing gonads and adrenal glands. In contrast to the pattern seen in mouse embryos (Ikeda et al 1994), *SF1* is more highly expressed in developing ovaries compared to testes in both the chicken alligator embryo. Greater expression in female chick embryos is seen from stage 30 (day 6.5 of embryogenesis) and is maintained up until at least stage 35 (day 8.5), at which time there is strong expression in both female gonads, but weaker expression in the developing male gonads.

In the alligator, *SF1* was expressed in the developing gonad/mesonephros/adrenal complex during stages 20–23 and in the gonad during stages 24–27 throughout male and female sex determination. *SF1* expression appeared to be at least as high or a higher level in the developing ovary than the testis from early on (stage 22) in the TSP (Western et al 2000). This result needs to be interpreted with caution since *SF1* is also expressed in the male and female adrenal gland, which was included in the stage 20–23 samples. However, stage 24–27 samples included only gonadal tissue. In the alligator the level of aromatase expression and activity in the developing gonad increases after the TSP (Smith et al 1995) corresponding with the presence of ovarian *SF1* expression (Western et al 2000). Paradoxically, *SF1* is strongly expressed in the testis but down-regulated in the developing ovary of *Trachemys scripta*, a turtle with TSD (Fleming et al 1999). *Aromatase* expression and oestrogen synthesis have been strongly implicated in ovarian differentiation in this species (Wibbels et al 1998). The significance of the different gonadal expression patterns of *SF1* in non-mammalian vertebrates is yet to be determined.

It is possible that the higher *SF1* expression in the ovary during chick/alligator gonadogenesis reflects a role of SF1 in steroidogenesis, particularly aromatase regulation. Strong ovarian expression of *SF1* may be required to ensure sufficient oestrogen production for normal ovarian differentiation. In males, the lower level of *SF1* expression may nevertheless play a role in regulating *AMH* expression in chick (Oreal et al 1998) and alligator (Western et al 2000). *AMH* is also expressed in embryonic ovaries of both species (Smith et al 1999a, Western et al 1999a). SF1 may regulate *AMH* in both sexes.

DAX1

In humans, *DAX1* is located on a portion of the X chromosome which, when abnormally duplicated, results in male-to-female sex reversal (Bardoni et al 1994). Loss-of-function mutations in *DAX1* cause hypoplastic adrenal development (Zanaria et al 1994). Hence the acronym, *DAX1*: Dosage sensitive sex reversal, Adrenal hypoplasia congenita, on the X chromosome, number 1. Consistent with a role in gonadal development, *Dax1* is expressed in embryonic mouse gonads at the time of sexual differentiation. In mouse, expression declines in males at the time of *Sry* activation, while expression increases in females at around the same time (Swain et al 1996). Strains of transgenic mice carrying weak alleles of *Sry* together with extra copies of *Dax1* can show male-to-female sex reversal (Swain et al 1998). The human sex reversal and mouse transgenic data indicate that *DAX1* can act as an 'anti-testis' factor, antagonizing SRY function. Interestingly, gonads develop normally in *Dax1* null mutant mice, with the exception of impaired spermatogenesis in males and compromised endocrine cell development in both sexes (Yu et al 1998). *DAX1* may be associated with gametogenesis and endocrine development of the gonads, under normal conditions. The mammalian *DAX1* gene encodes a novel orphan nuclear receptor. The C-terminus of the protein includes a conserved region homologous to the ligand-binding domain of the ligand-activated receptors. However, instead of a typical zinc finger motif, an unusual tandem repeat region is present at the N-terminus. This repeat region is thought to represent a novel DNA-binding domain.

We cloned a chicken *DAX1* homologue from an embryonic urogenital ridge cDNA library and compared the deduced protein with that found in mammals. The chicken DAX1 protein shows 63% amino acid identity with the human protein over the region of the conserved ligand-binding domain. However, the chicken protein lacks the unusual tandem repeat motif seen at the N-terminus in mammals, although it has weak homology to one of the repeats (Smith et al 2000). This suggests that the chicken and mammalian proteins may bind DNA via different motifs. An alternative possibility is that chicken DAX1 does not actually bind DNA. In the mammals, some studies have demonstrated direct DNA-binding by DAX1 (Zazopoulos et al 1997), while others have not (Nachtigal et al 1998). Fluorescence *in situ* hybridization analysis shows that chicken *DAX1* is autosomal, located on the long arm of chromosome 1 (Smith et al 2000). Using RNase protection assays, we have shown that *DAX1* is expressed in embryonic chicken gonads. Expression is up-regulated during sexual differentiation, with somewhat higher expression in female gonads than in the male.

In the alligator, *DAX1* was expressed in the developing gonad/mesonephros/adrenal complex during stages 20–23 and in the gonad during stages 24–27 throughout male and female sex determination. There appeared to be little difference between the male and female expression patterns of *DAX1* (Western et al 2000). These expression patterns for *DAX1* are broadly similar to those seen in mouse embryos. However, *DAX1* expression is not down-regulated at the onset of testis differentiation in the chicken and alligator, as occurs in the mouse (Swain et al 1996). Thus, in the chicken and alligator, *DAX1* may have a role in both sexes, probably involved with gametogenesis and steroidogenesis, as the mammalian studies suggest.

DMRT1

DMRT1 (*DM*-Related Transcription factor, number *1*) is a putative sex-determining gene in mammals, identified through its homology with two genes involved in male sexual development in *Drosophila melanogaster* and *Caenorhabditis elegans* (Raymond et al 1999). These genes encode known or putative transcription factors, characterized by a DNA binding motif called the DM domain. In humans, *DMRT1* is located within the minimal region of chromosome 9p shown to be deleted in several patients with XY male-to-female sex reversal. In human embryos, *DMRT1* is expressed in male but not female gonads at the time of sexual differentiation (Moniot et al 2000). Similarly, *DMRT1* is expressed specifically in the gonads of mouse embryos, showing stronger expression in males than in females after the onset of sexual differentiation (Raymond et al 1999). These lines of evidence suggest that *DMRT1* plays a role in male sexual differentiation. To date no mutations have been identified in the *DMRT1* gene of human XY sex-reversed patients, although gene knockout studies in mice have shown that *Dmrt1* is required to maintain normal testis development. More recently, several different DM genes have been identified but their role, if any, in sexual development has yet to be defined.

The chicken *DMRT1* homologue is located on the Z sex chromosome (Nanda et al 1999). We have studied *DMRT1* expression in the chicken embryo during gonadogenesis, using whole mount *in situ* hybridization (Smith et al 1999b). As in the mouse, *DMRT1* is expressed specifically in the urogenital system of developing embryos. In the chicken, *DMRT1* expression is significantly stronger in male gonads compared to female gonads prior to and during the period of gonadal sex differentiation. This sexual dimorphism is apparent from at least developmental stages 25–28 (day 4.5–5.5). In male embryos, expression is localized in the medullary cords of the developing gonads, consistent with an organizational role for the gene in testis formation. The male Müllerian ducts also show stronger *DMRT1* expression than the female ducts. In tissue sections, expression is confined to the mesenchyme surrounding the Müllerian duct. Even though birds do appear to exhibit Z chromosome inactivation (McQueen et al 2001), the higher expression of *DMRT1* in male gonads may reflect that it escapes Z-inactivation. In the alligator embryo, *DMRT1* showed a very similar expression profile to that observed in birds, being expressed early in both the developing ovary and testis but becoming higher in the developing testis than the ovary. The spatial expression pattern of *DMRT1* suggests that its male-determining role in mammals has been conserved in birds and reptiles. However, the expression of chicken *DMRT1* well before the onset of sexual differentiation suggests that other factors are also necessary to initiate testis formation.

SOX9

The *SRY*-related gene, *SOX9*, appears to have a male-specific role in mammals, birds and reptiles. In chicken embryos, *SOX9* begins to be expressed only in male gonads from stage 30 (day 6.5). Expression is not seen in female embryonic gonads (Kent et al 1996, Smith et al 1999a). The exact role of *SOX9* in avian gonadal development is unclear, although mammalian studies indicate that it is involved in Sertoli cell development. In mouse embryos, one of the functions of *Sox9* appears to be the activation of *anti-Müllerian hormone* (*Amh*, also known as Müllerian inhibitory substance, *Mis*) gene expression (Arrango et al 1999). In chicken embryos, however, the onset of *AMH* expression precedes the onset of *SOX9* expression (Oreal et al 1998, Smith et al 1999a). In the developing alligator testis *SOX9* expression was first observed very close to the end of the TSP (at stage 23.5) and its expression appeared to be confined to the *AMH*-expressing medullary cells. These cells were organizing into testis tubules, a behaviour consistent with Sertoli cell development. Therefore it appears that within the developing alligator testis, medullary cells begin to proliferate and an increasing number of these cells express *AMH* (Western et al 1999a,b). This is followed by the onset of a low-level expression of *SOX9* in all the *AMH*-expressing cells by the end of the TSP. After the TSP both *SOX9* and *AMH* are strongly expressed in the cells aligned within the testis tubules. At no stage was *SOX9* or *AMH* expression observed by *in situ* hybridization studies in the developing gonads of alligators raised at either female determining temperature (30 °C or 34.5 °C) (Western et al 1999a,b).

AMH and *SOX9*

However, in the alligator and chick *SOX9* expression appears to be initiated and up-regulated after the testis-specific expression of *AMH* (Oreal et al 1998, Western et al 1999a,b). This implies that at least the initiation of *SOX9* expression is not required for AMH production in the alligator and chick. The temporal and spatial expression of *AMH* in the medullary cells of the developing alligator testis and the timing of the TSP relative to *SOX9* expression also imply that pre-Sertoli cell differentiation and sex determination precede *SOX9* expression in the alligator gonad (Western et al 1999a,b).

Considering the high level of *SOX9* sequence conservation and the sex-specific expression of this gene it is tempting to assume that it functions at the same points in the testis pathway of mammals and non-mammals. Paradoxically the expression patterns observed in the alligator and chick suggest otherwise. Can these differences be explained by suggesting that changes in SOX9 function have occurred during evolution? Considering the evident plasticity of sex determining mechanisms in various vertebrates this may be a reasonable explanation of the current data. It has been shown that SOX9 is likely to perform multiple functions during mammalian sex determination and testis differentiation. For example: mutations in *SOX9* cause campomelic dysplasia and XY female sex reversal; strong evidence supports a role for SOX9 in *AMH* control and *SOX9* expression continues in the developing testis during the final stages of fetal development and testis differentiation. Multiple functions for SOX9 may help explain the difference in testicular *SOX9* expression observed between the alligator/chick and that of the mouse. It is possible that SOX9 is required at different stages of testis development in the different vertebrates and that only some of these functions have been conserved through evolution.

To date, all attempts to clone an orthologue of *SRY* from non-mammalian vertebrates (and also from the monotremes) have failed. It has been suggested that the initiation of testis development by *SRY* is a recently-evolved mammalian specific process. This evolutionary change may have occurred in conjunction with the *SOX9* gene, thus replacing an evolutionary precursor with the *SRY*/*SOX9*-initiated testis pathway present in today's mammals. In mouse it seems probable that *Sox9* functions in both early and late during testis development. Similarly, the initiation of testis-specific *AMH* expression prior to *SOX9* expression during alligator and chick testicular development suggests that SOX9 is not required to initiate *AMH* expression (at least) and probably not for the initiation of Sertoli cell differentiation in these species. However, the strong testis-specific up-regulation of alligator/chick SOX9 in the later stages of testis differentiation strongly suggests that *SOX9* has important testis-specific functions in these stages. We suggest that the latter function(s) of *SOX9* in alligator and chick (and probably other non-mammalian vertebrates) testis development have been conserved throughout evolution and are likely to be important in all higher vertebrates. In mammals, we suggest that *SOX9* functions at multiple levels during testis determination, including a recently evolved function during the very early stages of Sertoli cell commitment and other function(s) during the later stages of testis differentiation.

Conclusions

Sex determination in the chicken and alligator embryo shows some conserved and some divergent features when compared to the mammalian system. The sex chromosomes are different to those of mammals, and no *SRY* gene has been found in the chicken or alligator (or any other non-mammal). However, other genes implicated in mammalian gonadal development are also expressed in embryonic chicken and alligator gonads. *DMRT1* is more highly expressed in males from the earliest stages examined. In birds, *DMRT1* may represent the postulated dose-dependent Z-linked factor underlying avian sex determination. In chickens and alligators *SF1* is expressed in both sexes prior to gonadal differentiation, as in the mouse, but expression becomes higher in females than in males after the onset of differentiation (dissimilar to the mouse). *DAX1* expression is up-regulated in both (chicken and alligator) sexes, but is higher in females than in males. Higher *DAX1* and *SF1* expression in the chicken and alligator females may be correlated with the high levels of hormone production in the embryonic ovary. In the chicken and alligator embryo, some sexually dimorphic gene expression occurs prior to histological differentiation of the gonads. *DMRT1*, for example, is more strongly expressed in males than in females from early stages, when the gonads are morphologically undifferentiated. Similarly, *AMH* gene expression at early stages precedes histological differentiation (Oreal et al 1998, Western et al 2000). These observations suggest that sexual differentiation at the molecular level is initiated prior to overt morphological differentiation of the gonads. The chicken and alligator embryos therefore serve as useful models for the analysis of vertebrate sex determination in general.

Acknowledgements

This work is supported by a National Health and Medical Research Council (NH&MRC) grant to AHS & CAS and an Australian Research Council grant to AHS.

References

Arrango NA, Lovell-Badge R, Behringer R 1999 Targeted mutagenesis of the endogenous mouse *Mis* gene promoter: *in vivo* definition of genetic pathways of vertebrate sexual development. Cell 99:409–419

Bardoni B, Zanaria E, Guioli S et al 1994 A dosage sensitive locus at chromosome Xp21 is involved in male to female sex reversal. Nat Genet 7:497–501

Fleming A, Wibbels T, Skipper JK, Crews D 1999 Developmental expression of steroidogenic factor 1 in a turtle with temperature-dependent sex determination. Gen Comp Endocrinol 116:336–346

Graves JAM 1995 The origin and function of the mammalian Y chromosome and Y-borne genes — an evolving understanding. Bioessays 17:311–320

Ikeda Y, Shen WH, Ingraham HA, Parker KL 1994 Developmental expression of mouse steroidogenic factor-1, an essential regulator of the steroid hydroxylases. Mol Endocrinol 8:654–662

Kent J, Wheatley SC, Andrews JE, Sinclair AH, Koopman P 1996 A male-specific role for SOX9 in vertebrate sex determination. Development 122:2813–2822

Lang JW, Andrews HV 1994 Temperature-dependent sex determination in crocodilians. J Exp Biol 270:28–44

Luo X, Ikeda Y, Parker KL 1994 A cell-specific nuclear receptor is essential for adrenal and gonadal development and sexual differentiation. Cell 77:481–490

McQueen HA, McBride D, Miele G, Bird AP, Clinton M 2001 Dosage compensation in birds. Curr Biol 11:253–257

Moniot B, Berta P, Scherer G, Sudbeck P, Poulat F 2000 Male specific expression suggests role of *DMRT1* in human sex determination. Mech Dev 91:323–325

Nachtigal MW, Hirokawa Y, Enyeart-VanHouten DL, Flanagan JN, Hammer GD, Ingraham HA 1998 Wilms' tumor 1 and Dax-1 modulate the orphan nuclear receptor SF-1 in sex-specific gene expression. Cell 93:445–454

Nanda I, Shan Z, Schartl M 1999 300 million years of conserved synteny between chicken Z and human chromosome 9. Nat Genet 21:258–259

Oreal E, Pieau C, Mattei MG et al 1998 Early expression of *AMH* in chicken embryoinc gonads precedes testicular *SOX9* expression. Dev Dyn 212:522–532

Raymond CS, Kettlewell JR, Hirsch B, Bardwell VJ, Zarkower D 1999 Expression of *Dmrt1* in the genital ridge of mouse and chicken embryos suggests a role in vertebrate sexual development. Dev Biol 215:208–220

Smith CA, Elf P, Lang JW, Joss JMP 1995 Aromatase enzyme activity during gonadal sex differentiation in alligator embryos. Differentiation 58:281–290

Smith CA, Smith MJ, Sinclair AH 1999a Gene expression during gonadogenesis in the chicken embryo. Gene 234:395–402

Smith CA, McClive PJ, Western PS, Reed KJ, Sinclair AH 1999b Conservation of a sex-determining gene. Nature 402:601–602

Smith CA, Smith MJ, Sinclair AH 1999c Expression of chicken Steroidogenic factor-1 during gonadal sex differentiation. Gen Comp Endocrinol 113:187–196

Smith CA, Clifford V, Western PS, Wilcox SA, Bell KS, Sinclair AH 2000 Cloning and expression of a *DAX1* homologue in the chicken embryo. J Molec Endocrinol 24: 23–32

Swain A, Zanaria E, Hacker A, Lovell-Badge R, Camerino G 1996 Mouse *Dax1* expression is consistent with a role in sex determination as well as in adrenal and hypothalamus function. Nat Genet 12:404–409

Swain A, Narvaez V, Burgoyne P, Camerino G, Lovell-Badge R 1998 *Dax1* antagonises *Sry* action in mammalian sex determination. Nature 391:761–767

Western PS, Harry JL, Graves JAM, Sinclair AH 1999a Temperature-dependent sex determination in the American alligator: AMH precedes SOX9 expression. Dev Dyn 216:411–419

Western PS, Harry JL, Graves JAM, Sinclair AH 1999b Temperature-dependent sex determination: upregulation of SOX9 expression after commitment to male development. Dev Dyn 214:171–177

Western PS, Harry JL, Graves JAM, Sinclair AH 2000 Temperature-dependent sex determination in the American alligator: expression of SF1, WT1 and DAX1 during gonadogenesis. Gene 241:223–232

Wibbels T, Cowan J, LeBoeuf R 1998 Temperature-dependent sex determination in the red-eared slider turtle, *Trachemys scripta.* J Exp Zool 281:409–416
Yu RN, Masafumi I, Saunders TL, Camper SA, Jameson JL 1998 Role of Ahch in gonadal development and gametogenesis. Nat Genet 20:353–357
Zazopoulos E, Enzo L, Stocco DM, Sassone-Corsi P 1997 DNA binding and transciptional repression by DAX-1 blocks steroidogenesis. Nature 390:311–315
Zanaria E, Muscatelli F, Bardoni B et al 1994 An unusual member of the nuclear hormone receptor superfamily responsible for X-linked adrenal hypoplasia congenita. Nature 372:635–641

DISCUSSION

Short: It always used to be said that if the unilateral ovary was removed, then the contralateral gonadal rudiment would hypertrophy as a testis. Is this true?

Sinclair: It doesn't form a complete testis. It looks a little bit like a testis, but it isn't functional. Some people claim that they have seen germ cells in these testes.

Mittwoch: I thought that if the ovary is removed at an early stage, the germ cells will still be in the right gonad, and the amount of testicular differentiation will vary. You may get a few sperm cells. If the ovary is removed later than one month after hatching, the germ cells in the right gonad will have disappeared and there will be no spermatogenesis, but testosterone production may occur (Domm 1939, King 1975). Do you have any explanation at all about the difference between left and right?

Sinclair: Unfortunately not. We would like to do subtractions between the left and right ovaries to see which genes are being turned on and off. This is one of the fascinating things about the chick. One of the other nice features is that you can also reverse sex using aromatase blockers. You can take a ZW individual, and by using an aromatase blocker you can induce a bilateral testis to form if you intervene at the right time. Timing is crucial.

Short: Isn't it true that in some species of birds, such as budgerigars, the side on which the ovary occurs is reversed?

Sinclair: I don't recall this.

Graves: In rattites I think both sides develop into an ovary.

Renfree: Many birds have both sides. What is interesting is that in monotremes, in the platypus there is only one functional ovary, and it is the right one. But in the echidna, there are two functional ovaries. This is not unique to chickens.

Sinclair: Professor John Hudson is a paediatric surgeon (Royal Children's Hospital, Melbourne), who has operated on lots of children with intersex disorders. He says that in cases of hermaphrodites, he often sees development of the left gonad into an ovary, and the right gonad is usually a testicle.

Vilain: If I remember correctly it is the same asymmetry in mouse. True hermaphrodite mice more commonly have the right gonad as an ovary. One possible reason for this is anatomical, because of the anatomic level and the embryonic development of the renal vein that differs on both sides.

McLaren: In female mice, where the right ovary sheds more eggs than the left (McLaren 1963), the ovarian artery comes off the dorsal aorta more cranially on the right side than on the left. Perhaps blood pressure is slightly higher.

Renfree: In males, John Hudson notices that in failure of testicular descent it is uniformly the right side that is retained.

Wilkins: There's a lot known about the molecular biology of left–right differences. I remember Gail Martin saying that birds and mammals differ in the placement of some components. Many of these differences can be traced back to early embryogenesis.

Koopman: Andrew Sinclair, do you have any feel for whether in chickens it is *AMH* that is expressed early or *SOX9* that is expressed late, compared with mouse, or both? This is important, because if *SOX9* is expressed late, it challenges the idea that *SOX9* is the common sex-determining gene in vertebrates.

Sinclair: It is hard to make this comparison between chick and mouse. My best guess would be that *SOX9* is coming on a bit late.

Capel: Do you see testis cord structures before you see *SOX9*? This is what appears to be the case in your *AMH in situs*.

Sinclair: Yes, it does seem that this is happening. Pre-Sertoli cells are appearing before strong up-regulation of *SOX9*.

Koopman: We find that *Sox8* is male-specific during gonad development in mice. Do you have any evidence that *SOX8* is male-specific in chickens?

Sinclair: The *in situs* on the chick gonad do not show any difference.

Capel: Have you tried to do migration experiments in the chick yet?

Sinclair: No, but we intend to start doing this.

Capel: We have tried them in turtle, without any success so far. We are having temperature problems, trying to culture a genetically labelled mouse mesonephros with a turtle gonad at a temperature that produces males in turtle gonads (26 °C). Culture temperatures would be more compatible in chick. Since turtles are seasonal breeders we only have one shot at it a year.

Short: One says glibly that in mammals, sex is determined at fertilization; in birds it is determined at ovulation. We know that we can now separate mammalian X- from Y-bearing sperm, but can we distinguish between avian Z-bearing and W-bearing eggs?

Charlesworth: There's a good deal of convincing evidence coming out that birds can regulate their sex ratio in response to environmental factors. This has been shown in parrots and Seychelles warblers among others.

Wilkins: Are you sure that this is interference with the sex-determining mechanism and not differential survival?

Charlesworth: My guess is that it is some kind of directed disjunction of the chromosome, but the mechanism is not known.

Renfree: In the Seychelles warbler it is differential survival.

Short: You would think that with such a vast gamete, with a basic sex difference, there ought to be some distinguishing feature.

Graves: It's hard enough to separate X- and Y-bearing sperm.

Harley: Does anyone know whether *SOX9* is involved in regulating aromatase, perhaps via SF1 in some way?

Sinclair: No.

Wilkins: I'm confused about SF1. Yesterday it was being discussed as something that is important for testis determination in mammals, but you are saying that in birds its main function is to boost aromatase.

Sinclair: It could act at many different points in the pathway. The ovary is more active in birds compared with mammals, and it requires SF1 to up-regulate aromatase to produce more oestrogen, which is necessary for normal ovarian development.

Mittwoch: In mammals that develop in female-hormonal environment, the testis must develop early, but in birds this may not be so necessary.

Capel: When do germ cells in bird ovaries enter meiosis? This might give the male pathway longer to work.

Sinclair: I don't know.

Short: I know we are supposed to be discussing sex determination and not differentiation, but I can't help adding one thing that always amazes me, that in birds if you ovariectomize the female, it develops the male plumage. If you take the ovaries out of a peahen it turns into a peacock (Owens & Short 1995). It looks as if in birds the male plumage is the neutral state on which the female plumage is superimposed as a cryptic defence strategy by the action of ovarian oestrogen. Is this true of anything other than plumage?

Lovell-Badge: This is occasionally seen in gynandromorphs. These are amazing birds in which one side is male and the other side is female.

Scherer: Andrew Sinclair, you mentioned that *DMRT1* is a candidate sex-determining gene. Do you know anyone who has tried to work out whether *DMRT1* is dosage compensated?

Sinclair: We haven't. Mike Clinton at the Roslin Institue, Edinburgh, has just published a paper showing that there is dosage compensation of Z-linked genes in birds, but *DMRT1* was not examined (McQueen et al 2001).

Zarkower: In my paper (Zarkower 2002, this volume) I mention some possible mechanisms by which *Dmrt1* expression is adjusted to allow it to avoid dosage compensation.

Short: I have always been fascinated by the old work of F. A. E. Crew (1927), who reported a complete functional female–male sex reversal in a chicken. This bird started life as a hen laying eggs and then became a cockerel. This is common enough, but in this case the cockerel was fertile.

Vilain: This is reminiscent of 5α-reductase deficiency in humans. Although they are not fertile, they start their lives as girls and become boys at puberty.

Short: The amazing thing about this avian case of sex reversal is that both sexes were fertile. I can't think of another example of a functional hermaphrodite. It suggests an interesting lability of the avian germline, switching from female gamete production to male gamete production depending on the environment of the soma of the gonad.

References

Crew FAE 1927 The genetics of sexuality in animals. Cambridge University Press, Cambridge

Domm LV 1939 Modifications in sex and sexual characters in birds. In: Allen E (ed) Sex and internal secretions, 2nd edn. Ballière Tindall, London, p 227–327

King AS 1975 Aves urogenital system. In: Getty R (ed) Sisson and Grossman's the anatomy of the domestic animals, 5th edn. Saunders, Philadelphia, p 1919–1964

McLaren A 1963 The distribution of eggs and embryos between sides in the mouse. J Endocrinol 27:157–181

McQueen HA, McBride D, Miele G, Bird AP, Clinton M 2001 Dosage compensation in birds. Curr Biol 11:253–257

Owens IPF, Short RV 1995 Hormonal basis of sexual dimorphism in birds: implications for new theories of sexual selection. Trends Ecol Evol 10:44–47

Zarkower D 2002 Invertebrates may not be so different after all. In: The genetics and biology of sex determination. Wiley, Chichester (Novartis Found Symp 244) p 115–135

Invertebrates may not be so different after all

David Zarkower

Department of Genetics, Cell Biology, and Development, University of Minnesota, Minneapolis, MN 55455, USA

Abstract. Sex determination is widespread, but uses highly varied molecular mechanisms. A possible case of conservation between phyla is that of *doublesex* (*dsx*) from *Drosophila* and *mab-3* (male abnormal 3) from *Caenorhabditis elegans*, genes related in sequence and some elements of function. *mab-3* controls multiple aspects of male development, including sense organ formation in the tail and yolk transcription in the intestine, both similar to functions of *dsx*. Indeed, the male isoform of DSX can replace MAB-3 in *C. elegans*. Do related genes control sexual development in vertebrates, despite great differences in the biology of sex determination? We have identified several *dsx*-related genes in mouse and human. One, *Dmrt1*, appears to play a conserved role in vertebrate male gonad development. In humans, *DMRT1* maps to a short interval required for testis differentiation. In all vertebrates examined, including mammals, birds, fish, and reptiles, *Dmrt1* is expressed early in the genital ridge, in most cases with higher expression in future male gonads. A null mutation in murine *Dmrt1* causes severe defects in testis differentiation, resembling those associated with human deletions removing the gene. Mutant females are unaffected. Other DM domain genes are expressed in embryonic gonad and are currently under study.

2002 The genetics and biology of sex determination. Wiley, Chichester (Novartis Foundation Symposium 244) p 115–135

Genetic approaches have identified many genes that control the establishment of sexual dimorphism, particularly in the model organisms *Caenorhabditis elegans* and *Drosophila melanogaster*. Surprisingly, however, the cloning of these genes has revealed almost no molecular similarity in the regulatory pathways that determine sex in these two species, or indeed between sex determining genes in species of any two phyla. This contrasts with other major developmental processes, such as patterning of the primary body axes, where homologous genes play highly conserved roles in many highly distantly related phyla.

Why is sex determination not obviously conserved? Conceivably the answer could be that sex determining mechanisms have, in fact, arisen independently multiple times. This seems unlikely, and a more plausible explanation is that sex

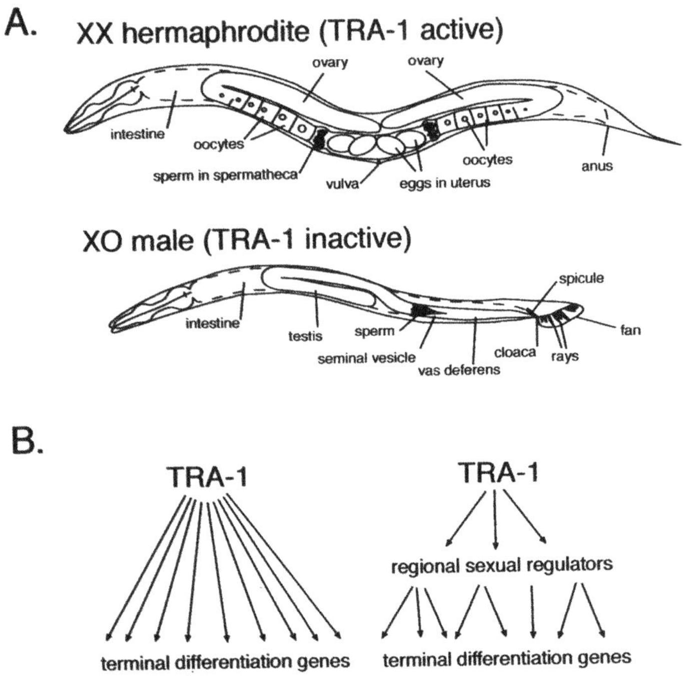

FIG. 1. Sexual dimorphism in *C. elegans*. (A) The sexes of the nematode worm *C. elegans*. Top: XX hermaphrodite, in which the TRA-1 protein is active. The *C. elegans* hermaphrodite is somatically female, but briefly undergoes male differentiation of the germline before switching to oogenesis, generating a mixed sex germ line. Bottom: XO male, in which TRA-1 is inactive. There are many differences between the two sexes. In addition to those indicated, there is extensive sexual dimorphism in the nervous system and musculature. (B) Alternative models of *tra-1* action. Left: TRA-1 could directly repress all genes that are expressed dimorphically. Right: TRA-1 could directly control a smaller number of downstream genes that in turn regulate the rest of the dimorphically expressed genes.

determination mechanisms evolve quickly, with little conservation recognizable over the time span that separates current phyla. Our work suggests that some similarities do remain, at least among invertebrates.

Invertebrate sex determination is a useful paradigm for regulation by genetic switches with major developmental consequences (reviewed by Cline & Meyer 1996). In *C. elegans*, for example, more than 30% of cells are sexually specialized (Fig. 1), and sexual development requires the differential control of cell lineages, cell migrations, programmed cell death, morphogenesis, and other processes of fundamental biological importance (Hodgkin 1988, Hodgkin et al 1989). Determining how the sex determination pathway causes these events to occur sex-specifically will not only explain sexual dimorphism, but also will help illuminate how these processes are controlled and executed in other contexts.

Nematode sexual development is controlled by a regulatory cascade that reads the number of X chromosomes (in the form of the ratio of X chromosomes to sets of autosomes, or X:A) and sets the activity of the *transformer-1* (*tra-1*) gene. Accordingly, in XX animals *tra-1* is active and promotes female somatic development, whereas in XO animals *tra-1* is inactive, permitting male somatic development to occur (Hodgkin 1987, Schedl et al 1989). *tra-1* is genetically epistatic to all of the other globally-acting sex determination genes in the soma, and therefore these genes can be viewed as serving primarily to ensure that *tra-1* activity is appropriately controlled in the two sexes (Hodgkin 1987, Schedl et al 1989). Genetic analysis demonstrates that TRA-1 can regulate, directly or indirectly, all genes required for somatic sexual differentiation (Hodgkin 1987). *tra-1* encodes a zinc finger transcription factor, TRA-1 (Zarkower & Hodgkin 1992), and the identification of the genes whose transcription TRA-1 regulates will be crucial to an understanding of how sexual dimorphism is established.

One can envision at least two general models for the control of sexual dimorphism by TRA-1 (Fig. 1). In principle TRA-1 might directly regulate the transcription of all genes that must be differentially expressed in the two sexes. Alternatively, TRA-1 might 'delegate' its regulatory authority to a suite of downstream sexual regulators. Each of these downstream genes, directly controlled by TRA-1, would then regulate a subset of sexually dimorphic genes responsible for sexual differentiation. The latter model appears to be more accurate, based in part on the study of one direct TRA-1 target gene called *mab-3*.

Similarity between worm and fly sexual regulators

mab-3 (male abnormal 3) was identified by Jonathan Hodgkin in a genetic screen for males incapable of mating (Shen & Hodgkin 1988). Mutant hermaphrodites are unaffected, but males have at least two very different defects (Fig. 2). In the tail, *mab-3* males lack sense organs of the peripheral nervous system called V rays, because the ray neuroblasts fail to differentiate properly. In the intestine *mab-3* mutant males fail to repress transcription of vitellogenin (yolk protein) genes. Thus the tail is defective and the intestine is sex-reversed.

Cloning of *mab-3* revealed that it is related to the *doublesex* (*dsx*) gene of *D. melanogaster* (Raymond et al 1998). In particular, both genes encode proteins containing a novel zinc finger DNA binding domain that we named the DM domain (after *dsx* and *mab-3*). This motif was functionally identified by Burtis and colleagues in *dsx* on the basis of its ability to bind DNA *in vitro* (Erdman & Burtis 1993). Subsequent database searches and degenerate PCR approaches have identified a number of additional DM domain-containing genes in a variety of

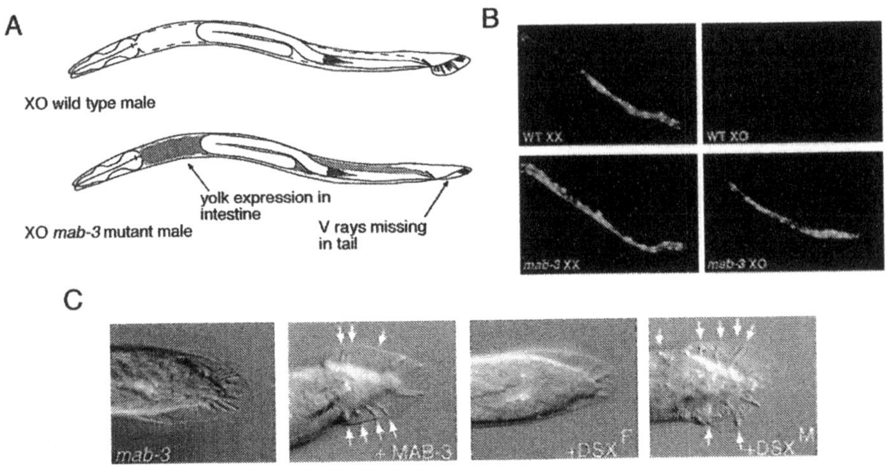

FIG. 2. Functions of *mab-3* in male development. (A) Cartoon of *mab-3* male phenotype showing loss of V rays in the tail and ectopic expression of yolk (vitellogenin) in the intestine. (B) Transcriptional regulation of vitellogenins by *mab-3*. A *vit-2* promoter fragment fused to green fluorescent protein (GFP) accurately recapitulates vitellogenin expression and regulation by *mab-3*. Thus *mab-3* regulates *vit-2* expression at the level of transcription. Mutation of a MAB-3 binding site (not shown) eliminates sex-specific regulation, indicating that the regulation is direct (adapted from Yi & Zarkower 1999, with permission). (C) Dsx^M can replace MAB-3 *in vivo*. *mab-3* (*null*) mutant male tail (far left) lacks V rays but retains T rays. Expressing MAB-3 by heatshock (second from left) restores V ray development. Expressing DsxF has no effect (third from left), but expressing Dsx^M (far right) can restore ray development nearly as well as MAB-3 (adapted from Raymond et al 1998, with permission).

species (Ottolenghi et al 2000, Raymond et al 1999b). The DM domain has an unusual intertwined structure and, uniquely among zinc finger motifs, binds DNA by interaction with the minor groove (Zhu et al 2000).

Of what significance is the molecular similarity between *mab-3* and *dsx*? The less interesting possibility is that this is a case of either convergence or coincidence. Indeed, *C. elegans* has at least twelve DM domain genes and *Drosophila* has four. Moreover, DSX and MAB-3 resemble one another at the protein sequence level no more closely than other pairs of DM domain proteins in the two organisms. However, functional similarities between the two genes suggest a more interesting alternative: that *dsx* and *mab-3* are, in fact, descended from a common ancient ancestral sexual regulator (Fig. 2). There are four lines of evidence. First, the two genes both function downstream in their respective pathways, acting in parallel with other downstream regulators, and controlled by upstream global regulators. Second, the two genes control related sex-specific processes, including sense organ differentiation and yolk transcription. Third, the two genes encode proteins that bind related DNA sequences (Yi & Zarkower 1999).

Fourth, the male isoform of DSX (but not the female isoform) can replace MAB-3 in *C. elegans* sensory ray differentiation (Raymond et al 1998). These multiple similarities have led us to suggest that *dsx* and *mab-3* may well be an example of evolutionary conservation (Raymond et al 1998, Yi et al 2000, Yi & Zarkower 1999).

How does *mab-3* fit into the *C. elegans* sex determination pathway? Reporter gene analysis has shown that the *mab-3* promoter contains neuron-specific and intestine-specific regulatory elements (Yi et al 2000). Transgenic experiments reveal that, in the intestine, TRA-1 directly represses *mab-3* in XX animals and MAB-3 directly represses vitellogenins in XO animals. In this tissue, therefore, the pathway appears to be completely connected, from the X chromosome to products of terminal differentiation. In the nervous system TRA-1 appears to regulate *mab-3* indirectly. *mab-3* serves to potentiate the function of the neurogenic bHLH transcription factor *lin-32* to promote sensory ray neuroblast formation (Yi et al 2000). *mab-3* also is required for normal interaction of males with hermaphrodites and for expression of at least two genes in male sensory neurons that may mediate this interaction (Yi et al 2000). It is not known whether *dsx* performs similar functions in the *Drosophila* nervous system. We have identified several genetic suppressors that can restore sensory ray differentiation to *mab-3* null mutants (J. Ross & D. Zarkower, unpublished results). It will be of interest to see whether genes related to these suppressors interact with *dsx* in the fly.

A human DM domain gene linked to testis dysgenesis

The similarities between *mab-3* and *dsx* have raised the possibility that DM domain genes might be conserved in vertebrate sexual development. Before considering the evidence, it is important to note two factors that complicate the issue. First, all species we have examined have multiple DM domain genes, and there is evidence that not all are involved in sexual development. Thus the DM domain on its own provides no clue as to biological function. Second, vertebrate sexual development is very different from that of invertebrates. In the former, the key events of sex determination occur in the early embryonic gonad, while in the latter, sex determination occurs throughout the body, in most or possibly all cells. As a consequence, there is no expectation that a vertebrate *dsx* or *mab-3* counterpart should perform analogous functions, such as regulating yolk expression or sex-specific sense organ differentiation. We have sought instead DM domain genes expressed in the genital ridge (the gonad primordium) of vertebrates with diverse sex-determining systems.

Our searches for vertebrate DM domain genes, both *in silico* and by degenerate PCR, have so far identified six genes. One of these, *Dmrt1*, is involved exclusively in sexual development; another, *Dmrt2*, is required for patterning of the somatic mesoderm (K. Seo, J. R. Kettlewell, H. Kokubo, D. Zarkower & R. Johnson, unpublished results); and the rest are currently under investigation. We first found the human *DMRT1* gene in a database search that identified a testis cDNA clone containing a DM domain. Hybridization of this cDNA to a multi-tissue dot blot with mRNA from 50 tissues only detected expression in the testis (Raymond et al 1998). Mapping of *DMRT1* by fluorescence *in situ* hybridization placed the gene on the distal short arm of chromosome 9 (9p24.3) (Raymond et al 1998). This region when hemizygous is associated with defective testis differentiation, severe enough in some cases to cause feminization of non-gonadal tissues (Bennett et al 1993, Crocker et al 1988, Hoo et al 1989).

Sequencing of the *DMRT1* coding exons from a large number of sex-reversed individuals, both XY females and XX males, failed to identify an unambiguous point mutation (Raymond et al 1999b). One possibility is that *DMRT1* is not involved in the 9p deletion syndrome. However its embryonic expression (Moniot et al 2000), combined with expression and functional data from the mouse, as described below, suggests an important role in human testis development. An alternative explanation is that the 9p deletions affect another gene in addition to *DMRT1*, and the compound hemizygosity of these two genes results in the observed phenotype. Intriguingly, the DM domain gene *DMRT3* is the nearest neighbour of *DMRT1* in both mouse (C. S. Raymond & D. Zarkower, unpublished results) and human (Ottolenghi et al 2000), and is expressed in the embryonic testis in the mouse (C. S. Raymond, S. Kim & D. Zarkower, unpublished results).

Conserved *Dmrt1* expression in diverse vertebrates

Studies of *DMRT1* homologues in other vertebrates suggest a widely conserved role in male gonad development. In birds the sex chromosomes are denoted Z and W, with females (ZW) the heterogametic sex. The Z chromosome has extensive conserved synteny with human chromosome 9, including the presence of *Dmrt1* (Nanda et al 1999). Avian *Dmrt1* is expressed in the genital ridge at higher levels in ZZ than ZW embryos, starting prior to sexual differentiation (Fig. 3; Raymond et al 1999a, Smith et al 1999). Non-coding RNAs transcribed female-specifically from a tightly linked region (MHM) accumulate on the Z chromosome adjacent to the *Dmrt1* locus, possibly helping explain the reduced expression of *Dmrt1* in ZW embryos (Teranishi et al 2001). The sex linkage of *Dmrt1* in birds is quite ancient, as the gene is located on the Z chromosome of the emu (S. Shetty & J. A. M. Graves, personal communication).

In many reptiles sex is determined by the ambient temperature during a critical period of embryonic development. In the Red-Eared Slider turtle, we found that *Dmrt1* is expressed in the genital ridge at higher levels in embryos incubated at the

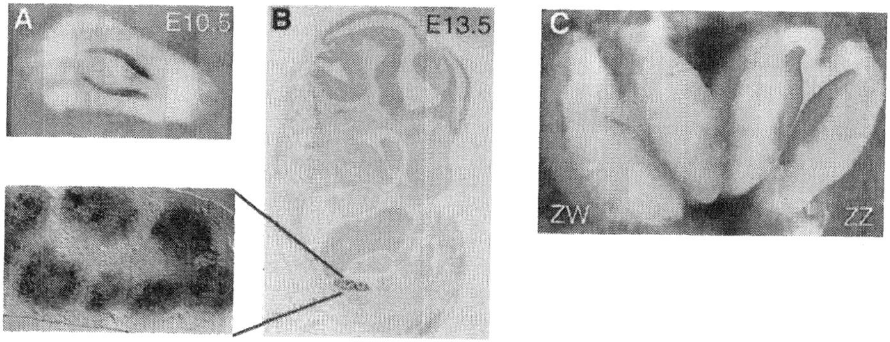

FIG. 3. Conserved expression of *Dmrt1* in the embryonic gonad. (A) *In situ* hybridization of *Dmrt1* probe to dissected E10.5 mouse embryo shows expression in genital ridges. (B) *In situ* hybridization of *Dmrt1* probe to sectioned E13.5 XY embryo shows expression only in testis (enlargement at right shows expression in pre-Sertoli cells and germ cells but not in interstitial cells of the testis). Testis-specific expression has been confirmed by RNase protection and RT-PCR experiments (not shown). (C) *In situ* hybridization to embryonic stage 31 chicken mesonephros/genital ridge complexes showing higher *Dmrt1* mRNA expression in ZZ (male) than ZW (female) genital ridges. (Adapted from Raymond et al 1999a, with permission.)

male promoting temperature (Kettlewell et al 2000). As in chickens, differential expression is evident prior to the onset of sexual differentiation. Similar results have been observed in the American Alligator (Smith et al 1999). Lastly, others have found that *Dmrt1* is expressed male-specifically in the early genital ridge in fish (Marchand et al 2000).

The fact that dimorphic *Dmrt1* expression precedes sexual differentiation in so many vertebrate taxa is particularly striking, and suggests that the gene has been functionally maintained during the evolution of different vertebrate primary sex determining mechanisms for at least 300 million years. This is apparently not the case with other vertebrate sexual regulators that have been examined. *Sry*, for example, does not exist outside the mammals. The related gene *Sox9* is widely conserved and male-enriched among vertebrates, but in birds and reptiles its expression does not become male-enriched until after the onset of testis differentiation. Thus it is possible that *Dmrt1* acts at an earlier step and in a greater variety of vertebrates than other sexual regulators that have been identified. Functional studies are needed to test this possibility.

Dmrt1 is required for testis differentiation in the mouse

As outlined above, the widespread conservation of *Dmrt1* sequence and expression among vertebrates is highly suggestive of a conserved role for *Dmrt1* in testis

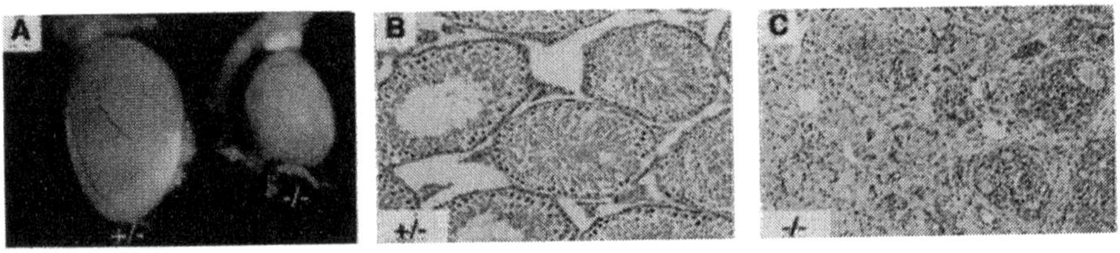

FIG. 4. *Dmrt1* null phenotype. (A) Adult testes from heterozygous (+/−) and homozygous (−/−) *Dmrt1* mutant mice. (B) Section of testis from heterozygous mutant showing normal morphology. (C) Section from homozygous mutant showing severely dysmorphic phenotype. A few cord remnants are present; germ cells are missing; Sertoli cells have immature morphology and are dying, and there is infiltration by macrophage-like cells. (Adapted from Raymond et al 2000, with permission.)

development. To test the function of *Dmrt1* we disrupted the gene in the mouse by homologous recombination (Raymond et al 2000). In the targeted allele, the basal promoter and first exon (encoding the DM domain) of *Dmrt1* are flanked by recognition sites for Cre recombinase. Excision of these sequences is predicted to render *Dmrt1* non-functional, and as expected no protein is made from the deleted allele.

Dmrt1 null mutant XX animals are unaffected by the mutation, with normal ovary development and fertility, but homozygous mutant XY animals have severely dysmorphic testes. Surprisingly, despite the early genital ridge expression of *Dmrt1* (Fig. 3), embryonic testis development is normal and extragonadal development is male. Postnatally, however, there are multiple defects in testis differentiation. The first morphological defect is apparent at about 7 days postnatally, when germ cells should move from the centre to the margin of the seminiferous tubules and differentiate into spermatogonia. This does not happen in the *Dmrt1* mutant testis. Instead, between 7 and 10 days postnatally, when meiosis normally begins, germ cells in the mutant testis die, leaving seminiferous tubules containing only immature Sertoli cells. The Sertoli cells also fail to complete differentiation, as judged by morphology and gene expression, and later the testis becomes highly disorganized, with few remaining seminiferous tubules, extensive cell death, and invasion by macrophages (Fig. 4).

Does the *Dmrt1* mutant phenotype account for the testis defects seen in humans with 9p deletions? Certainly there are differences. Most notably, 9p deletions can cause embryonic testis defects leading to feminization outside the gonad, while the mouse mutants show only postnatal testis defects. Also, 9p deletions are haploinsufficient, while murine *Dmrt1* is recessive. There are, however, striking similarities between the mouse and human phenotypes, suggesting that loss of *DMRT1* is at least one important component of the 9p deletion phenotype. 9p

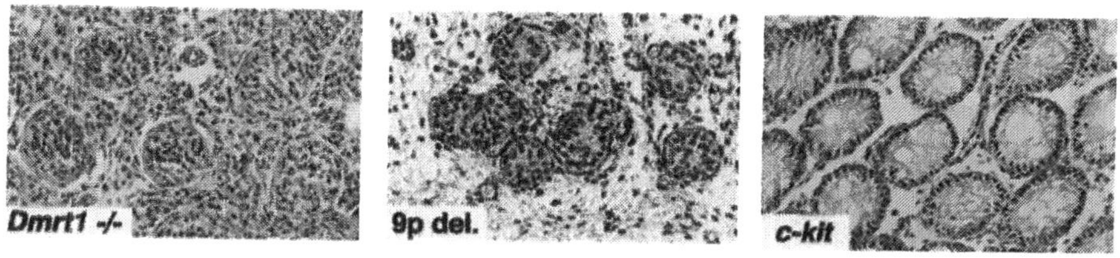

FIG. 5. Human 9p deletions and murine *Dmrt1* mutation have similar phenotypes. Left panel: Section of testis from 6 week old *Dmrt1* mutant mouse. Seminiferous tubules are still present, but lack germ cells and contain uniformly distributed undifferentiated Sertoli cells. Middle panel: Section of immature dysgenic testis from a 12 year old 46,XY patient with deletion of 9p24.3. Seminiferous tubules are present but, as in the mouse, lack germ cells and a central lumenal space, and contain undifferentiated Sertoli cells. (Left and right panels from Raymond et al 2000, and middle panel from Ion et al 1998, with permission.)

deletions, like the *Dmrt1* mutation, affect testis but not ovary development, and despite little published histology from 9p-deleted humans, young *Dmrt1* mutant mice and 9p-deleted humans do appear similar in testis morphology (Fig. 5). In both cases seminiferous tubules, if present, are deficient in germ cells and contain evenly distributed immature Sertoli cells. This contrasts with the effect of simple germ cell loss, such as in a c-*kit* mutant. In that case, Sertoli cells complete differentiation and are found at the margins of the seminiferous tubules, with only Sertoli cell cytoplasm present in the centre.

Comparison of expression and mutant phenotype in the mouse raises as yet unanswered questions concerning what, if any, is the role of *Dmrt1* in the embryonic testis. The reasons to suspect an early function for *Dmrt1* are primarily its conserved early expression in diverse vertebrates and the XY feminization that can occur in humans with 9p24.3 deletions. In the mouse any such role must be genetically redundant, at least in the strain background in which the mutant was made.

There are several possible explanations for the differences between 9p deletions in human and *Dmrt1* mutations in mouse. First, *Dmrt1* may simply function later in mouse than in human. Second, *Dmrt1* may act redundantly in the early gonad in mouse but non-redundantly in human. Third, as discussed above, 9p deletions may remove additional genes involved in testis development, leading to a more severe phenotype than a mutation in *Dmrt1* alone. Fourth, genetic background is likely to be important, since only a minority of 9p-deleted humans show signs of sex reversal. We are currently testing the latter two possibilities.

What is the relationship of *dsx*, *mab-3* and *Dmrt1*?

Returning to the original question, do *dsx*, *mab-3* and perhaps *Dmrt1* derive from the evolutionary conservation of an ancestral sex-determining gene? Alternatively, is this a case of convergent evolution or coincidence, with flies, worms and vertebrates choosing independently to regulate aspects of male development with DM domain-containing transcription factors? The case for conservation is strongest between insects and nematodes, where *dsx* and *mab-3* perform some analogous biological functions and are at least partially interchangeable. Likewise, among the vertebrates the apparently universal early gonad expression of *Dmrt1* suggests a longstanding role in testis development. Between the invertebrates and the vertebrates, however, agnosticism currently seems safest, as the fundamentally different biology of vertebrate sex determination confounds simple comparisons. The study of intermediate taxa should clarify how widely DM domain genes are involved in sexual development and in what capacities, and will help determine the evolutionary relationships of these genes.

Assuming for the moment that the similarity of *mab-3* and *dsx* does reflect evolutionary conservation, why are these genes conserved while the genes that regulate them are not? Two factors particularly deserve mention (for further discussion, see Marin & Baker 1998, Zarkower 2001). First, Wilkins has proposed that sex determining regulatory pathways evolve by accretion of regulators in a 'bottom-up' fashion (Wilkins 1995). In this model, new regulators, which can be of any sort, are recruited to the top of the pathway as needed to correct imbalances of sex ratio by regulating downstream genes in one sex or the other. As a result, the ancient genes are found downstream, whereas the upstream genes are more recent additions. In addition, it has been suggested that downstream genes in any regulatory pathway are subject to greater constraint due to pleiotropy. This is because they regulate multiple target genes and the upstream genes mainly do not (Waxman & Peck 1998). How generally this principle applies to sex determination is unclear, as, for example, *tra-1* is both highly pleiotropic and exceedingly rapid in its evolution (de Bono & Hodgkin 1996). Again assuming that *mab-3* and *dsx*, and perhaps *Dmrt1*, are the result of evolutionary conservation, is this a unique example or are there other cases of conservation of sexual regulators between these phyla? Aided by genome sequencing and new molecular genetic tools, efforts are under way to identify large numbers of genes involved in sexual regulation in worms, flies and mice. These screens will eventually help settle this intriguing question.

Acknowledgements

The author thanks P. Koopman and the Novartis Foundation for the opportunity to attend this Symposium, and J. Dempster for much logistical assistance. The work from the author's laboratory described here has been supported by the University of Minnesota Center for Developmental Biology, the University of Minnesota Graduate School, Minnesota Medical Foundation and the National Institutes of Health.

References

Bennett CP, Docherty Z, Robb SA, Ramani P, Hawkins JR, Grant D 1993 Deletion 9p and sex reversal. J Med Genet 30:518–520

Cline TW, Meyer BJ 1996 Vive la différence: males versus females in flies versus worms. Ann Rev Genet 30:637–702

Crocker M, Coghill SB, Cortinho R 1988 An unbalanced autosomal translocation (7;9) associated with feminization. Clin Genet 34:70–73

de Bono M, Hodgkin J 1996 Evolution of sex determination in *Caenorhabditis*: unusually high divergence of tra-1 and its functional consequences. Genetics 144:587–595

Erdman SE, Burtis KC 1993 The Drosophila doublesex proteins share a novel zinc finger related DNA binding domain. EMBO J 12:527–535

Hodgkin J 1987 A genetic analysis of the sex-determining gene, tra-1, in the nematode *Caenorhabditis elegans*. Genes Dev 1:731–745

Hodgkin J 1988 Sexual dimorphism and sex determination. In: Wood WB (ed) The nematode *Caenorhabditis elegans*. Cold Spring Harbor, New York, p 243–279

Hodgkin J, Chisholm AD, Shen MM 1989 Major sex-determining genes and the control of sexual dimorphism in *Caenorhabditis elegans*. Genome 31:625–637

Hoo JJ, Salafsky IS, Lin CC 1989 Possible localisation of a recessive testis forming gene on 9p24. Am J Hum Genet 45:A73

Ion R, Telvi L, Chaussain J-L et al 1998 Failure of testicular development associated with a rearrangement of 9p24.1 proximal to the *SNF2* gene. Hum Genet 102:151–156

Kettlewell JR, Raymond CS, Zarkower D 2000 Temperature-dependent expression of turtle Dmrt1 prior to sexual differentiation. Genesis 26:174–178

Marchand O, Govoroun M, D'Cotta H et al 2000 DMRT1 expression during gonadal differentiation and spermatogenesis in the rainbow trout, *Oncorhynchus mykiss*. Biochim Biophys Acta 1493:180–187

Marin I, Baker BS 1998 The evolutionary dynamics of sex determination. Science 281:1990–1994

Moniot B, Berta P, Scherer G, Sudbeck P, Poulat F 2000 Male specific expression suggests role of DMRT1 in human sex determination. Mech Dev 91:323–325

Nanda I, Shan Z, Schartl M et al 1999 300 million years of conserved synteny between chicken Z and human chromosome 9. Nat Genet 21:258–259

Ottolenghi C, Veitia R, Quintana-Murci L et al 2000 The region on 9p associated with 46,XY sex reversal contains several transcripts expressed in the urogenital system and a novel doublesex-related domain. Genomics 64:170–178

Raymond CS, Shamu CE, Shen MM et al 1998 Evidence for evolutionary conservation of sex-determining genes. Nature 391:691–695

Raymond CS, Kettlewell JR, Hirsch B, Bardwell VJ, Zarkower D 1999a Expression of *Dmrt1* in the genital ridge of mouse and chicken embryos suggests a role in vertebrate sexual development. Dev Biol 215:208–220

Raymond CS, Parker ED, Kettlewell JR et al 1999b A region of human chromosome 9p required for testis development contains two genes related to known sexual regulators. Hum Mol Genet 8:989–996

Raymond CS, Murphy MW, O'Sullivan MG, Bardwell VJ, Zarkower D 2000 *Dmrt1*, a gene related to worm and fly sexual regulators, is required for mammalian testis differentiation. Genes Dev 14:2587–2595

Schedl T, Graham PL, Barton MK, Kimble J 1989 Analysis of the role of tra-1 in germline sex determination in the nematode *Caenorhabditis elegans*. Genetics 123:755–769

Shen MM, Hodgkin J 1988 mab-3, a gene required for sex-specific yolk protein expression and a male-specific lineage in *C. elegans*. Cell 54:1019–1031

Smith CA, McClive PJ, Western PS, Reed KJ, Sinclair AH 1999 Conservation of a sex-determining gene. Nature 402:601–602
Teranishi M, Shimada Y, Hori T et al 2001 Transcripts of the MHM region on the chicken Z chromosome accumulate as non-coding RNA in the nucleus of female cells adjacent to the *DMRT1* locus. Chromosome Res 9:147–165
Waxman D, Peck JR 1998 Pleiotropy and the preservation of perfection. Science 279:1210–1213
Wilkins AS 1995 Moving up the hierarchy: a hypothesis on the evolution of a genetic sex determination pathway. BioEssays 17:71–77
Yi W, Zarkower D 1999 Similarity of DNA binding and transcriptional regulation by *Caenorhabditis elegans* MAB-3 and *Drosophila melanogaster* DSX suggests conservation of sex determining mechanisms. Development 126:873–881
Yi W, Ross JM, Zarkower D 2000 *mab-3* is a direct tra-1 target gene that regulates diverse aspects of *C. elegans* male sexual development and behavior. Development 127:4469–4480
Zarkower D 2001 Establishing sexual dimorphism: conservation amidst diversity? Nat Rev Genet 2:175–185
Zarkower D, Hodgkin J 1992 Molecular analysis of the *C. elegans* sex-determining gene tra-1: a gene encoding two zinc finger proteins. Cell 70:237–249
Zhu L, Wilken J, Phillips NB et al 2000 Sexual dimorphism in diverse metazoans is regulated by a novel class of intertwined zinc fingers. Genes Dev 14:1750–1764

DISCUSSION

Graves: Are there any *tra-1* homologues in mammals?

Zarkower: Yes, the *GLI* genes. Also, *tra-2* of *C. elegans* looks a bit like *Patched.* We think that the worm pathway may have been at least partially formed by recruitment from an unrelated signalling pathway. As far as I know, there's no evidence that the *GLI* genes in vertebrates are involved in any meaningful way in sex determination. It would appear that *mab-3* may have been a more general sexual regulator early on in evolution, and *tra-1* may have been one of the genes that was then recruited upstream. It is intriguing that the family of genes that includes *mab-3* is a relatively large one. It turns out that one of the other DM domain genes is also involved in male differentiation.

Behringer: Could you expand on the preliminary results you have on sex reversal?

Zarkower: These are very preliminary. The initial experiment was quite simple and poorly controlled. We took advantage of the Y chromosome from the *Mus domesticus poschiavinus* strain, which will quite nicely sex reverse on a B6 background but not on most other backgrounds (the DBA is the one that has been most widely looked at). Eva Eicher's lab has used this effect to map loci responsible for the different effect of this Y chromosome on B6 compared with DBA, and thereby identified at least three autosomal loci. We have put our mutant on a mixed background that should not sex reverse, although we need to demonstrate this more clearly, in the presence of the *poschiavinus* Y chromosome. We see sex reversal that segregates perfectly with the *Dmrt1* mutant allele. There are a couple of controls missing. We need to track the B6 chromosomes:

statistically we can argue at this point that it is highly unlikely that the sex reversal we see is due to B6 autosomal alleles, because there is too much of it and it is too perfectly correlated with the *Dmrt1* mutant allele, but we need to prove this. We also need to show that we haven't done something unrelated to *Dmrt1* elsewhere on chromosome 19. It could be that it is not actually caused by the *Dmrt1* mutation but something horrible that happened to the embryonic stem cell line we used. We need to use the targeted but not deleted allele of the gene to show this doesn't cause sex reversal.

Behringer: What do you mean by sex reversal?

Zarkower: At the moment, we mean that externally the animals appear to be female. We need to open them up to see what is inside.

Wilkins: Did you say that there are two other *Dmrt* genes in the mouse?

Zarkower: There are either six or seven, including *Dmrt1*.

Wilkins: In humans there are three that are closely linked.

Zarkower: We found the same group of half a dozen in both human and mouse. We don't know much about the linkage in the mouse, except that *Dmrt1*, *2* and *3*—the ones you are referring to—are linked in mouse as they are in human.

Renfree: You said that *Dmrt1* in mouse was up-regulated or strongest from about 15.5 days, which is when testosterone production begins and there is differentiation as distinct from determination. What role is *Dmrt1* playing in sex determination/differentiation? Is it really a differentiating gene and not a determining gene?

Zarkower: The evidence suggests this at the moment. Depending on what we find when we open up the sex-reversed mice, we may feel differently. One of several possible roles for *Dmrt1* in the mouse is to act as the genital-ridge-specific activator of *Sry*, since there needs to be one. DMRT1 is the only transcription factor that has been identified that is expressed in genital ridge and not elsewhere at the time that *Sry* switches on. Unfortunately there's no evidence for this. I'm hoping that if we can get the sensitized background working well, we will be able to generate *Dmrt1* mutant embryos that are feminized, and test what other genes are affected. The conserved male-specific expression in embryos of other vertebrates would suggest that *Dmrt1* is doing something early in a lot of vertebrates, but not necessarily in the mouse. There is also a report that early expression is sex specific in human, and so what one would like to think is that while the RNA expression isn't sex specific in the early mouse gonad, the function may well be. Interaction of some sort with a gene such as *Sry* could explain this.

Koopman: Wouldn't it appear that *Dmrt1* represents a more ancient gene, and would therefore be further down the pathway than *Sry*?

Zarkower: *Dmrt1* may be more ancestral, as it occurs more widely than *Sry*. Thus one might expect it to be downstream. But when you say 'downstream', a gene that is downstream in a regulatory pathway can nevertheless act at quite an upstream biological step. Let me give you an example. Is *tra-1* an upstream gene or downstream gene? There are 10 genes upstream of it in the genetic pathway, so by that criterion it is a downstream gene. On the other hand, you can ablate all of those genes and artificially turn *tra-1* on and off, and get fertile male/female strains that will mate with each other, proving that *tra-1* can control the whole process. On this basis one could argue that *tra-1* is an upstream gene, and yet the genetics and molecular biology suggest that it is downstream.

Wilkins: With genetic manipulation you can convert a downstream gene into an upstream gene. It is always context dependent.

If the *Dmrt* genes really are early downstream and conserved genes, this poses the interesting question of how the *Drosophila* pathway relates to this whole business. In particular, in *Drosophila* there is differential splicing that looks very different, but I think it is interesting that the *DMRT* genes in human also have differential splicing. The product that is highly expressed in testis is very similar in its exon structure to the *DSX* male copy.

Zarkower: We have no evidence that *Drmt1* is alternatively spliced in any meaningful way. *Drmt2* is alternatively spliced, but it doesn't appear to be expressed in embryonic gonad. We have collaborated with Randy Johnson's lab to knock it out, and it doesn't have a sex-determining phenotype.

Wilkins: I was referring to a paper by Ottolenghi et al (2000), in which they showed differential splicing of what I think was *Drmt2*, but there was one form that was heavily expressed in the testis relative to everything else.

Zarkower: If I remember correctly, their expression analysis was all done in adult tissue. We have also looked at late expression, and those isoforms are very highly expressed in many tissues. In the embryonic gonad in the mouse, we can't detect convincing expression. We would very much like to think that something analogous to *Dsx* is taking place in vertebrates, and maybe in *C. elegans* also. But we haven't seen evidence for this. I think what may be more likely is that the sex-specific splicing in *Drosophila* is a late evolutionary adaptation. The default splice mode for *Dsx* is male-specific, and if splicing regulators were recruited to adjust the sex ratio, your model could explain the very different pathways that exist in these species today.

Harley: Is *Dmrt1* expressed at all germ cell stages? What stage is it arrested at in your knockouts? Have you looked at male motility syndromes?

Zarkower: *Dmrt1* is expressed in germ cells from as early as we have looked, which is 10.5 days. We could and should look earlier. In some *in situ* hybridizations we have seen *Dmrt1* expressed in cells that are just outside of the genital ridge that look like they might be germ cells migrating in at 10.5 days. It may turn on in germ cells before they enter the genital ridge. In terms of the germ cell phenotype in the knockout, the first defect we see in the germ cells is at about

7 d postnatally. In sibling animals that are heterozygotes, those germ cells have migrated peripherally and inserted themselves among the Sertoli cells. The normal cells have begun to differentiate into spermatogonia, whereas in the mutant most of the cells haven't. I'm not sure that we have actually looked the next day, but if we look a couple of days later they are mostly gone. We have been trying to figure out what happened to them. We don't see any convincing difference in apoptosis between the mutant and wild-type. The problem is that this is also the stage when proliferation is picking up again. It could be that there is a steady rate of apoptosis that is unaffected by the mutation, and that due to reduced proliferation the cell population disappears. This is something we intend to test.

Greenfield: Are you able specifically to ablate the pre-Sertoli cells independently of the germ cells, and vice versa?

Zarkower: We have made the mutation as a conditional knockout, and we are currently doing the germ-cell-specific targeting. As you know, Sertoli-cell-specific targeting is a bit more difficult and we haven't done that yet. I should stress that we think that there are probably autonomous defects in both cell types. We know that there has to be a problem with Sertoli cells, because they don't differentiate and they die. The germ cell phenotype could be due to problems with Sertoli cells. We are suspicious, however, that there may be an autonomous requirement for *Dmrt1* in germ cells, which is what we are testing by the germ-cell-specific targeting. The reason for this is that the protein expression of *Dmrt1* in the germ cells goes from relatively low levels to very high levels at about the same stage as the mutant defect becomes apparent. If we look in adults, we see a cycling of expression of *Dmrt1* in early spermatogonia in the adult testis.

Swain: Do you think *Dmrt1* works as a repressor?

Zarkower: We don't know, but Vivian Bardwell's lab has some preliminary data suggesting that it may act as one. If you fuse it to a heterologous DNA-binding domain and do a standard transfection assay, it represses. This is in a heterologous cell system, so it is suggestive but not convincing. Her lab has also done a yeast two-hybrid screen and pulled out a couple of interacting proteins that also interact strongly in a mammalian two-hybrid assay. One of these is related to a protein that has been found in co-repressor complexes.

Swain: If *mab-3* works in *Drosophila*, do you think that these genes are just DM domains with a repressor domain attached?

Zarkower: Not in the same sense that *Sox* genes appear to be relatively non-specific. MAB-3 and DSX are quite highly sequence-specific DNA binding proteins, and we have *in vivo* targets for them. They appear to act as enhancer-blocking proteins involved in short range transcriptional repression, as does *tra-1*. There are reasons to think that this is a particularly good way to evolve a regulator of many genes. In terms of *Dmrt1* we don't know because it has proved difficult to define the binding site. The human DMRT1 protein is extremely oxidation sensitive and very hard to work with.

Capel: Is the expression in the early mouse gonad specific to germ cells, or is it in the somatic cells of both sexes?

Zarkower: In the early gonad we see identical expression in both sexes in germ cells and somatic cells.

Lovell-Badge: We have also looked at this, and the expression is in somatic cells and germ cells at those early stages in both sexes.

Harley: Are your male mice 'aloof'?

Zarkower: They don't plug. We need to do a lot more with them. We only know that they are infertile, and since they have no germ cells it is quite obvious why. Also, their steroid hormone levels are presumably not quite what they could be, since the gonad is severely dysgenic.

Greenfield: Is it inconceivable that *Mab3* or *Dsx* could rescue your mouse mutant?

Zarkower: It is not inconceivable. We have tried the reciprocal experiment, putting *Dmrt1* into a worm, and this didn't work.

Behringer: Do *mab-3* mutant worms have gonad defects?

Zarkower: They don't have gonad defects that we know of, but the gonad doesn't have the same sort of hallowed position in *C. elegans* sex determination that it does in mammals.

Behringer: In the worm, is *mab-3* a sex-determining gene or a sex-differentiation gene?

Zarkower: In the worm this becomes a semantic problem. *mab-3* sex reverses the intestine but the neuroblasts that require *mab-3* in the tail have their presence controlled by *tra-1* and their differentiation by *mab-3*. One can view *mab-3* as a sex-determination gene in one tissue and a sex-differentiation gene in another. This suggests that *mab-3* is at the border between determination and differentiation.

Behringer: It is the same for *Dsx* in the fly?

Zarkower: Yes.

Behringer: If you go on conservation, would *Dmrt1* in vertebrates be determining or differentiating?

Zarkower: It could be either. One can become the other during evolution.

Koopman: Since *Dmrt1* seems to be acting as a repressor, could we fit it into some double repressor model of testis determination?

Zarkower: We could. I'd like to know that it definitely is a repressor, first.

Wilkins: This may seem a little egocentric, but I have been very gratified by the discovery of the *Dmrt* genes. We still don't fully understand their significance, but two colleagues and I have been developing a model for how one can build up the *Drosophila* sex determination pathway through a sequence of mutational steps. We have a workable scheme that involves building up the pathway from the downstream element, *dsx*, not by simple recruitment of repressors but through a sequence of mutations. Each of these mutations reverses the actions of the previous upstream controlling step. From my perspective the pattern of evolution is at least somewhat similar to what I proposed in 1995, where one can begin to describe many of these pathways, starting with an ancestral downstream element and building up the pathways in complex and different patterns in the different animal lineages.

Zarkower: Yes. As I mention in my paper, our results, while limited so far, are quite consistent with the model you proposed.

Capel: Dmrt1 appears to be working as a differentiation factor, at least in the mouse. On the other hand, I think that your evidence in turtles and chickens is fairly strong that it is a very early gene in the decision making process in sex determination. Combined with the data that *Amh* and *Sox9* can reverse their order of expression in different species, I find this whole idea that genes can occupy different positions in the pathway very strange.

Zarkower: If it is reversal of order of action, it is strange. If it is multiple roles for *Sox9* in some species, then it is less strange: you could just lose an early role in some evolutionary lineages.

Capel: So we have many genes with overlapping functions that can shift their order?

Schedl: That certainly holds true for the *Pax* gene family. In mouse mesonephric development *Pax2* is expressed before *Pax8*. In contrast, in *Xenopus* the onset of *Pax2* and *Pax8* expression is swapped. Something like this could happen in the chick, with *Sox8* or *Sox10* being expressed earlier than *Sox9*.

Harley: I'd like to return to your hypothesis about *Dmrt1* being a genital ridge-specific activator for *Sry*. Can you measure *Sry* levels in your knockout model?

Zarkower: We could, but we haven't bothered to do those experiments because there is no phenotype in the background that we made the initial knockout on. If we have a background that will show us the gonad defect, this should be possible. I should stress that this is just one option, and that it might not apply in all mammals.

Capel: If *Dmrt1* is an activator of *Sry*, why didn't this affect the initial function of *Sry*?

Zarkower: Presumably this was because of genetic redundancy, which would also be why there is no early phenotype in a normal genetic background.

Greenfield: I think it will be important to get all these mutants onto microarrays so we can look at the transcriptional consequences of mutating every relevant gene.

Zarkower: We are trying to do this.

Poulat: In human we have seen *DMRT1* expressed only in male. This goes against the idea that *DMRT1* could be upstream of *SRY*. It is expressed at approximately the same time as *SRY* and not in germ cells.

Zarkower: That observation of yours is the main argument against the possibility in humans. Mice may be different, or *Dmrt1* may not activate *Sry*.

Graves: Because *Dmrt1* is an old gene and *Sry* is a very new gene, would you like to speculate whether *Dmrt1* might have taken control of *Sry* or vice versa?

Zarkower: We assume that the ground state in early vertebrates is *Dmrt1* doing something important. This role will depend on what the sex-determining system is. In reptiles it may be a temperature-sensitive allele of *Dmrt1* that gives differential expression at different temperatures; in birds it may be linkage to the Z chromosome, together with a W-dependent methylation difference as a later embellishment. The simplest way of introducing *Sry* in mammals would be for *Sry* to arise as a dominant mutation that controls *Dmrt1*. This would argue against the possibility that the relationship is the other way round.

Lovell-Badge: If *Dsx* and *Mab3* are really conserved in this way and at this level, what about the other genes that are at the same level, such as *fruitless* (*fru*) and *dissatisfaction* (*dsf*) in *Drosophila*? Are they also conserved in *C. elegans*?

Zarkower: No, there are genes related to *Drosophila fru* and *dsf* in *C. elegans*, but we haven't been able to find evidence that these are involved in anything interesting that is sex specific. One thing to stress in the fly is that *dsx* really controls most of sexual differentiation. The other genes known to act at that level in the pathway are doing relatively minor things in a small number of cells. These are probably functions that the worm doesn't have.

Short: Does anyone have any insight into the sex-determination process in plants?

Charlesworth: The ancestral state in flowering plants is to be hermaphrodite. Usually there is very little difference between males and females in dioecious species, except that the males lack female function and vice versa. The identity of the genes responsible for this is unknown, but it is probably just a matter of incorporation of male sterility and female sterility mutations.

Graves: There are, of course, plants that have sex chromosomes. *Silene* evolved an X and Y chromosome system that has no homology to the animal equivalents. It is unclear what the genes are on those chromosomes.

Charlesworth: In *Silene latifolia*, there are now three X and Y genes known, and they have nothing to do with sex determination.

McLaren: Is it not also true that within a single individual plant, some of the flowers will be male and some will be female? This introduces interesting developmental problems.

Mittwoch: There is also an interesting developmental problem in birds. In chick embryos, at day 5.5 (stage 28), there is a definite difference between left and right gonad. The left gonad is bigger than the right (Mittwoch et al 1971) and on day 6 has more DNA and protein (Gasc 1978). It is morphologically distinct in having an incipient ovarian cortex. The left gonad in both sexes has some ovarian potential, whereas the right gonad, in most cases, has only testicular potential in both sexes (Domm 1939, King 1975). This raises two questions. Could this difference between left and right at this stage be due to a difference in vascularization on the two sides? Second, if this were to be due to vascularization, what would it tell us about sex determination and differentiation?

Short: You obviously have an idea in mind!

Mittwoch: I know nothing about vascularization, but a connection between the degree of vascularization and the level of cell proliferation seems likely.

Short: Anne McLaren, you were defending vascularization earlier as one of the things that might account for the asymmetry of ovarian function in the mouse.

McLaren: It could be related to small differences in timing during development, if one side became vascularized slightly earlier than the other.

Capel: Vascularization is one of the most obvious things that *Sry* controls, and it is one of the earliest steps in testis formation.

Short: I guess we are back to the gynandromorphs again. How do we explain the gynandromorph, with a complete bilateral asymmetry?

Wilkins: In *Drosophila* it is easy—a chromosomal difference in segregation—but in the vertebrates it is completely mysterious.

I believe David Zarkower generalized and said that sex is determined cell autonomously in invertebrates. This is not true in crustacea, from which the insects derived. I have read that in crustacea it is hormonal.

Charlesworth: Males have something called the androgenic gland.

Zarkower: I didn't mean to make that a sweeping generalization. There are signalling molecules in even the *C. elegans* pathway, so if you look at that level you see non-autonomy. I only meant to suggest that, at least in some invertebrate species where it has been possible to test it, individual cells throughout the body do read the X chromosome ratio.

Wilkins: We have a small number of model systems in which this is true, but we have to be careful not to over-generalize.

Charlesworth: The interesting thing in crustacea is that infectious agents can override the sex determination mechanism by knocking out the androgenic glands, and feminize males.

Short: I couldn't help thinking that we were reworking R. A. Fisher's '*Genetical theory of natural selection*' (1930). In his chapter on sexual differentiation he argues how insect sex could be cell autonomous. And then in birds and mammals, when you want a greater degree of sexual diversity, you confine your genetic sexual dimorphism to the gonads, which then produce sexually dimorphic hormones that can open up the entire autosomal complement of genes for dimorphic expression. He argues that this is a great advance over the cell autonomous, very constricting mechanism of sex determination.

Zarkower: I am not sure that it is so constricting. What have evolved are a number of interesting genetic interactions between genes involved in other processes and sex-determining genes allowing these other pathways to be deployed sex-specifically in certain tissues. Data are appearing in flies and *C. elegans* that there is an interaction between the *Hox* system and *dsx/mab-3* to allow sex-specific posterior patterning to occur. Also, in flies there are some nice recent papers showing that the male and female isoforms of *dsx* will sex-specifically modulate the response to more than one signalling pathway, to cause those pathways to act sex specifically in the genital disc.

Short: One thing no one has touched on is the constraint that viviparity imposes on sexual differentiation, once sex hormones are also used for controlling some aspects of gestation. One thinks in particular of oestrogen. Do you then have to start protecting the fetus from the sex hormones made by the placenta?

Renfree: There are so many viviparous animals, ranging from invertebrates through to vertebrates, and they all manage to have their sex allocated appropriately. One presumes that it is either not a problem, or there are many different mechanisms to solve it. I guess it depends where in the pathway the hormones become critical.

McLaren: As far as eutherian mammals go, the prenatal protection of the fetus from extraneous hormones is only partial. There is good evidence that in the mouse a male fetus with female fetuses either side will be feminized and vice versa. Of course, this is sex differentiation and not sex determination. I have often wondered how those hormones get across from one fetus to the next.

Wilkins: Does this mean that there is more vascular connection between sibling fetuses than between the individual fetuses and the maternal blood circulation?

McLaren: I doubt that it is a vascular connection. It could be just seepage. This is what has puzzled me.

Short: A female mouse sandwiched between two male fetuses is still fertile, presumably.

McLaren: Yes, but there are behavioural and anatomical differences. The anogenital distance is modified, for example.

Josso: If there were vascular exchanges it would cause real problems.

Capel: What do yolk proteins do in worms and flies?

Zarkower: It isn't known. There's a model in flies that yolk proteins have an affinity for ecdysone and act as a timer for its release. I'm not sure there is any evidence for this. In worms we don't know; there are no mutants. These proteins are made in the intestine and secreted in the body cavity. There is a specific import system in the gonad that brings them in and puts them in oocytes. They are the most abundant things that the adult hermaphrodite makes. One presumes that they are there for nutritional value and possibly other things.

Capel: This doesn't happen early in development, presumably.

Zarkower: It only happens in the adult intestine, because this is the only time that oocytes are present.

References

Domm LV 1939 Modifications in sex and sexual characters in birds. In: Allen E (ed) Sex and internal secretions, 2nd edn. Ballière Tindall, London, p 227–327

Fisher RA 1930 The genetical theory of natural selection. Clarendon Press, Oxford

Gasc J-M 1978 Growth and sexual differentiation in the gonads of chick and duck embryos. J Embryol Exp Morphol 44:1–13

King AS 1975 Aves urogenital system. In: Getty R (ed) Sisson and Grossman's the anatomy of the domestic animals, 5th edn. Saunders, Philadelphia, p 1919–1964

Mittwoch U, Narayanan TL, Delhanty JDA, Smith CAB 1971 Gonadal growth in chick embryos. Nat New Biol 231:197–200

Ottolenghi C, Veitia R, Quintana-Murci L et al 2000 The region on 9p associated with 46,XY sex reversal contains several transcripts expressed in the urogenital system and a novel doublesex-related domain. Genomics 64:170–178

The hormonal control of sexual development

Marilyn B. Renfree*, Jean D. Wilson*† and Geoffrey Shaw*

**Department of Zoology, The University of Melbourne, Victoria 3010, Australia and †Department of Internal Medicine, University of Texas Southwestern Medical Center, 5323 Harry Hines Boulevard, Dallas TX 75390-8857, USA*

Abstract. The formation of the testis or ovary is a critical step in development. The pioneering studies of Professor Alfred Jost showed that the hormones produced by the embryonic rabbit testis are essential for development of the male phenotype. Sexually dimorphic hormones play a key role in the transition from an undifferentiated gonad into the mature testis and ovary. Marsupials, with their altricial young, provide an accessible model for the study of sexual differentiation because most of these events occur postnatally, while the young are attached to teats within their mothers' pouches. The relatively long time-course for the marsupial sexual differentiation has provided an excellent opportunity to correlate morphological changes with the genes and hormones that control them. Using this model species we have demonstrated that not all sexual dimorphisms are controlled by hormones. Virilization of the prostate and phallus is androgen dependent but appears to rely on circulating 5α-androstane-3α,17β-diol which is converted to dihydrotestosterone in these target tissues. Collectively these studies have led to the development of new paradigms to explain the hormonal mechanisms mediating sexual differentiation.

2002 The genetics and biology of sex determination. Wiley, Chichester (Novartis Foundation Symposium 244) p 136–156

The formation of the testis or ovary is a critical step in development and for the continuation of species. The cloning and characterization of the testis-determining gene *SRY* and several genes 'downstream' from it has reawakened interest in the pathways regulating gonadal differentiation but there has been relatively little attention paid to other aspects of phenotypic sexual differentiation. Gonadal hormones play a critical role in the translation of gene expression into phenotypic differentiation, but the fields of molecular development and endocrinology have only recently begun to come together to investigate the control of sexual differentiation. In all mammals, hormones have profound effects on sexual differentiation, none more dramatic than those that occur at puberty. However, the hormonal control of early sexual differentiation has been

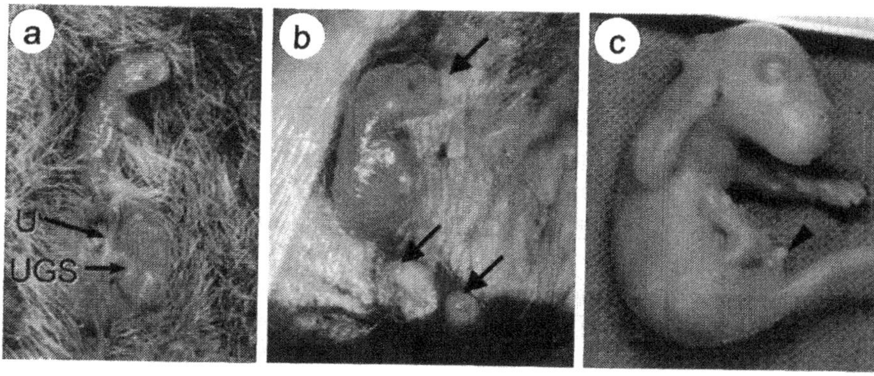

FIG. 1. The developing tammar wallaby. At birth (A) the tammar is altricial (average crown-rump length 16 mm). Some features, such as forelimbs and mouth are relatively well developed, whilst others are poorly developed, such as the hind limbs and gonads. The newborn young, with its umbilicus (U) trailing behind it, climbs unaided from the opening of the urogenital sinus (UGS), which is not sexually dimorphic, to the pouch where (b) it attaches to one of the four available teats (arrows). The young is therefore readily accessible for experimental study while most of sexual differentiation occurs. By day 25 post partum (c) male young have a differentiated Wolffian duct system, regressed Müllerian duct and prostatic buds are forming, but the phallus (arrowhead) does not become sexually dimorphic until about day 100.

difficult to examine because it occurs *in utero* in eutherians and many assumptions have been based on the endocrine responses of neonatal, pubertal and adult mammals. Marsupials, on the other hand, with their altricial young, provide a new model for the study of sexual differentiation. The advantages of using marsupials to study sexual differentiation are many. To begin with, the entire process occurs after birth when the young are accessible in their mother's pouch (Fig. 1). Hormones and inhibitors can be administered directly to the pouch young overcoming the issue as to whether they cross the maternal–fetal placental barrier. The fact that pouch young can be removed and replaced onto the teat makes it possible to perform surgery on the neonates. Perhaps the greatest advantage is that the process of phenotypic development is slower than in eutherian mammals, occurring in distinct phases so that it is possible to study each process.

Marsupial and eutherian mammals diverged from a common ancestor about 100 million years ago, but retained many common mechanisms directing sex determination and differentiation. In marsupials the Y chromosome is testis-determining and contains a homologue of the eutherian *SRY* gene (reviewed in Renfree et al 1995). As in eutherians, the fetal testis produces anti-Müllerian hormone (AMH, also known as Müllerian inhibitory substance, MIS) (Hutson et al 1988) and androgens (Renfree et al 1992, Wilson et al 1999, Shaw et al 2000) that direct subsequent development of the male urogenital sinus and phallus. However, unlike eutherians, the development of the scrotum and mammary primordia does not depend on gonadal hormones but instead is determined by a gene or genes on the X chromosome (O et al 1988, Renfree & Short 1988).

The overriding importance of testicular hormones was established by the pioneering studies of Professor Alfred Jost begun almost 50 years ago. His work on the embryonic rabbit established the paradigm that the testis was essential for development of the male phenotype (reviewed in Jost 1970) (Fig. 2, Table 1). In the rabbit, gonadal sex is recognizable on the fifteenth day of pregnancy, but the remainder of the genital tract remains identical in males and females until day 20 (Jost 1961). Females remain undifferentiated until day 23, but in males Müllerian duct regression occurs and prostatic anlagen appear between days 20 and 22. Jost's remarkable experiments on rabbits castrated before day 20 of pregnancy *in utero* resulted in the development of a female phenotype in male fetuses. Jost suggested that there is a window of time during which male development can be prevented but after which has irreversible consequences (Jost 1961).

Testicular and ovarian differentiation

The altricial state of the marsupial neonate means that most of sexual differentiation in marsupials takes place postnatally (Fig. 1). The marsupial urogenital system develops from an indifferent stage at birth, when both Wolffian and Müllerian ducts are present, to the phenotypically distinct male or female condition during early pouch life. The marsupial is born with a fully functional mesonephros, and the Wolffian (mesonephric) duct is patent to the urogenital sinus, whereas the metanephric kidney does not become functional until about two weeks after birth (Renfree et al 1996).

Gonadal differentiation follows a typical pattern except for its timing in relationship to birth. In the tammar wallaby, testicular differentiation commences around the time of birth, with clearly defined seminiferous cords, by day 2 pp (post partum), but ovarian development is not evident before days 6–8 pp (Renfree et al 1996). No difference is seen in gonadal mass during the first eight weeks of pouch life, but testicular weights begin to diverge from ovarian weights around day 60 pp, and are significantly heavier by day 80 pp.

As the Sertoli cells differentiate they interact with and modulate Leydig cell differentiation, germ cell proliferation and seminiferous tubule formation. Leydig cells produce the steroid hormone testosterone that stimulates the development of the Wolffian ducts into the vasa deferentia and epididymides, and virilization of the urogenital sinus and phallus. The Sertoli cells produce the protein hormone AMH and also secrete androgen binding protein (ABP). The primary role of AMH is to induce regression of the Müllerian ducts that would

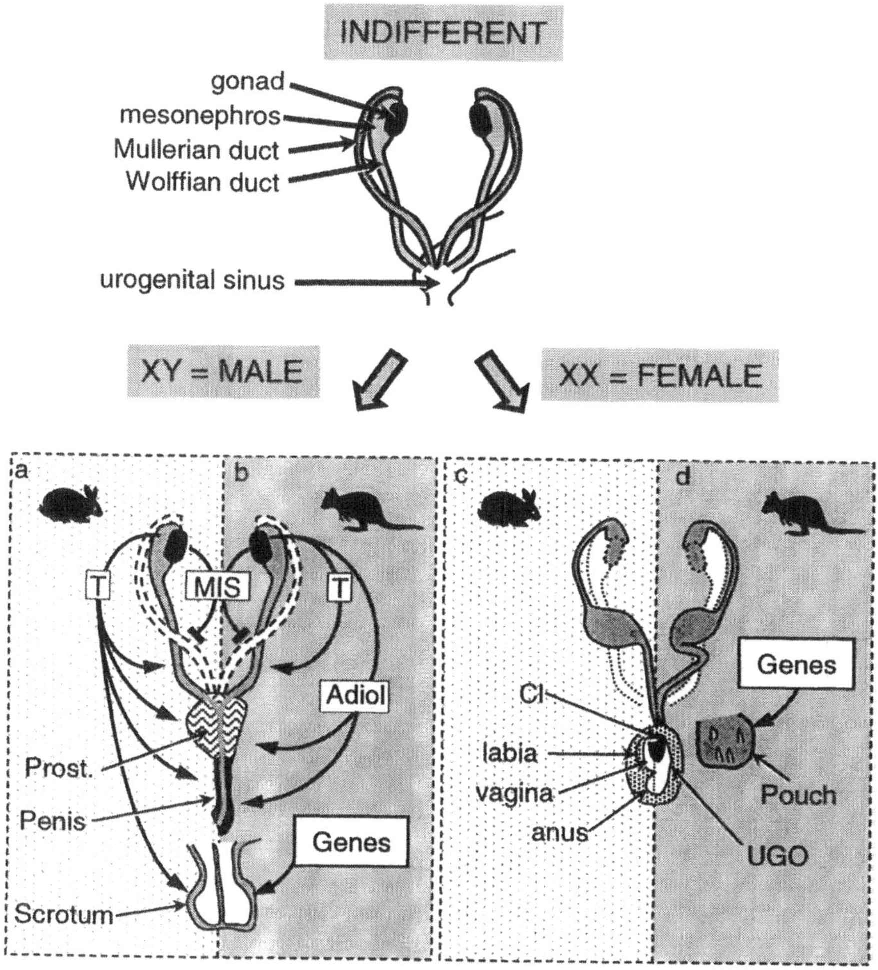

FIG. 2. Control of sexual differentiation in eutherian (a, c) and marsupial (b, d) mammals. Unlike in eutherians, in wallabies development of some dimorphisms, notably the scrotum, pouch and mammary glands, is controlled independent of testicular hormones by an X-linked gene or genes. MIS, Müllerian inhibitory substance (AMH); T, testosterone; Adiol, 5α-androstane-3α,17β-diol; Prost, prostate; P, penis; Cl, clitoris; UGO, urogenital opening.

otherwise form the oviducts, uterus and upper vagina. GATA4, a transcription factor which has a sexually dimorphic expression pattern (Viger et al 1998) may enhance *AMH* gene transcription through a direct interaction with the nuclear receptor SF1 (Tremblay & Viger 1999). *GATA4* and *AMH* have similar expression patterns. In males, germ cells apparently play no role in the early

TABLE 1 Sexual differentiation of rabbit fetuses and effects of castration

	Normal differentiation		
Day	*Females*	*Males*	*Castration effects on male genital tract*
19	Indifferent	Indifferent	Feminine organogenesis
20	Indifferent	Involution of Müllerian ducts begins	± complete feminine organogenesis
21	Indifferent	Analgen anterior Prostate	Anterior prostate present
22	Indifferent	Definite masculine trends	Uterine sections present; hypospadias
23	Fusion of posterior part of Müllerian ducts	Anlagen posterior and lateral prostate	Deferent duct absent; otherwise masculine
24	Involution of Wolffian ducts begins	Differences in genital tubercle	Masculine organogenesis
25 to 26	Definite feminine features	Definite masculine features	

From Jost (1961).

differentiation of the testis, since seminiferous cords, Sertoli cells and Leydig cells can develop in the absence of germ cells. However, loss of germ cells from ovaries leads to the formation of Sertoli cells that organize into seminiferous-like tubules in marsupials (Whitworth et al 1996) and in eutherians (reviewed in McLaren 1991, Whitworth 1998). This suggests that in females, an interaction between the germ cells and the supporting cell lineage inhibits Sertoli cell formation. Sertoli and granulosa cells are thought to be derived from a common progenitor supporting cell line, and both express AMH in the adult. Gonadal sex reversal can be induced in both female to male and male to female directions by gonadal transplantation, by administration of AMH in culture, or by exogenous oestradiol (Burns 1961, Whitworth et al 1996, Coveney et al 2001, Renfree et al 2001).

Wolffian and Müllerian ducts

Differentiation of the Wolffian and Müllerian ducts takes place under the influence of gonadal hormones in both groups of mammals (Fig. 2). In the rat, the fetal testis first becomes distinguishable at 13 days 15 h (Jost 1970), with well-organized seminiferous cords in the anterior part of the gonad by 14 days 14 h, and commences testosterone secretion around day 15.0–15.5 (Jost 1970). The first effects of gonadal androgen on the morphology of the rat Wolffian duct do not appear before day 15.5, at about the time of morphological differentiation of the fetal Leydig cells (Eusterschulte et al 1992). On the day of birth in the tammar wallaby *Macropus eugenii* and the grey opossum *Monodelphis domestica*, the testes contain very little testosterone (Renfree et al 1992, Fadem & Harder 1992, Xie et al 1998). However, by days 5–10 pp in the tammar in males testicular testosterone content rises to around 1 ng/mg and the Wolffian ducts begin to differentiate, but in females ovarian testosterone is undetectable (Renfree et al 1992).

Mammary gland and scrotum

Most mammals of both sexes possess mammary glands, even if only transiently during development. The majority of marsupials are exceptions to this rule. In tammars and other Australian marsupials, males never have mammary development, even the first rudiments (O et al 1988, Renfree & Short 1988, Renfree et al 1996). In opossums and other American marsupials, males have fewer mammary primordia than females (Renfree et al 1990, Robinson et al 1991). The scrotum in eutherian mammals is caudal to the penis: in marsupials it is cranial. In the tammar, scrotal bulges are first seen in the male fetus and mammary primordia in the female fetus on day 22, 4 or 5 days before birth and before the gonads differentiate, and 6 or 7 days before gonadal steroids are detectable (O et al 1988, Renfree & Short 1988, Renfree et al 1992, 1996). A similar pattern of development occurs in the brushtail possum, *Trichosurus vulpecula* (Ullmann 1993). Treatment of neonates with exogenous steroids has no influence on mammary, pouch or scrotum development (Shaw et al 1988) (Table 2). Burns, in his pioneering studies on sexual differentiation in the North American opossum, *Didelphis virginiana,* also found no effects of androgen or oestrogen treatment on the presence of the pouch or scrotum, although the internal genitalia were affected just as Jost had shown for the rabbit. However, the Jost experiments were so persuasive that Burns (1961) concluded that the Jost model applied in its entirety to marsupial as well as eutherian mammals, despite Burns' own results to the contrary (reviewed in Wilson et al 1995).

The pouch is just visible in female tammar neonates on days 5 or 6 pp, and clearly evident by days 7 or 8 pp. However, primordia of the folds can be identified histologically in females at day 24 or gestation, many days before gonadal differentiation. Since XXY marsupial males with testes have a pouch and mammary glands, whereas XO marsupials do not, the conclusion from the collective observations is that both these structures are under the control of X-linked genes (reviewed in Renfree et al 1995) (Fig. 2).

Early workers believed that the labio-scrotal folds of eutherians are homologous to the scrotal/pouch folds of marsupials. However, the external opening of the

TABLE 2 Effect of hormone treatment on sexual dimorphisms in tammars

(A) Hormone-dependent dimorphisms

Treatment	*Control*		*Oestradiol*	*Androgen*
Sex	*Male*	*Female*	*Male*	*Female*
Gonad	Testis	Ovary	Abnormal testis	Ovary
Gonad position	Scrotal	Abdominal	Abdominal	Abdominal
Gubernaculum & processus vaginalis	Extends to scrotum	Small and disappears	Terminates outside scrotum	Small and disappears
Wolffian duct	Normal	Regressed	Regressed	Hypertrophic
Müllerian duct	Regressed	Normal	Stimulated	Developed
Urogenital sinus	Normal	Normal	Hypertrophic	Hypertrophic
Prostate	Normal	Normal	—	Normal

(B) Hormone-independent dimorphisms

Treatment	*Control*		*Oestradiol*	*Androgen*
Sex	*Male*	*Female*	*Male*	*Female*
Mammary gland	Absent	Present	Absent	Present
Pouch	Absent	Present	Absent	Present
Scrotum	Normal	Absent	Normal	Absent

Data from O et al (1988), Shaw et al (1988), Coveney et al (2001).

urogenital system is not sexually dimorphic in marsupials, and there are no labia (Renfree 1992, 1994) (see Fig. 1). Since the penis is caudal to the scrotum this places the pouch and scrotal primordia in close proximity. E. J. McCrady reported that the scrotum and pouch arose from common anlagen in the American opossum, beginning as paired folds just cranial to the phallus (see Renfree et al 1992). However, we have shown in the tammar that the scrotal primordia arise as paired bulges in the groin region at about day 21 of gestation, while the pouch primordia are slightly more cranial and develop at around day 24 of gestation. We conclude that pouch and scrotum arise from different anlage in the same, or closely adjacent, morphogenetic fields (Robinson et al 1991, Renfree 1994), and in Australian marsupials at least, are developmentally mutually exclusive.

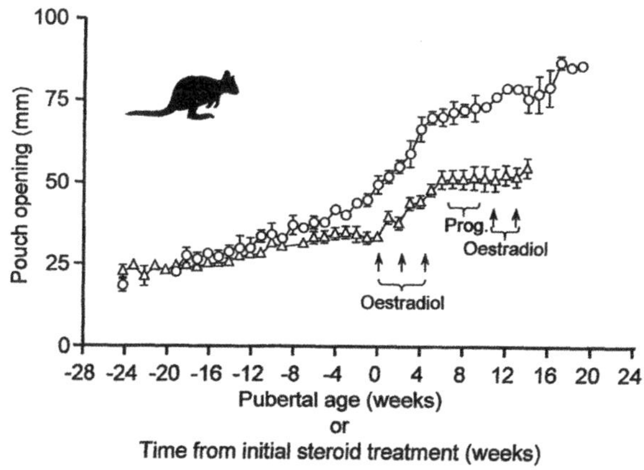

FIG. 3. Pouch growth in female tammar wallabies through puberty. There is a spurt of pouch growth around puberty, in intact females (circles) which is inhibited in ovariectomized females (triangles) unless these are treated with oestradiol. Progesterone has no effect on pouch growth (Redrawn from Nurse & Renfree 1994).

The pouch, a secondary sexual structure characteristic of female marsupials, is one of the largest sexually dimorphic structures in mammals (Wilson et al 1995). Early experimental treatment with massive doses of oestrogen apparently induced the formation of a pouch from a scrotum in castrated male brushtailed possums (see Nurse & Renfree 1994). These early findings have never been confirmed, and neither androgen nor oestrogen have any effect on pouch or scrotal development of opossums or tammars, even in massive does (see Burns 1961, Fadem & Tesoriero 1986, Moore & Thurstan 1990, Shaw et al 1988). However, the pouch is responsive to steroids during sexual maturation at puberty, and oestrogen mediates pouch growth and the eversion of the teats at puberty (Nurse & Renfree 1994) (Fig. 3). Progesterone has no effect on teat eversion or pouch growth, but ovariectomy disrupts pouch maturation.

The genes that might control pouch and scrotal development have not been identified. *SOX3* is an X-linked gene and was a possible candidate gene to control differentiation of the scrotum and mammary glands. However, no transcripts can be detected in the scrotum, mammary primordia or pouch folds throughout development (Pask et al 2000). In contrast, autosomal *SOX9* (located on tammar chromosome 2) is expressed in the scrotum and mammary glands before birth, but is down-regulated by the day of birth in both tissues (J. L. Harry, A. J. Pask, G. Shaw & M. B. Renfree unpublished results). We are currently investigating the other candidate X-linked genes in pouch and scrotum.

Androgens and virilization

It is well established that androgens play a critical developmental role in the maturation of the Wolffian duct system and the virilization of the urogenital sinus and external genitalia. Circulating androgens virilize the urogenital sinus and the external genitalia, but the Wolffian ducts appear to be virilized ipsilaterally, either by lymphatic transport, diffusion, or secretion of androgen through the lumen of the Wolffian ducts (Jost 1970). However, the androgen(s) that actually perform these functions in eutherians have never been identified, because male phenotypic differentiation takes place so early in embryogenesis that it has not been possible to obtain blood for hormone measurements until after phenotypic sexual development is complete. On the basis of studies of mutations in humans and animals it was widely assumed that androgens virilize the embryo in a fashion similar to the process in adults, namely that the testicular hormone testosterone is secreted into the circulation and acts via the androgen receptor in target tissues either as testosterone itself or as its 5α-reduced metabolite 5α-dihydrotestosterone (DHT) (Wilson & George 1994).

In the marsupial it is possible to examine the mechanism of male phenotypic development in a way that cannot be done in any eutherian mammal (see Fig.1). Virilization of the tammar pouch young takes place in three phases. Formation of the epididymis starts before day 20 and the prostate between days 25 and 35 (Shaw et al 1988, Renfree et al 1996). Sexually dimorphic development of the male phallus does not occur until after day 100 (Butler et al 1999). This time difference makes it possible to study prostatic and penile virilization to be studied independently.

As in eutherian mammals, the developing marsupial testes, but not the ovaries, produce AMH and testosterone (Hutson et al 1988, George et al 1985, Renfree et al 1992, Wilson et al 1999). Since testosterone is the principal androgen in the testis at the time of marsupial sexual differentiation it was tacitly assumed that testosterone is the key hormone in virilization. However, virilization of the marsupial male urogenital tract begins after the onset of androgen synthesis (George et al 1985, Renfree et al 1992). Gonadal testosterone concentrations are low in male and female tammar gonads at birth, but in males they rise around day 2, coinciding with the formation of the seminiferous tubules (Renfree et al 1992). However, at this stage plasma androgens are not sexually dimorphic (Wilson et al 1999). Androgen transport in the plasma of the tammar wallaby (and some other marsupial species) differs from that in most mammals in that there is no high-affinity transport protein in plasma analogous to sex hormone binding globulin (SHBG). Consequently, testosterone and DHT in plasma are transported bound to low affinity, non-saturable carriers, principally serum albumin. Virilization of the urogenital sinus is androgen dependent (Shaw et al 1988, Tyndale-Biscoe & Hinds 1989, Lucas et al 1997, Ryhorchuk et al 1997, Butler et al 1998), but the first signs of prostatic development, the appearance of prostatic buds in the urogenital sinus, does not commence until 3 weeks after the onset of testosterone production. Similarly, the phallus does not become sexually dimorphic until about day 100 pp (Butler et al 1999), after the fall in testicular testosterone concentration about day 45–50 after birth. The androgen receptor (AR), is expressed in the urogenital sinus of the fetus of both sexes from as early as day 19 (early head-fold), 7 days before birth and the first rise in testicular testosterone, therefore AR is not rate-limiting for virilization (Butler et al 1998). Despite the overwhelming evidence from our laboratory and by others (Tyndale-Biscoe & Hinds 1989) that testicular androgens are required for male phallic development in the wallaby, we have been unable to demonstrate sexual dimorphism after day 50 in the plasma levels of androgens, notably during the phase of pouch life when differential phallic growth occurs (days 75–200). This presents an enigma in that the tissues are clearly androgen sensitive, but differentiate well after the initiation of testosterone synthesis.

The delay in virilization of the urogenital sinus and phallus cannot be due to the lack of the more potent androgen DHT, since the urogenital sinus and phallus both contain the enzyme necessary for its synthesis, 5α-reductase, in high concentrations by at least day 10–11 pp (Renfree et al 1992). The precise mechanism by which virilization is initiated in the developing male marsupial is not entirely understood, but we have identified another androgen, 5α-androstane-3α,17β-diol (5α-adiol), that appears to have a key role in this process (Shaw et al 2000) (Fig. 4). 5α-adiol is synthesized in testes and secreted into the plasma of pouch young (Shaw et al 2000), and is present in higher concentrations in male than female young, unlike testosterone which circulates in similar concentrations in the two sexes (Wilson et al 1999). Oral administration of 5α-adiol to female tammar pouch young from days 20–30 induces development of a prostate (Shaw et al 2000) and administration of 5α-adiol to female pouch young from days 70–150 induces prostate and phallic growth similar to that in males (Leihy et al 2001). Similarly, administration of small doses of 5α-adiol enanthate from day 20–45 causes virilization of the female urogenital sinus (Leihy et al 2001). Exogenous testosterone and DHT can also induce prostatic development (Shaw et al 1988, Ryorchuk et al 1997, Leihy et al 2001). Both testosterone (Renfree et al 1992) and 5α-adiol (Shaw et al 2000) are produced in high amounts in the testes, but not the ovaries of tammar pouch young, however any DHT that is formed in the pouch young testes is quickly converted into 5α-adiol (Shaw et al 2000). Because 5α-adiol is the predominant androgen in the tammar testis during the period of virilization and is the only known androgen that is higher in the male plasma than the female plasma at the time of prostate formation (Shaw et al 2000). We therefore conclude that this hormone plays a unique role as a circulating hormone to control the formation of the male phenotype. In the urogenital sinus and phallus 5α-adiol is

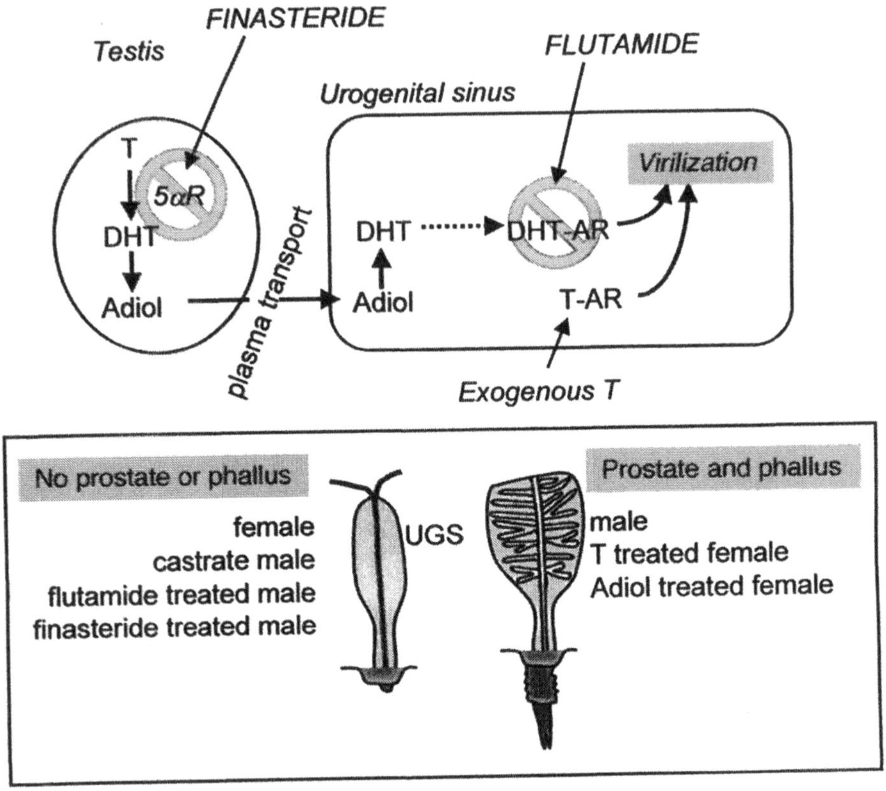

FIG. 4. Model for androgenic control of prostatic and phallic development in tammars. Prostate development is dependent in 5α-reduced androgens because treatment with an androgen receptor blocker, flutamide or a 5α-reductase inhibitor, finasteride, inhibits prostate development. 5α-androstane-3α,17β-diol (5α-adiol) is a major androgen produced by the testis that is sexually dimorphic in plasma and treatment with low does of 5α-adiol induces prostate and phallus development. T, testosterone; DHT, dihydrotestosterone; AR, androgen receptor; 5α-R, 5α-reductase; UGS, urogenital sinus.

converted back to DHT (Shaw et al 2000), which is thought to be the active intracellular androgen in target tissues. Since steroid 5α-reduction is irreversible, and neither testosterone nor DHT are sexually dimorphic in the circulation, we conclude that 5α-adiol is the androgen that is secreted by the tammar testis into plasma to initiate virilization of the urogenital sinus and phallus.

Since 5α-adiol is responsible for prostate formation in the tammar and is a major androgen in the immature rat and rabbit testes, it is the leading candidate for a universal role for this function in these mammals. In the rabbit, testosterone synthesis begins in the fetal testis between days 17 and 17.5 of gestation, and virilization of the urogenital sinus requires exposure to androgens for only four days, namely days 19 to 23 (Jost 1961). It is possible that the role of androgen in phallic development occurs during a critical window of time and that some other factor or factors then take over to cause differential growth. Precedent exists for such a phenomenon in both the rabbit prostate (androgen is required only for 4 days [19–23], and growth of the tissue is androgen independent thereafter) and in the wallaby (growth of the phallus and prostate continued in the female after apparent atrophy of transplanted testes in the experiment of Tyndale-Biscoe & Hinds 1989). There is also a critical window of time in male tammars sometime between day 25 and 13 months pp, when exposure to androgens imprints the response of the hypothalamopituitary axis to oestradiol challenge (see below). In contrast to the urogenital sinus and external genitalia which are virilized by circulating androgens, the Wolffian ducts virilize by an ipsilateral process in which androgen is delivered directly to the tissue, presumably via the lumen of the Wolffian ducts. Because of low levels of 5α-reductase in the Wolffian ducts, it has been widely assumed that the hormone that mediates this process is testosterone itself, but our demonstration that the major testicular androgens in the early tammar pouch young are 5α-adiol and DHT (Shaw et al 2000) and that the metabolic sequence in the epididymis favours the formation of 5α-adiol from testosterone suggest that steroid 5-reduction may be critical for this process as well. The epididymis is well developed before the time of commencement of masculinization of the urogenital sinus around day 25. In mature eutherian mammals androgen is transported from the lumen of the Wolffian duct into the epithelial cell bound to androgen-binding protein (ABP or prostatein) which is synthesised in Sertoli cells and secreted into the lumen of the Wolffian ducts (Joseph 1994). This molecule is a leading candidate for mediating the virilization of this tissue. Further characterization of the formation and endocrine effects of 5α-adiol are underway.

Brain sex and hormonal control of puberty

Androgens masculinize the brain either by their conversion to oestradiol in that tissue or directly via the androgen receptor. Some of the effects of the androgens permanently masculinize the brain during a critical period in early development. In eutherian mammals, sex differences in male-type sexual behaviour can be attributed to both organizational and activational effects of testicular hormones acting on the central nervous system. In contrast, in all primates and in male and female tammar wallabies, the expression of male-type sexual behaviour appears to be completely dependent on the adult steroid hormone environment (Rudd et al 1996). Male behaviour can be induced in female tammars with testosterone implants, and is lost in castrated males.

Although there appears to be no permanent organizational imprinting of the male or female tammar brain as in many eutherians, there are sex differences in the positive feedback response of luteinizing hormone (LH) to oestradiol, similar to the preovulatory surge of LH in oestrous females (Rudd et al 1999). Ovariectomized and intact female tammars both respond to an oestradiol challenge with an LH surge, whereas castrated males or intact males do not. However, if the males are castrated at 26 days of age they respond like females, but if castrated (pre-pubertally) at 14 months they respond like males (Fig. 5). These results suggest that the positive feedback mechanisms in the male tammar are permanently suppressed by an organizational action of testicular hormones acting sometime after 26 days but before the only other time point studied, 14 months (Rudd et al 1999). It is interesting to note that the early castrations were done in the middle of the 45 day period when testosterone is elevated in the testes (Renfree et al 1992, Wilson et al 1999), so we predict that the critical period of androgen exposure coincides with these high testicular testosterone concentrations and that the brain is effectively imprinted by day 45.

There are also differences in males and female with respect to puberty. Female tammars are seasonal breeders, but puberty can occur at any time of the year once the young female attains a body weight of around 2.0 kg, much smaller than the average adult female weight of 5 kg (Williams et al 1998). This usually occurs at round 9–10 months of age when the female young first leaves the pouch (Williams et al 1998). In contrast, males mature later, and both testicular growth and maturation of the hypothalamic–pituitary–testicular axis begins at 19 months. Puberty is complete with the appearance of mature sperm in the testes by 25 months of age (Williamson et al 1990).

Conclusions

Sexual differentiation in marsupials, as in eutherians, is a sequential process beginning with the establishment of chromosomal sex at the time of fertilization. The sex chromosomes exert extra-gonadal and gonadal effects, the former being particularly pronounced in marsupials since they involve the scrotum, pouch, mammary gland, gubernaculum and processus vaginalis, and these effects precede gonadal differentiation. Gonadal hormones drive the subsequent sexual differentiation of the Wolffian and Müllerian ducts, as in eutherians. Prostatic, urogenital sinus and phallic development in the wallaby are not temporally related to the production of androgens.

The relatively long time-course for marsupial sexual development provides an excellent opportunity to correlate phenotypic changes with gene expression patterns and hormone synthesis. Similarly, the long time lag in marsupial development between the production of androgens and their action on target

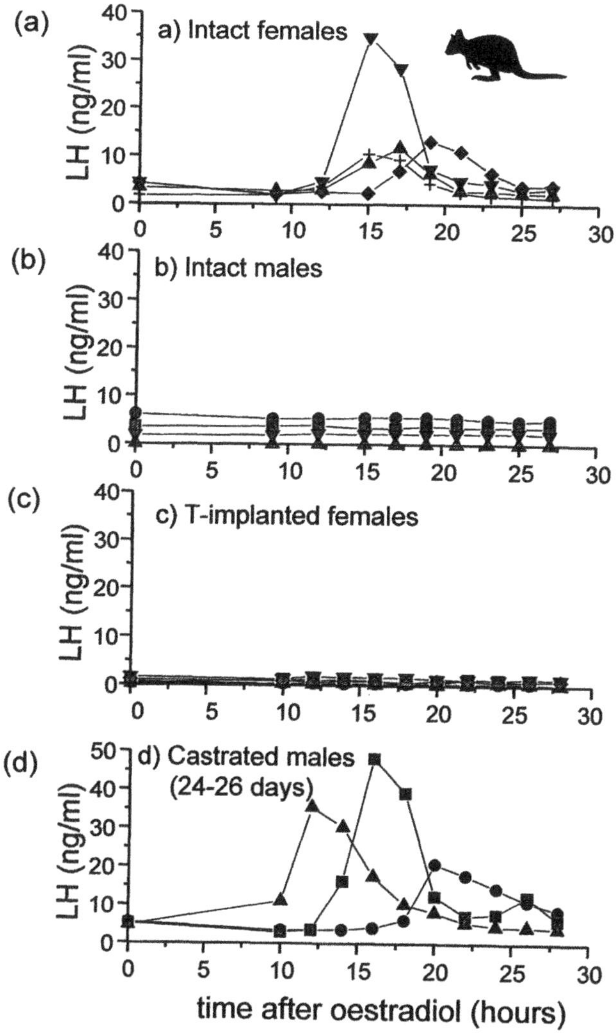

FIG. 5. Effect of early castration on the LH surge response to oestradiol challenge. Intact females (a) and males castrated (d) at d 24–26 post partum show a marked surge in LH 15–20 hours after an injection of 17β-oestradiol, but intact adult males (b) and testosterone-implanted females (c) do not respond. (Redrawn from Rudd et al 1999.)

tissues to induce virilization has led to the development of new paradigms to explain the mechanisms mediating the process of sexual differentiation. Virilization of the urogenital sinus and phallus appears to depend on a testosterone metabolite, 5α-adiol which is the predominant circulating androgen produced by the testis and which is converted to the potent androgen DHT in the target tissues. 5α-adiol has been recognized to be a potent androgen since the 1930s and known to be formed in immature rabbit, human and rat testes, but a specific physiological role for the hormone has never before been identified. A role for 5α-adiol in male phenotypic development explains the previous conundrum, namely the need for testicular hormones and the prevention of virilization by inhibitors of 5α-reductase and binding of DHT to the androgen receptor. This discovery makes it possible to approach the major unresolved issues in male phenotypic development in a new way.

Acknowledgements

We thank our students for their numerous contributions to these studies, in particular Dr Carl Rudd, Dr Serena Williams (Nurse), Dr Deanne Whitworth, Dr Christopher Butler, Douglas Coveney and Michael Leihy. We especially thank Professor Roger Short, Professor John Hutson, Dr Wai-Sum O and Dr Fred George for their collaboration to the early phases of this research. The work was supported by grants from the National Health and Medical Research Council of Australia and the University of Texas, Southwestern Medical Center, Dallas, USA, and the Perot Family Foundation.

References

Burns RK 1961 Role of hormones in the differentiation of sex. In: Young WC (ed) Sex and internal secretions, vol 1. Williams & Wilkins, Baltimore, p 76–158

Butler CM, Harry JL, Deakin JE, Cooper DW, Renfree MB 1998 Developmental expression of the androgen receptor during virilization of the urogenital system of a marsupial. Biol Reprod 59:725–732

Butler CM, Renfree MB, Shaw G 1999 Development of the penis and clitoris in the tammar wallaby, *Macropus eugenii*. Anat Embryol 199:451–457

Coveney D, Shaw G, Renfree MB 2001 Oestrogen induced gonadal sex reversal in the tammar wallaby. Biol Reprod 65, 613–621

Eusterschulte B, Reisert I, Pilgrim C 1992 Absence of sex differences in size of the genital ducts of the rat prior to embryonic day 15.5–16.0. Tissue Cell 24:483–489

Fadem BH, Tesoriero JV 1986 Inhibition of testicular development and feminization of the male genitalia by neonatal estrogen treatment in a marsupial. Biol Reprod 34:771–776

Fadem BH, Harder JD 1992 Evidence for high levels of androgen in peripheral plasma during postnatal development in a marsupial: the gray short-tailed opossum (*Monodelphis domestica*). Biol Reprod 46:105–108

George FW, Hodgins MB, Wilson JD 1985 The synthesis and metabolism of gonadal steroids in pouch young of the opossum, *Didelphis virginiana*. Endocrinology 116:1145–1150

Hutson JM, Shaw G, O WS, Short RV, Renfree MB 1988 Müllerian inhibitory substance production and testicular migration and descent in the pouch young of a marsupial. Development 104:549–556

Joseph DR 1994 Structure, function, and regulation of androgen-binding protein/sex hormone-binding globulin. Vitam Horm 49:197–280

Jost A 1961 The role of fetal hormones in prenatal development. The Harvey Lectures 53: 201–226

Jost A 1970 Hormonal factors in the sex differentiation of the mammalian foetus. Philos Trans R Soc Lond B Biol Sci 259:119–130

Leihy M, Shaw G, Wilson JD, Renfree MB 2001 Virilization of the urogenital sinus of the tammar wallaby is not unique to 5α-androstane-3α,17β-diol. Mol Cell Endocrinol 181: 111–115

Lucas JC, Renfree MB, Shaw G, Butler CM 1997 The influence of the anti-androgen flutamide on early sexual differentiation of the male marsupial. J Reprod Fertil 109:205–212

McLaren A 1991 Development of the mammalian gonad: the fate of the supporting cell lineage. Bioessays 13:151–156

Moore HD, Thurstan SM 1990 Sexual differentiation in the grey short-tailed opossum, *Monodelphis domestica,* and the effect of oestradiol benzoate on development in the male. J Zool (Lond) 221:639–658

Nurse SC, Renfree MB 1994 Pubertal development of the pouch and teats in a marsupial, *Macropus eugenii.* J Reprod Fertil 101:279–285

O WS, Short RV, Renfree MB, Shaw G 1988 Primary genetic control of somatic sexual differentiation in a mammal. Nature 332:716–717

Pask A, Harry JL, Renfree MB, Graves JAM 2000 Absence of *SOX3* in the developing marsupial gonad is not consistent with a conserved role in mammalian sex determination. Genesis 27:145–152

Renfree MB 1992 The role of genes and hormones in marsupial sexual differentiation. J Zool (Lond) 226:165–173

Renfree MB 1994 Sexual dimorphisms in the gonads of marsupial mammals. In: Short RV, Balaban E (eds) The differences between the sexes. Cambridge University Press, Cambridge, p 213–230

Renfree MB, Harry JL, Shaw G 1995 The marsupial male: a role model for sexual development. Phil Trans Roy Soc Lond B 350:243–251

Renfree MB, Robinson ES, Short RV, Vandeberg JL 1990 Mammary glands in male marsupials. 1. Primordia in neonatal opossums, *Didelphis virginiana* and *Monodelphis domestica.* Development 110:385–390

Renfree MB, Wilson D, Short RV, Shaw G, George FW 1992 Steroid hormone content of the gonads of the tammar wallaby during sexual differentiation. Biol Reprod 47:644–647

Renfree MB, O WS, Short RV, Shaw G 1996 Sexual differentiation of the urogenital system of the fetal and neonatal tammar wallaby, *Macropus eugenii.* Anat Embryol 194:111–134

Renfree MB, Short RV 1988 Sex determination in marsupials: evidence for a marsupial-eutherian dichotomy. Philos Trans R Soc Lond B Biol Sci 322:41–53

Renfree MB, Coveney D, Shaw G 2001 The influence of oestrogen on the developing male marsupial. Reprod Fert Dev, in press

Robinson ES, Renfree MB, Short RV, VandeBerg JL 1991 Mammary glands in male marsupials, 2. Development and regression of mammary primordia in *Monodelphis domestica* and *Didelphis virginiana.* Reprod Fert Dev 3:295–301

Rudd CD, Short RV, Shaw G, Renfree MB 1996 Testosterone control of male type sexual behavior in the tammar wallaby (*Macropus eugenii*). Horm Behav 30:446–454

Rudd CD, Short RV, McFarlane JR, Renfree MB 1999 Sexual differenitation of oestradiol-LH positive feedback in a marsupial. J Reprod Fertil 115:269–274

Ryhorchuk AR, Shaw G, Butler CM, Renfree MB 1997 Effects of the 5α-reductase inhibitor finasteride on the developing prostate and testis of a marsupial. J Androl 182:123–130

Shaw G, Renfree MB, Short RV, O WS 1988 Experimental manipulation of sexual differentiation in wallaby pouch young treated with exogenous steroids. Development 104:689–701

Shaw G, Renfree MB, Leihy MB, Shackleton CHL, Roitman E, Wilson JD 2000 Prostate formation in a marsupial is mediated by the testicular androgen 5α-androstane-3α, 17β-diol. Proc Natl Acad Sci 97:12256–12259

Tremblay JJ, Viger RS 1999 Transcription factor GATA-4 enhances Mullerian inhibiting substance gene transcription through a direct interaction with the nuclear receptor SF-1. Mol Endocrinol 13:1388–1401

Tyndale-Biscoe CH, Hinds LA 1989 Influence of the immature testis on sexual differentiation in the tammar wallaby *Macropus eugenii* (Macropodidae: Marsupialia). Reprod Fert Dev 1:243–254

Ullmann SL 1993 Differentiation of the gonads and initiation of mammary gland and scrotum development in the brushtail possum *Trichosurus vulpecula* (Marsupialia). Anat Embryol 187:475–484

Viger RS, Mertineit C, Trasler JM, Nemer M 1998 Transcription factor GATA-4 is expressed in a sexually dimorphic pattern during mouse gonadal development and is a potent activator of the Müllerian inhibiting substance promoter. Development 125:2665–2675

Whitworth DJ, Shaw G, Renfree MB 1996 Gonadal sex reversal of the developing marsupial ovary in vivo and in vitro. Development 122:4057–4063

Whitworth DJ 1998 XX Germ Cells: the difference between an ovary and a testis. Trends Endocrinol Metab 9:2–6

Williams SC, Fletcher TP, Renfree MB 1998 Puberty in the female tammar wallaby. Biol Reprod 58:1117–1122

Williamson P, Fletcher TP, Renfree MB 1990 Testicular development and maturation of hypothalamic-pituitary-testicular axis in the male tammar, *Macropus eugenii*. J Reprod Fertil 88:549–557.

Wilson JD, George FW 1994 Sex Determination and Differentiation, In: Knobil E, Neill JF, Greenwald GS, Markert CL, Pfaff DW (eds) The physiology of reproduction, 2nd edn. Lippincott Williams & Wilkins, Philadelphia, PA p 3–28

Wilson JD, George FW, Renfree MB 1995 The endocrine role in mammalian sexual differentiation. Recent Prog Horm Res 50:349–364

Wilson JD, George FW, Shaw G, Renfree MB 1999 Virilization of the male pouch young of the tammar wallaby does not appear to be mediated by plasma testosterone or dihydrotestosterone. Biol Reprod 61:471–475

Xie Q, Mackay S, Ullmann SL, Gilmore DP, Payne AP, Gray C 1998 Postnatal development of Leydig cells in the opossum (*Monodelphis domestica*): an immunocytochemical and endocrinological study. Biol Reprod 58:664–669

DISCUSSION

Short: What do you think about the evolution of testicular descent? Do you think that eutherians and marsupials ended up with the same result, but did it differently? Or is this difference in genetic versus hormonal control of the scrotum just incidental?

Renfree: I think the scrotum has evolved many times in mammals. You only have to look at the variety of locations of the testis in the eutherian mammals — from the abdominal testis of the elephant to the inguinal testis of the mole — to see this. Likewise in marsupials there are a variety of testis locations. Presumably the evolutionary pressure was to get the testis outside, for whatever reason. Monotremes don't have a scrotum; they have abdominal testes. Presumably somewhere between monotremes and the eutherian line there was some drive to get the male gametes in a cooler location.

Lovell-Badge: Is there any indication of what INSL3 does in marsupials?

Renfree: Testicular descent is inhibited after oestradiol treatment, and we get a failure of closure of the inguinal canal. We think this is because of the down-regulation of INSL3. We are chasing this at the moment, but we haven't pulled the marsupial gene out yet.

McLaren: You see precocious entry into meiosis in hormonally disrupted testes. From what one knows about germ cell development in other animals, I would predict that this is female meiosis and not male meiosis, and that those germ cells would go on into oogenesis and develop into oocytes. Have you kept the animals long enough to know?

Renfree: Of these ones that are born early, we only get a very small number. We have been trying for the whole of this breeding season to get some born naturally on day 25. We have only had two this year. Our plan is to let them grow up, because it would be very interesting to see what happens. Ideally, we would like to look at them at all the different stages. I agree with you; I think it is female meiosis.

Short: Anne McLaren, was the thought running through your mind that the oestrogen might just damage the testis and hence interfere with the ability of the seminiferous tubule to inhibit meiosis?

McLaren: Yes, indeed. The testis was clearly developing quite abnormally. In other contexts if cord formation is disrupted, the normal inhibitory effect is lost and the germ cells go into meiosis.

Renfree: I should emphasize that the oestradiol effect is not physiological but a pharmacological effect. It is not a normal part of the sex differentiation pathway.

Harley: What converts DHT to androstandiol?

Renfree: There are several isoforms of the 3α-hydroxysteroid dehydrogenases. Some of them oxidize, others reduce. The different tissues have their own specific isoform.

Harley: Which cells make those?

Renfree: The prostate, the urogenital sinus and the penis among others. It is a fairly widespread enzyme, but it can be switched on and off at certain times in development. If the hormone is circulating around in the blood you can get the differential timing by synthesis of the appropriate enzyme in the appropriate target tissue.

Swain: Is that what explains the timing difference in development of the penis?

Renfree: We think so.

Josso: Does the oestradiol treatment influence the Müllerian regression?

Renfree: The Müllerian ducts appear to be enlarged, but it depends on which part of the duct you look at. It is very difficult to get the exact location to compare the sizes. Müllerian duct volume is increased as measured from one edge of the mesenchyme to the other, and the lumen diameter is increased.

Vilain: I would like to comment on the issue of the postnatal peak of testosterone you showed. It is interesting because in humans there is also a postnatal peak of testosterone: it peaks at about one week of age and then goes down slowly until about one year where it reaches a very low prepubertal level. No one really understands the significance of this peak. One could see it as a rehearsal of puberty, but this is probably not true. For instance, mutations in *DAX1* in humans that result in delayed puberty because of hypogonadotrophic hypogonadism do not affect this minipuberty during the first year of age. Do you think this could have anything to do with brain imprinting? Is there any way to disrupt just this postnatal peak without disrupting puberty, and look at the sexual behaviour of marsupials?

Renfree: I don't know how you could just disrupt the postnatal peak. If you do something to inhibit it, this will almost certainly have an effect further on in development. We think there is a whole new story to be found for fetal androgens in eutherian mammals. It has always been tacitly assumed that testosterone is doing the virilization. In human fetuses this starts at about 8 weeks and goes on from there, but there are very few measurements. It looks like some of these effects could be due to differential synthesis of the appropriate enzymes in the target tissues. As you would know, we can't explain some 70–80% cases of pseudohermaphroditism. 5α-reductase explains a small proportion. Jean Wilson thinks that differences in enzyme synthesis might well explain some of these cases.

Capel: It looks from your data as though the surge of testosterone is resetting sensitivity to oestradiol in the male. An important feature of this resetting might be to lower the sensitivity to oestradiol in the male, so that they are unaffected by other influences (e.g. environmental oestrogens).

Short: The studies of boys born after stilboesterol exposure *in utero* showed that they were remarkably normal in terms of spermatogenesis and fertility. Massive exposure of the fetuses to stilboesterol seemed to have remarkably little effect later in life (Wilcox et al 1995).

Capel: What is known about the influence of maternal oestrogens on the fetus? I know there is a huge literature on the placenta.

Renfree: There have been many attempts to give oestrogen to pregnant females. Almost all of the studies I know of in which oestrogen has been given to neonatal eutherians give the same results as we get with the full-term marsupial treatments: a disorganized testis, but still a testis, and perhaps also some deleterious effects on other parts of the genital tract. Everyone has always said that there is no effect when oestrogen is given to the pregnant mother because there are protection mechanisms via the placenta. I don't see how that can be true: if you give inappropriate hormone treatments to women, or there are abnormal concentrations, as in congenital adrenal hyperplasia or after diethylstilbestrol, there are effects on the fetus. In tammars we have given oestrogen to the mother to see whether we could affect on the fetus, but they all aborted.

McLaren: What is the basis of the protection against oestrogen in marmoset twins? They always have twins, and 50% of the twins are one male, one female. But there is no freemartin effect in spite of the fact that they share a common blood circulation (Benirschke et al 1962).

Josso: Could it be because the blood circulation is established later? After all, freemartins are not really virilized. They appear normal; the problem is that AMH crosses to the male fetus, but to have an effect it must cross very early. If one finds anastamoses at birth, this gives no information as to the time the blood was exchanged.

Short: There is certainly a vascular anastomosis fairly early on in marmoset pregnancies. I always just assumed that this was because AMH didn't enter the fetal circulation.

Josso: In humans, the critical stage ends at 8 weeks of fetal life. This is very early. After this time it is no longer possible to induce Müllerian regression.

Short: Blanche Capel, I was wondering from your question whether you were resurrecting the idea about endocrine disruptors in the environment and the declining male sperm count. The feeling at the moment is that the evidence for the declining sperm count is very shaky. It is an artefact of a meta analysis, and there isn't any compelling evidence that oestrogens are adversely affecting human sperm counts.

Renfree: Recent work showing that there are important roles for oestrogen in normal male sexual development and function, and that males have oestrogen receptors α and β, puts a whole new complexion on the idea that oestrogens are solely female hormones.

Capel: In my other question asking about the delivery of maternal hormones to the fetus, I was trying to establish whether fetal development happens against a background of oestrogens delivered from the circulation.

Renfree: Ursula Mittwoch has suggested that the reason marsupials are born so altricial is because they cannot tolerate oestrogens. However, there is at least one species, the swamp wallaby, that has a prepartum oestrus. It has an oestrus before birth and ovulates, conceives and then gives birth a few days later. That fetus is exposed to high levels of oestrogen.

Josso: Are you saying that there are two fetuses?

Renfree: Yes, there is a developing blastocyst in one uterus, and a full-term fetus in the other.

Short: The human fetus must be incredibly good at metabolizing and conjugating any oestradiol that is around, hence all the conjugated oestriol, a very weak oestrogen, that is present in fetal blood. I have always thought that this must be the mechanism by which the fetus protects itself from oestrogenization.

Graves: We know exactly where the tammar wallaby X chromosome is homologous to the human X chromosome. Is it now time to go through the genes on the human X and pick out which is likely to be the scrotum-determining factor?

Renfree: Yes, as you know we are trying to do that now. We have selected some candidates. When it is found, it will be interesting to see whether the same gene is expressed in eutherian mammals.

Graves: Saifi & Chandra (1999) have published a list of syndromes on the X that affect the gonads. Most of those are just syndromes and they are vaguely mapped, and so there is no candidate gene.

Renfree: The Aarskog syndrome is one that is a good candidate, because it has something called a shawl scrotum around the penis. The gene responsible is *FGD1*.

References

Benirschke K, Anderson JM, Brownhill LE 1962 Marrow chimerism in marmosets. Science 138:513–515

Saifi GM, Chandra HS 1999 An apparent excess of sex- and reproduction-related genes on the human X chromosome. Proc R Soc Lond B Biol Sci 266:203–209

Wilcox AJ, Baird DD, Weinberg CR, Hornsby PP, Herbst A 1995 Fertility in men exposed prenatally to diethystilbestrol. New Eng J Med 332:1411–1416

Genetic studies of MIS signalling in sexual development

Soazik P. Jamin, Nelson A. Arango, Yuji Mishina and Richard R. Behringer[1]

Department of Molecular Genetics, University of Texas M. D. Anderson Cancer Center, 1515 Holcombe Blvd, Houston, TX 77030, USA

Abstract. The Müllerian ducts are composed of an epithelium and surrounding mesenchyme that have the potential to differentiate into female reproductive organs, including the oviducts, uterus and upper vagina. In eutherian mammals, Müllerian inhibiting substance/anti-Müllerian hormone (MIS/AMH) secreted by the fetal testis causes the regression of the Müllerian ducts to prevent the differentiation of female reproductive organs in males. MIS signalling in the Müllerian duct is mediated by the MIS type II receptor (MISRII) that is expressed in the mesenchyme surrounding the epithelium. MIS signalling alters the Müllerian duct mesenchyme, leading to the elimination of the ductal epithelium. Loss of MIS signalling, by mutation of *MIS* or *MISRII*, leads to the differentiation of female reproductive organs in males that can cause cryptorchidism and infertility. We have exploited the mouse *MisrII* locus to express heterologous genes in the cellular target of MIS signalling, the Müllerian duct mesenchyme. This approach can be used with conditional genetic strategies to identify factors that are required for the regression of the female genital duct system.

2002 The genetics and biology of sex determination. Wiley, Chichester (Novartis Foundation Symposium 244) p157–168

In eutherian mammals, the developing male and female fetuses both form two pairs of genital ducts, the mesonephric ducts (Wolffian ducts) and paramesonephric ducts (Müllerian ducts). The Wolffian ducts have the potential to form male reproductive organs, including the seminal vesicles, vas deferens and epididymides. The Müllerian ducts have the potential to differentiate into the oviducts, uterus and upper portion of the vagina. To realize the differentiated sexual phenotypes, one of these duct systems must differentiate and the other must be eliminated. In males, this switch is mediated by hormones produced by the fetal testis. Initially, Müllerian inhibitory substance (MIS; also known as anti-Müllerian hormone, AMH) is secreted by the Sertoli cells to induce the regression

[1]This chapter was presented at the symposium by Richard Behringer, to whom correspondence should be addressed.

of the Müllerian ducts, thereby eliminating the formation of female genital duct-derived tissues in males. Testosterone secreted by Leydig cells leads to the differentiation of the Wolffian duct derivatives. In females, this switch occurs in the absence of MIS and testosterone. Without MIS, the Müllerian ducts differentiate but in the absence of testosterone, the Wolffian ducts degenerate. In human males, the absence of MIS, caused by mutations in the *MIS* gene, leads to the differentiation of Müllerian duct derivatives, a condition known as persistent Müllerian duct syndrome (Belville et al 1999). A similar phenotype is observed in *Mis* mutant mice generated by gene targeting in embryonic stem (ES) cells (Behringer et al 1994). These findings demonstrate that MIS is an important regulator of male sexual differentiation (for review see Josso et al 1993).

MIS is a member of the transforming growth factor β (TGFβ) superfamily of growth and differentiation factors. TGFβ family ligands bind to membrane bound serine/threonine kinase type II receptors that complex with membrane bound serine/threonine kinase type I receptors causing phosphorylation of Smad proteins that regulate downstream gene transcription (Massagué et al 2000). The MIS type II receptor has been cloned (Baarends et al 1994, di Clemente et al 1994, Teixeira et al 1996, Mishina et al 1997) and interestingly, is expressed in a highly tissue-specific pattern. Expression is detected in the mesenchyme surrounding the Müllerian duct epithelium, and in Sertoli, Leydig and granulosa cells (Baarends et al 1994, di Clemente et al 1994, Teixeira et al 1996, Racine et al 1998, Lee et al 1999). Humans and mice with mutations in the MIS type II receptor gene have phenotypes that are identical to MIS ligand gene mutations (Imbeaud et al 1995, Mishina et al 1996). The specificity for MIS signalling for Müllerian duct regression is most likely to be due to the restricted expression of the MIS type II receptor in the mesenchyme. Indeed, MIS type II receptor knockout mice that overexpress a human *MIS* transgene, do not develop any of the reproductive abnormalities of transgenic mice that overexpress human *MIS* (Behringer et al 1990, Mishina et al 1999a). In addition to the MIS type II receptor, there should be a MIS type I receptor to mediate MIS signals for Müllerian duct regression. The TGFβ family type I receptors are called activin-like kinases (ALK). Generally, these ALKs are widely expressed and *Alk* mutations usually lead to very early embryonic lethal phenotypes, precluding conclusions on their potential roles in MIS signalling (Mishina et al 1995, 1999b, Gu et al 1998, 1999, Oh et al 2000). These observations suggest that the identification of the MIS type I receptor requires a specialized *in vivo* approach.

Over the last decade, conditional genetic strategies in mice, notably the use of the Cre/*loxP* system (for review, see Nagy 2000), have matured. Cre is a DNA recombinase that recognizes *loxP* sites that are 34 bp DNA sequences. *LoxP* sites can be used to flank a segment of DNA. The *loxP*-flanked DNA segment is said to be 'floxed'. Cre will mediate a deletion of the floxed DNA segment if the flanking *loxP* sites are in direct orientation. This simple outcome has many potential applications. One application of this technology that has become very useful is the mutation of genes in tissue-specific patterns, so-called 'tissue-specific knockouts'. In its simplest form, a gene of interest is floxed using gene targeting in mouse ES cells. When Cre acts on the floxed allele, the gene should be deleted, leading to the generation of a null allele. For tissue-specific knockouts a second strain of mouse is needed, one that expresses Cre in a tissue-specific manner. These Cre-expressing mice can be generated by gene targeting or traditional transgenic mouse methods. A series of crosses between mice carrying the floxed allele and the Cre transgenes will finally result in a mouse in which tissue-specific Cre expression deletes the gene in that tissue but not in other tissues. Thus, one can determine the required role of the gene specifically in that tissue. Here, we have devised a strategy to express heterologous genes in the cellular target of MIS-induced Müllerian duct regression, the mesenchyme surrounding the ductal epithelium. This strategy has been used to generate mice that express Cre in the Müllerian duct mesenchyme, providing a genetic tool for tissue-specific knockouts of genes that regulate Müllerian duct differentiation and regression.

Results and discussion

Because *MisrII* is expressed in the cellular target of MIS-induced Müllerian duct regression, we decided to exploit the regulation of this locus to express heterologous genes in the Müllerian duct mesenchyme. To investigate this, we introduced the *lacZ* gene into the endogenous *MisrII* locus by gene targeting in mouse ES cells (Arango et al 1999). The *lacZ* gene was introduced into exon 5 of the mouse *MisrII* locus using an *IRES-lacZ-pA* expression cassette (Fig. 1). This should lead to the production of a bi-cistronic mRNA that encodes β galactosidase (βgal) activity in a *MisrII*-specific pattern. Indeed, mice heterozygous for this *MisrII-lacZ* knock-in express βgal activity in the Müllerian ducts (Fig. 2). Furthermore, histological analysis showed that the βgal expression was restricted to the ductal mesenchyme (N. Arango & R. Behringer, unpublished observations). These findings suggest that this *MisrII* gene targeting strategy can be used to express heterologous genes in the cellular target for MIS-induced Müllerian duct regression.

We next introduced *Cre* into the *MisrII* locus using gene targeting in the identical manner described above for *lacZ* (Fig. 1). Mice heterozygous for the *MisrII-Cre* allele were then examined for Cre expression using a reporter mouse known as *Rosa26R* (*R26R*) (Soriano 1999). The cells of *R26R* mice will express βgal activity if they express Cre activity which deletes a segment of DNA that has been engineered to block *lacZ* expression. Therefore, *MisrII-Cre* mice were bred with *R26R* mice to generate fetuses heterozygous for both *MisrII-Cre* and *R26R*.

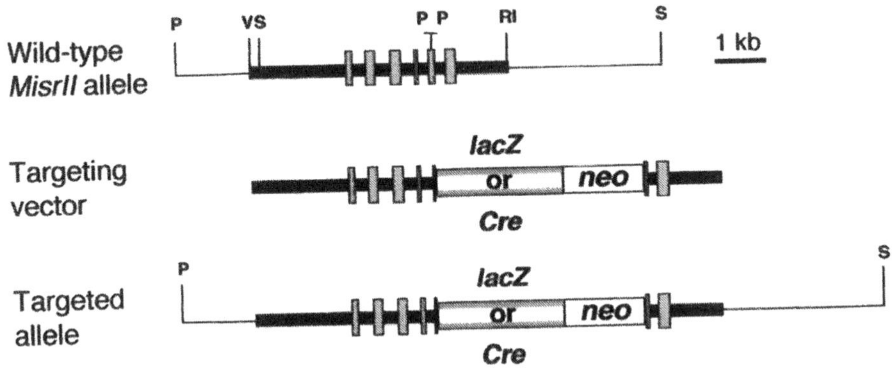

FIG. 1. Gene targeting strategy to introduce *lacZ* or *Cre* into the mouse *MisrII* locus. Partial structure of the wild-type *MisrII* locus showing the first 6 exons (shaded boxes). The region of chromosomal homology used to create the gene targeting vector is shown as a thick line. Targeting vectors were designed to introduce *lacZ-neo* or *Cre-neo* cassettes into the fifth exon of the *MisrII* gene.

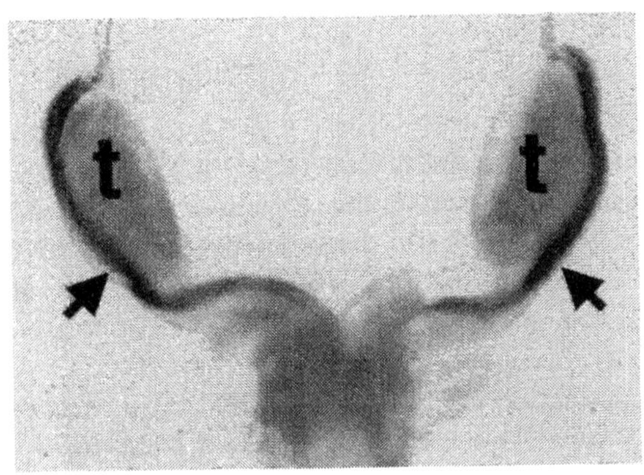

FIG. 2. Expression of βgal in the Müllerian ducts (arrows) of the reproductive tract of a 14.5 days post coitum *MisrII-lacZ* gene knock-in male mouse fetus. t, testis.

These *MisrII-Cre*;*R26R* fetuses were stained with Xgal to reveal βgal activity. At 12.5 days post coitum (dpc), βgal activity was detected only in male and female gonads and in the Müllerian ducts (Fig. 3). Histological analysis showed that βgal activity was detected in the somatic cells of the gonads and the mesenchyme cells of the Müllerian ducts (data not shown). These findings demonstrate that *MisrII-Cre* mice express Cre activity in the mesenchyme cells of the Müllerian ducts. Thus, we have generated a genetic tool for Müllerian duct mesenchyme-specific knockouts. Because *MisrII-Cre* mice express Cre in the somatic cells of the fetal gonads, they may also be useful for gonad-specific gene knockouts.

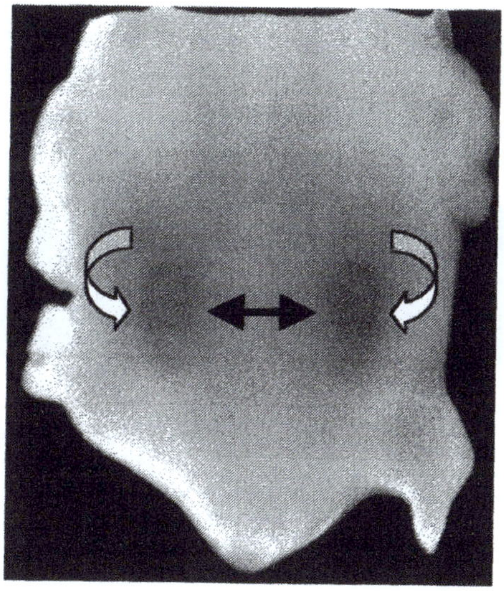

FIG. 3. Examination of Cre activity in 12.5 dpc *MisrII-Cre;R26R* double heterozygous mouse fetus. Dorsal view showing βgal staining, indicating Cre activity in the gonads (black arrows) and Müllerian ducts (white arrows).

The availability of the *MisrII-Cre* mice provided the opportunity to devise an unbiased strategy to genetically identify the MIS type I receptor. In this strategy, mice with floxed *Alk* genes are bred with mice carrying the *MisrII-Cre* transgene to generate males that carry both the floxed *Alk* gene and the *MisrII-Cre* transgene (Fig. 4). If the candidate *Alk* gene truly encodes the MIS type I receptor, then its mutation in the Müllerian duct mesenchyme should block MIS signalling, leading to the generation of males with a uterus.

We initially chose to test the role of ALK3 in MIS signalling because we had previously studied its function during mouse embryogenesis (Mishina et al 1995). ALK3 is a type IA bone morphogenetic protein (BMP) receptor (BMPR-IA). ALK3 mediates signals for BMP2, BMP4 and BMP7 (Massagué et al 2000). *Alk3* mutant mice die early during embryogenesis without forming mesoderm (Mishina et al 1995). Because *Alk3* is widely expressed and the mutants died so early during development, we decided to generate a conditional null allele by flanking exon 2 with *loxP* sites. When exon 2 is deleted by Cre, a null allele is generated that is indistinguishable in phenotype from the original *Alk3* knockout.

To determine whether *Alk3* encoded the MIS type I receptor, we interbred mice carrying the floxed *Alk3* allele with mice carrying the *MisrII-Cre* transgene. Males of the genotype *Alk3* flox/null; *MisrII-Cre* Cre/+ were found to be pseudohermaphrodites with a uterus and oviducts (Jamin et al 2002). This

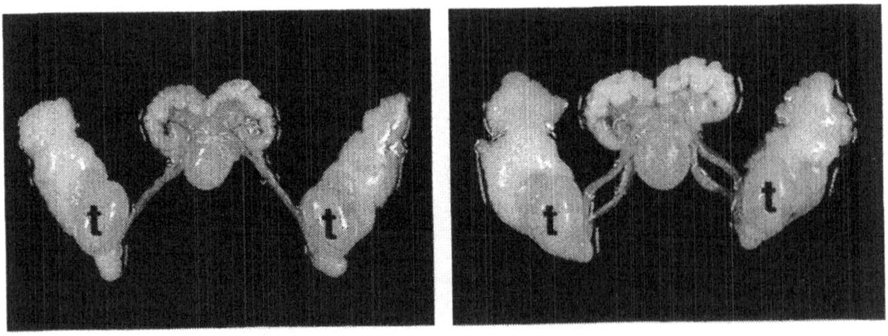

FIG. 4. Male pseudohermaphroditism in *Alk3* conditional knockout male mice. Gross morphology of 6 week old male reproductive tracts. (A) control male, (B) *Alk3* conditional mutant male (*Alk3* flox/null;*MisrII-Cre* Cre/+). The *Alk3* conditional mutant male has a uterus (arrowhead). Arrows, vas deferens; t, testis.

phenotype is identical to mice lacking the MIS ligand or the MIS type II receptor (Behringer et al 1994, Mishina et al 1996). Expression of *Mis* and *MisrII* was normal in these *Alk3* conditional knockout males, demonstrating that other essential components of the MIS signalling pathway were correctly expressed. These findings indicate that *Alk3* encodes the type I receptor for MIS-induced Müllerian duct regression.

Recent biochemical and tissue culture studies have pointed to ALK2 and ALK6 as MIS type I receptors (Gouédard et al 2000, Clarke et al 2001, Visser et al 2001). However, ALK6 cannot be the essential type I receptor for MIS-induced Müllerian duct regression because male *Alk6* knockout mice have normal Müllerian duct regression (Yi et al 2000, Clarke et al 2001). The role for ALK2 in Müllerian duct regression is not clear. *Alk2* knockout mice die early during embryogenesis prior to genital duct formation (Gu et al 1999, Mishina et al 1999b). In one study, female rat urogenital ridges cultured in the presence of MIS retained the Müllerian duct when exposed to antisense oligonucleotides for *Alk2* (Visser et al 2001). Unfortunately, mice with a floxed *Alk2* conditional allele do not yet exist to definitively determine a role for ALK2 in Müllerian duct regression.

Our *in vivo* findings demonstrate a required role for ALK3 in the Müllerian duct mesenchyme for the regression of the ductal epithelium. ALK3 can functionally interact with the BMP type II receptor to mediate BMP2, BMP4, and BMP7 signals and also with ActRII and ActRIIB to mediate BMP4 and GDF5 signals (Massagué et al 2000). Our studies show that ALK3 can also functionally interact with MISRII to mediate a different signalling pathway. Thus, one widely expressed type I receptor can interact with different type II receptors to mediate distinct signalling pathways. ALK3 is the orthologue of the type I decapentaplegic (DPP) receptor in

Drosophila known as thickvein (TKV). MIS-induced regression of the Müllerian ducts is found in reptiles, birds and mammals. Our findings indicate that a conserved TGFβ family signalling component has been co-opted during evolution for male sexual differentiation in amniotes.

Summary

The cellular target for the action of MIS on the regression of the Müllerian ducts is the mesenchyme adjacent to the ductal epithelium. We have devised a genetic strategy to modify this tissue, using gene targeting in mouse ES cells. Integration of heterologous genes into the *MisrII* locus leads to a pattern of expression that generally mimics the expression of the endogenous locus. The expression of Cre in the Müllerian duct mesenchyme opens up novel opportunities to generate tissue-specific mutations in this tissue to elucidate the factors that mediate MIS-induced Müllerian duct regression. Utilizing *MisrII-Cre* mice, we have identified ALK3/BMPR-IA as the type I receptor for MIS-induced Müllerian duct regression.

Acknowledgements

Supported by grants from the National Institutes of Health HD30284, AR42919, and the Barnts Family to R.R.B. S.P.J. was supported by a fellowship from the Lalor Foundation.

References

Arango NA, Lovell-Badge R, Behringer RR 1999 Targeted mutagenesis of the endogenous mouse *Mis* gene promoter: in vivo definition of genetic pathways of vertebrate sexual development. Cell 99:409–419

Baarends WM, van Helmond MJL, Post M et al 1994 A novel member of the transmembrane serine/threonine kinase receptor family is specifically expressed in the gonads and in mesenchymal cells adjacent to the Müllerian duct. Development 120:189–197

Behringer RR, Cate RL, Froelick GJ, Palmiter RD, Brinster RL 1990 Abnormal sexual development in transgenic mice chronically expressing Müllerian inhibiting substance. Nature 345:167–170

Behringer RR, Finegold MJ, Cate RL 1994 Müllerian-inhibiting substance function during mammalian sexual development. Cell 79:415–425

Belville C, Josso N, Picard J-Y 1999 Persistence of Müllerian derivatives in males. Am J Med Genet 89:218–223

Clarke TR, Hoshiya Y, Yi SE, Liu X, Lyons KM, Donahoe PK 2001 Mullerian inhibiting substance signaling uses a bone morphogenetic protein (BMP)-like pathway mediated by ALK2 and induces SMAD6 expression. Mol Endocrinol 15:946–959

di Clemente N, Wilson C, Faure E et al 1994 Cloning, expression, and alternative splicing of the receptor for anti-Müllerian hormone. Mol Endocrinol 8:1006–1020

Gouédard L, Chen YG, Thevenet L et al 2000 Engagement of bone morphogenetic protein type IB receptor and Smad1 signaling by anti-Mullerian hormone and its type II receptor. J Biol Chem 275:27973–27978

Gu Z, Nomura M, Simpson BB et al 1998 The type I activin receptor ActRIB is required for egg cylinder organization and gastrulation in the mouse. Genes Dev 12:844–857

Gu Z, Reynolds EM, Song J et al 1999 The type I serine/threonine kinase receptor ActRIA (ALK2) is required for gastrulation of the mouse embryo. Development 126:2551–2561

Imbeaud S, Faure E, Lamarre I et al 1995 Insensitivity to anti-Müllerian hormone due to a mutation in the human anti-Müllerian hormone receptor. Nat Genet 11:382–388

Jamin SP, Arango NA, Mishina Y, Hanks MC, Behringer RR 2002 BMPR-IA is a type I receptor that is required for male sexual differentiation in mammals. Submitted

Josso N, Cate RL, Picard JY et al 1993 Anti-Müllerian hormone: the Jost factor. Recent Prog Horm Res 48:1–59

Lee MM, Seah CC, Masiakos PT et al 1999 Müllerian-inhibiting substance type II receptor expression and function in purified rat Leydig cells. Endocrinology 140:2819–2827

Massagué J, Blain SW, Lo RS 2000 TGFβ signaling in growth control, cancer, and heritable disorders. Cell 103:295–309

Mishina Y, Suzuki A, Ueno N, Behringer RR 1995 *Bmpr* encodes a type I bone morphogenetic protein receptor that is essential for gastrulation during mouse embryogenesis. Genes Dev 15:3027–3037

Mishina Y, Rey R, Finegold MJ et al 1996 Genetic analysis of the Müllerian-inhibiting substance signal transduction pathway in mammalian sexual differentiation. Genes Dev 10:2577–2587

Mishina Y, Tizard R, Deng JM et al 1997 Sequence, genomic organization, and chromosomal location of the mouse Mullerian-inhibiting substance type II receptor gene. Biochem Biophys Res Commun 237:741–746

Mishina Y, Whitworth DJ, Racine C, Behringer RR 1999a High specificity of Mullerian-inhibiting substance signaling in vivo. Endocrinology 140:2084–2088

Mishina Y, Crombie R, Bradley A, Behringer RR 1999b Multiple roles for activin-like kinase-2 signaling during mouse embryogenesis. Dev Biol 213:314–326

Nagy A 2000 Cre recombinase: the universal reagent for genome tailoring. Genesis 26:116–117

Oh SP, Seki T, Goss KA et al 2000 Activin receptor-like kinase 1 modulates transforming growth factor-beta 1 signaling in the regulation of angiogenesis. Proc Natl Acad Sci USA 97:2626–2631

Racine C, Rey R, Forest MG et al 1998 Receptors for anti-Müllerian hormone on Leydig cells are responsible for its effects on steroidogenesis and cell differentiation. Proc Natl Acad Sci USA 95:594–599

Soriano P 1999 Generalized lacZ expression with the ROSA26 Cre reporter strain. Nat Genet 21:70–71

Teixeira J, He WW, Shah PC et al 1996 Developmental expression of a candidate Mullerian inhibiting substance type II receptor. Endocrinology 137:160–165

Visser JA, Olaso R, Verhoef-Post M, Kramer P, Themmen AP, Ingraham HA 2001 The serine/threonine transmembrane receptor ALK2 mediates Mullerian inhibiting substance signaling. Mol Endocrinol 15:936–945

Yi SE, Daluiski A, Pederson R, Rosen V, Lyons KM 2000 The type I BMP receptor BMPRIB is required for chondrogenesis in the mouse limb. Development 127:621–630

DISCUSSION

Josso: Our results identifying BMPRIB/ALK6 (a sister of ALK3) are not in complete opposition to Richard Behringer's. We clearly demonstrated that MIS/AMH acts through SMAD1 (Gouédard et al 2000); we didn't test SMAD5 and 8. Biochemically we showed that ALK6 was able to bind to the receptor and activate

BMP2-responsive genes, but we didn't really have a biochemical effect of MIS/AMH that we could relate to the type I receptor we had identified. In the conclusion of our paper, we said that there was certainly another receptor, because the receptor we had identified, BMPR1B, was not expressed in the cell lines of testicular origin which we had used, and it was hardly expressed at all in the testis. Therefore, I don't think it is surprising that another receptor has been found. My colleague Nathalie di Clemente has preliminary data showing that if she immunoprecipitates cells labelled with ^{32}P in testicular cell lines she gets a band that has the molecular weight of ALK3, not ALK6. This would be in perfect agreement with Richard Behringer's finding. However, I am still not completely satisfied as a clinician, because there is a syndrome called persistent Müllerian duct syndrome, which looks very much like the knockout mice Richard Behringer has described from the reproductive point of view. They are virilized, but they have a uterus. Some of these cases are due to *Mis/Amh* mutations, others are due to type 2 receptor mutations, and 16% are of unknown origin (Belville et al 1999). I find it difficult to believe that these unexplained persistent Müllerian duct syndrome cases in the human can be due to mutations of either one of the type I receptors that have been identified. These receptors are BMP-type I receptors, and the mutations, if they are not conditional, will lead to very early death or to the birth of individuals with skeletal abnormalities. I think therefore that there might be another type I receptor.

Behringer: Or it is another aspect of the pathway.

Josso: It couldn't be the SMADs. If you had a mutation of SMAD1 you would probably have an animal or patient that had cancer. SMAD1 not only works for BMPs but also for many other things.

Behringer: Further upstream, there could be a binding protein for MIS that transports or sequesters it.

Josso: Or a coactivator, perhaps.

Camarino: When you say that in 16% of cases the mutation is unknown, have you excluded by linkage that they are either type II or MIS mutations?

Josso: We have sequenced the promoter, but do you think that there could be other enhancer sequences?

Camarino: Yes, something like that. Do you have families?

Josso: Yes.

Camarino: It will be easy therefore to see whether you can exclude the loci by linkage.

Short: Earlier I mentioned the fascination of how it seems to be the oocyte and not the oogonium that is necessary for the induction of follicular cells in the ovary. Obviously you have lovely mutants in mice like the *W* allele where you can deplete germ cell populations. If you actually knock out the germ cell component of the ovary, what happens to those frustrated would-be follicular cells? Can they express

MIS, or do they need an oocyte there before there is any MIS production? If this is so, what a lovely model to study how the oocyte can switch on *Mis*!

Behringer: I don't think we have ever looked at a germ cell-deficient animal for *Mis* expression.

Lovell-Badge: We have. W^e homozygote males express *Mis* and females do not, but the embryos do not survive beyond about 15 days.

Josso: I thought if germ cells were deleted from the ovary, there was soon no ovary left at all.

McLaren: If the germ cells don't get into the ovary at all, or only a very few do, then the supporting cell lineage doesn't develop, and there is what the clinicians call a streak gonad. If the germ cells get in, but then later are lost, the granulosa cells tend to transdifferentiate into something more like Sertoli cells. There are a number of situations where this has been shown. In some cases it has been shown that they do secrete MIS.

Short: In an XO mouse, in which there is early oocyte atresia, what would happen with MIS production?

McLaren: In the mouse, a lot of germ cells still survive even in the XO ovary. The females are fertile even though their reproductive lifespan is shortened. I think in this case you wouldn't see anything unusual. It is more where the germ cells actually disappear, having induced granulosa cell formation.

Josso: You say that in the transdifferentiation the would-be granulosa cells are transformed into Sertoli-like cells. Sertoli cells make MIS/AMH, so this doesn't answer Roger Short's question. He would like to have MIS/AMH produced by a granulosa cell.

Short: Is the bottom line, then, that if there are no germ cells in the ovary, there is no MIS production?

Josso: There is no MIS/AMH production by granulosa cells, because there are no granulosa cells without germ cells. But there are cells that look like Sertoli cells, and these have a right to produce MIS/AMH. But what you are saying is probably of interest to clinicians. They are now asking us to measure MIS/AMH in the blood of women undergoing *in vitro* fertilization, because they believe that it might give them some indication as to the health of the granulosa cells.

Vilain: Richard Behringer, what was the phenotype of the transgenic mouse with *Mis* overexpression?

Behringer: In females it varies a little between the different lines of mice and the levels. Generally in females the oviducts and uterus are lost. Most of them will form an ovary. In one of the lines we have looked at carefully, they lose germ cells. Some of them started to form cord-like structures but then they degenerated. Most of the males seem OK, but some of them were not virilized. Nathalie Josso has looked at this more carefully, and she sees depressed Leydig cell function.

Vilain: So you would say that the transdifferentiation would be the consequence of the loss of germ cells, not a direct effect of MIS.

Greenfield: Have you ever observed any abnormalities with respect to mammary gland function? I am sure I saw a recent report saying that type II receptor expression in the mammary gland mediated apoptosis in a NF-κB-dependent fashion (Segev et al 2000).

Behringer: I don't know about that. I haven't looked specifically, but the *lacZ* fetuses that I have seen didn't look like they had mammary gland expression. Nathalie, have you ever looked in mammary glands for type II receptor expression?

Josso: No.

Behringer: Perhaps I can expand on the gonadal expression of the Cre reporter. I think it is a useful line. We see gonadal expression at 11.5 days in both sexes. In the female it looks like it is throughout. If you want to activate or delete expression conditionally, it might be very good for the female gonad. When we look at the postnatal gonad after these crosses, it is completely blue also. The testis is a little more variable. When Soazik Jamin did some sections of the fetal stages, she first saw the βgal activity in the interstitium, not in the cords, then when she looked after birth it was in the interstitial cells and probably also in the Sertoli cells. If we take this reporter as a readout of Cre activity, it looks like first there is activity in the interstitium and then finally it hits the Sertoli component. If this is true, then this Cre mouse might only be good for looking at the later stages if you want to alter Sertoli cell function. But we have been questioning the Rosa26 reporter. My student Akio Kobayashi decided to do a control test. He took Rosa26 heterozygous male and female fetuses, Xgal stained them and then sectioned them. At 14.5 days the cords are negative as are the equivalent structures in the females. There may be a problem with Rosa26 as a reporter in the gonad. I have talked a little to Blanche Capel about this, but she might not have seen this in her recombinations because she is using the recipient as a negative genital ridge. So we have to be a little bit careful with Rosa26. There are alternative Cre reporter lines. We are going to try one of these out.

Capel: We have done experiments using a blue gonad from Rosa26 and a white mesonephros, to see whether there was any back migration in the other direction that we weren't picking up. The whole gonad was blue in those experiments, so it might be a variant in the strain that you are using.

Behringer: We have maintained it on the typical B6 $\times$ 129 genetic background, but generated the fetuses by crosses with Swiss mice. Akio picked this up because he was doing chimera studies looking at the Müllerian duct epithelium, and he looked adjacent to the Müllerian ducts and saw all these pink gonads. This is what caused him to check this out.

Capel: It might be worth looking at your strain.

Short: Have you any evidence of the functional capacity of follicle cells that have not been able to express MIS? Can they, for example, produce a normal corpus luteum following ovulation?

Behringer: Axel Themmen reported alterations in follicle recruitment (Durlinger et al 1999). Anecdotally, I thought the *Mis*-deficient females were good breeders, that they bred frequently and had large litters. Perhaps it makes sense; Axel shows that the follicles are recruited at higher numbers and early, and in those crosses I was usually taking younger females and not older ones.

Short: How is MIS controlling follicle recruitment?

Behringer: I don't know how it is controlling it. But I find it very interesting that without MIS the animals are still fertile. What it appears to be doing is regulating the window of fertility. You can imagine that between different species you could play with this and alter the window of fertility.

Poulat: When you overexpress MIS you see transdifferentiation in the ovary in some cases. Do you think there is a link here with the double knockout of the oestrogen receptor, in which there is also transdifferentiation?

Behringer: The first time I saw the oestrogen receptor knockouts I thought they looked exactly like the MIS-overexpressing ovaries. But I think it may be that if the ovary is damaged and the germ cells are lost, the response is that the somatic cells become a bit Sertoli-like. I think of it more as a non-specific response.

McLaren: Is anything known about the effect of MIS on the mesenchyme surrounding the Müllerian duct, which messes up the epithelium of the duct?

Josso: Françoise Xavier in our lab has shown that in the Müllerian duct MIS/AMH causes the translocation of β catenin in the nucleus (Allard et al 2000), but this isn't really an answer to your question. I believe that the androgen receptor in the epididymis is also found in the mesenchyme. The fact that there is a receptor in the mesenchyme and the target cells are in the epithelium is not unique to the Müllerian duct.

McLaren: It seems to be a strange way of organizing things.

References

Allard S, Adin P, Gouedard L et al 2000 Molecular mechanisms of hormone-mediated Mullerian duct regression: involvement of beta-catenin. Development 127:3349–3360

Belville C, Josso N, Picard JY 1999 Persistence of Mullerian derivatives in males. Am J Med Genet 89:218–223

Durlinger AL, Kramer P, Karels B et al 1999 Control of primordial follicle recruitment by anti-Müllerian hormone in the mouse ovary. Endocrinology 140:5789–5796

Gouédard L, Chen YG, Thevenet L et al 2000 Engagement of bone morphogenetic protein type IB receptor and Smad1 signaling by anti-Müllerian hormone and its type II receptor. J Biol Chem 275:27973–27978

Segev DL, Ha TU, Tran TT et al 2000 Müllerian inhibiting substance inhibits breast cancer cell growth through an NFκB-mediated pathway. J Biol Chem 275:28371–28379

Social regulation of the brain: sex, size and status

Russell D. Fernald

Program in Neuroscience, Department of Psychology, Stanford University, Stanford, CA 94305, USA

Abstract. Fish comprise the largest group of extant vertebrates with approximately 25 000 known species. Some of these species are exceptional among vertebrates because they can change sex as adults. This observation raises ultimate questions about what selective forces led to the evolution of sex-changing ability and raises proximate questions about what mechanisms could account for this process. Sex change can be either from female to male (protogyny) or the reverse (protandry). In either case, the actual process of sex reversal requires reorganization of many critically important physiological systems from transformation of the gonads to modification of the neural and hormonal control systems. All of these changes require an individual animal to initiate the process based on information gleaned from the social situation. This is all the more remarkable because the information could be as simple as size discrimination or as complex as detecting subtle behavioural signals. Although it is self-evident that the brain controls behaviour, clearly behaviour can also 'control' the brain. How does behaviour cause changes in the brain? The work described here links molecular events with organismal behaviour by using an African cichlid fish model system in which social behaviours regulate reproduction. These animals have a complex social system based on the behaviour of two distinct classes of males, those with territories and those without. Changes in social status produced by behavioural interactions cause changes in neurons and endocrine responses. Surprisingly, growth rate is also regulated by social status and prior social history. Discovering how relevant social information is transduced into physiological processes requiring cellular and molecular action presents a major challenge.

2002 The genetics and biology of sex determination. Wiley, Chichester (Novartis Foundation Symposium 244) p 169–186

Among social animals, the behaviour of one individual depends on the behaviour of other individuals, as first described systematically by Konrad Lorenz (1935). The nature of such influential interactions depends on the species, the situation and the actual behavioural interaction. The most reliable predictor that behaviour will change due to an encounter is the social status of the individuals involved. For example, a dominant animal threatened by a non-dominant animal behaves differently than does a dominant animal threatened by another dominant individual. Similarly, behaviour by a female produces quite different reactions in males depending on their social status. It is fair to say that in every social system that has been observed, behaviour of individuals depends on their social status, on behavioural interactions and on the physical environment. This universal dependence of behaviour on social context is the primary scientific framework used to interpret behaviour during social interactions.

But how does an animal 'know' its own status and behave appropriately? And, how does an individual recognize an opportunity to change status upwards or acquiesce to an imposed change downwards? Clearly, in the short term, physiological processes allow the animal to act and, in the long term, cellular and molecular processes accommodate changes in its external reality or social status. Some of the required physiological and molecular changes must precede behavioural change but others are a consequence of that change. How are these internal changes regulated by social interaction? There must be a transduction of social information into internal change, but how? To a great extent, this must depend on how the animal perceives and interprets events in its own world.

von Uexkill (1909) first realized that every animal species experiences life differently, living in what he called its 'Umwelt', or unique perceptual world. A bat using sonic echoes to probe the world in darkness surely perceives its surroundings differently than a giraffe, which relies on its eyes, nose and ears, or a weakly electric fish that relies almost entirely on faint electrical signals for information. Each animal species has a particular complement of sensory capabilities that fundamentally restrict the physical stimuli it can use to make behavioural decisions. This constraint on the perceived world necessarily limits the possible behavioural responses of any animal. Writing at the turn of the last century, von Uexkill could not possibly have anticipated the discovery of magnetic, electric or pressure senses, nor could he imagine seeing into the infrared and ultraviolet, or even that light detection exists at some remarkable places other than the eye (Arikawa et al 1996). These discoveries make his writing all the more prescient, and the many interesting, unusual animal 'Umwelts' reveal the many ways that natural selection has shaped animal perceptions. The range of sensory capabilities are matched by variations in animal form and function that also reflect adaptations to the environment.

In evolutionary change, the ultimate arbiter of successful adaptations is behaviour. An animal that survives does so because it behaves successfully during the multitude of interactions with other animals and with its environment. Yet behaviour, in turn, depends on intricate physiological, cellular and ultimately molecular adaptations. A major challenge in biology is to understand the linkages across these levels of analysis as an animal interacts with its world. How is behaviour controlled via physiological processes and, correspondingly, how does behaviour influence physiological, cellular or molecular events? Here I will summarize evidence from our experiments designed to discover mechanisms that underlie the synergistic interactions between behaviour and physiology in a model system uniquely suited for this inquiry.

Model system

To understand how behaviour influences the brain and vice-versa, our laboratory studies a cichlid fish, *Haplochromis burtoni* native to Lake Tanganyika in central Africa. In its natural habitat, there are two kinds of males: those with territories and those without (Fernald & Hirata 1977a,b). Territorial males, which comprise only ~10–15% of the males, are brightly coloured, with a blue or yellow body colour, dramatic black stripe through the eye, vertical black bars on the body, a black spot on the tip of the gill cover and a large red patch just behind it (Fig. 1). In contrast, non-territorial males are cryptically coloured, making them difficult to

FIG. 1. Illustration of the body patterns for typical territorial (T) and non-territorial (NT) males. Top: NT males lack the robust markings of their territorial counterparts and are coloured to maximize camouflage. Bottom: the T male has distinctive anal fin spots, dark forehead and lachrymal (eye-bar) stripes and is brightly coloured, including orange humeral scales. The overall body colour may be either yellow or blue. (Modified from Fernald 1984.)

distinguish from the substrate and from females that are similarly camouflaged. That is, the non-territorial males appear nearly identical to females. The animals live in a lek-like social system in which the brightly coloured territorial males vigorously defend contiguous territories arrayed over a food supply (Fernald 1977). The number of territorial males is limited by the size of the available food supply.

This species has an elaborate social system that depends on signalling among animals. Social communication in *H. burtoni* depends primarily on visual signals (Fernald 1984). Territorial males are very active, performing at least 19 distinct behavioural acts during fast paced social encounters (Fernald 1977). They divide their time between digging a pit in the centre of their territory, fighting with neighbours at common territorial boundaries, chasing non-territorial animals away and soliciting and courting females. Solicitation and courtship behaviours are easily identified since the males display bright coloration patterns towards the courted female. Courtship includes 'leading' the female toward the territory and during 'courting' the male quivers his spread, brightly coloured anal fin in front of the female. Females led into the territory will feed by nipping at and sifting through the bottom cover.

Interestingly, non-territorial males mimic this female behaviour accurately enough so the territorial males will allow them to eat in the territory. Soon enough, however, the deception is discovered and the female impersonator is chased off. If a genetic female responds to the entreaties of a male, he will lead her into his pit and continue the elaborate courtship movements, swimming to the front of the female and rapidly quivering his entire body with his anal fin spread in her view. As the pair disappears into the spawning pit out of direct view of the territory, other animals exploit this opportunity to feed energetically. The spawning male repeatedly interrupts his courtship behaviour to chase intruders off his limited food supply. If physiologically ready and adequately stimulated, the female lays her eggs at the bottom of the pit, collecting them in her mouth almost immediately. After she lays several eggs, the male swims in front of her, again displaying the anal fins spots, his body quivering. The female then nips at the male's anal fin as though she mistakes his spots for uncollected ova. So, while attempting to 'collect' the spots, the female ingests the milt ejected near them by the male and ensures fertilization. After several bouts of this alternating behaviour, the female may go to the territory of another male to lay more eggs or depart from the territorial arena with the fertilized eggs to brood them (Fernald 1984).

This brief description of the natural behaviour of *H. burtoni* reveals the extensive role social interactions play in its daily life. Importantly, under the appropriate conditions, the behaviour of *H. burtoni* in the laboratory matches exactly that found in the field (Fernald 1977), making this a useful species for studying the influence of social behaviour on the brain. Clearly, the behaviour is guided by visual signals and the social scene largely governs the behaviour of individual animals. Each behavioural act influences the next, both in the observed individual and in the animals involved in the interaction. During these encounters, information is exchanged between individuals that influences the next behavioural interaction of these animals. How do animals exchange key information and what are the consequences of that exchange?

Differences between territorial and non-territorial males

As young *H. burtoni* grow, the social behaviour of conspecifics regulates their behavioural and gonadal development and even their growth in a differentiated fashion (Fraley & Fernald 1982). For the first seven to eight weeks of life, living in a group facilitates growth of males as compared to broodmates reared in total isolation with visual contact (Fig. 2). However, after this time, group-reared males that do not acquire and defend territories grow more slowly than those with territories. Males that do form territories develop their colour patterns faster, weigh more, and have larger and more highly developed gonads than animals reared under any other conditions (Fig. 2B). Concomitantly, group-reared fish show early developing agonistic/aggressive behavioural patterns (chase, tailbeat, fin spread) and chromatic patterns (eyebar, opercular spot) more than two weeks before these features appear in animals reared in physical isolation.

The absolute growth rate of *H. burtoni* under optimal conditions is dramatic and has resulted in novel developmental strategies over evolutionary time. These include the addition of new cells to the lens, retina and brain (Fernald & Wright 1983, Fernald 1983, 1989, Johns & Fernald 1981). Such social control of maturation and growth is found in many species (for example, Borowsky 1973, Schultz et al 1991) and takes a variety of forms. In *H. burtoni*, however, there are some unique effects of this social regulation of growth, most importantly that it is not limited to early development.

Juvenile males raised with adults present, as is the natural condition, show suppressed gonadal maturation relative to those reared without adults (Davis & Fernald 1990). As well as having smaller testes, these animals have smaller gonadotropin-releasing hormone (GnRH)-containing neurons in the preoptic area (POA), a region in the ventral telencephalon adjacent to the hypothalamus (Fig. 3). These neurons project to the pituitary (Bushnik & Fernald 1995) where they release GnRH. The somata sizes of GnRH-containing neurons differ eightfold in volume depending on the social conditions. Since GnRH is the main signalling peptide that regulates reproductive maturity, the social control of maturation acts by changing structures in the brain. Thus, the social control of maturation is reflected via changes in structures in the brain.

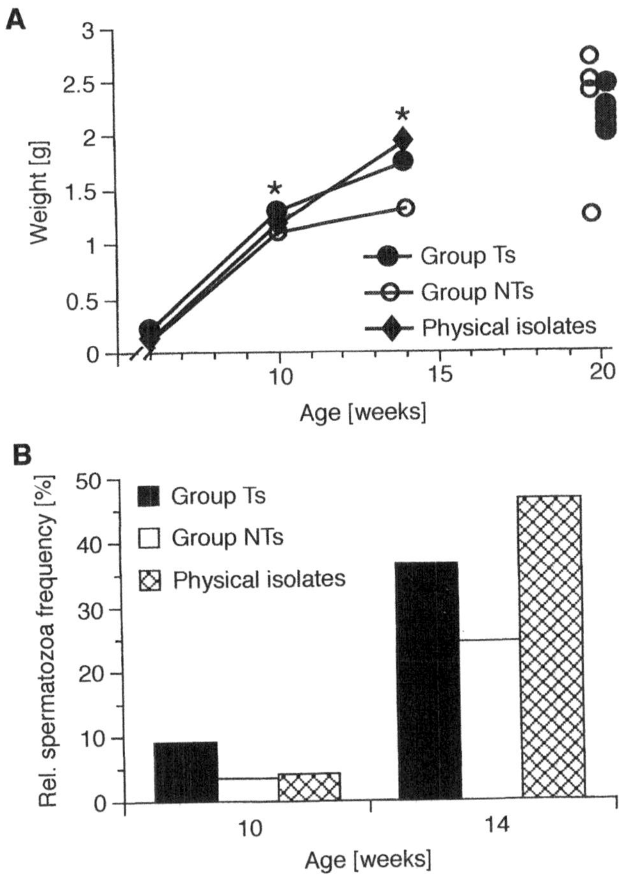

FIG. 2. Development and maturation in group-reared (open and filled circles) and physically isolated (diamonds) juvenile *H. burtoni*. (A) Growth rates expressed as body weight for the different categories. Asterisks indicate that group-reared territorial fish (Ts, filled circles) weigh significantly more after 10 and 14 weeks as compared to their non-territorial (NTs, open circles) tankmates. Differences in standard lengths are not significant (data not shown). Note that after 20 weeks size differences are no longer evident. (B) Relative estimates of mature spermatozoa in cross-sections of the central testicular lobule. Note the rapid increase in physically isolated males between week 10 and week 14. (After Fraley & Fernald 1982, Davis & Fernald 1990.)

What are the salient sensory cues that a juvenile male fish perceives which influence its initial social state? In the laboratory, if juvenile males are reared alone, they develop into territorial males with all of the defining characteristics from large gonads to prominent lachrymal stripes. This shows that every male has the potential for social dominance, that this is the default developmental

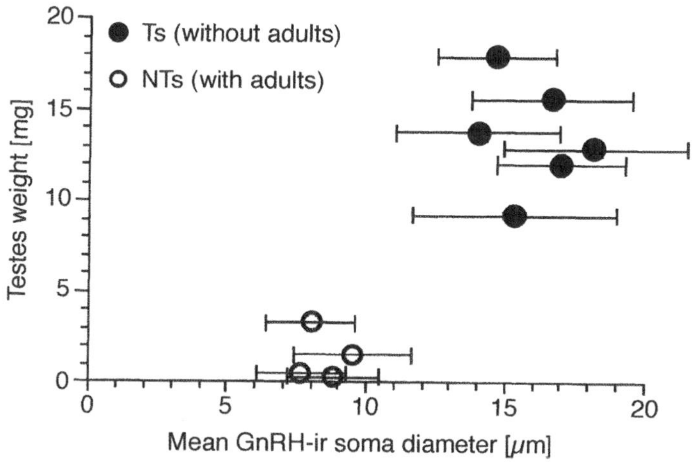

FIG. 3. Demonstration of social regulation of the reproductive axis in juvenile *H. burtoni*. Testes weights of 20 weeks old early-maturing (without adults present) territorial males (Ts; filled circles) and maturation-suppressed (with adults present) non-territorial males (NTs; empty circles) plotted against the respective average soma diameters for the largest 30% of preoptic GnRH-IR neurons (±SD). Neuron sizes are independent of body size in this experiment. Note the striking differences in cell size as well as testes weight between the two groups. (After Davis & Fernald 1990.)

pathway, and that any genetic influence on dominance is negligible in comparison to social cues.

We have begun to dissect these social cues by sensory modality to determine the ones responsible for suppressing non-territorial males and have discovered that, in addition to visual cues, tactile stimuli play a part (M. R. Davis & R. D. Fernald, unpublished observations). Thus, if a cohort of young fish are raised in the same aquarium as an older established community, the young males remain non-territorial, as stated above. If, however, the two groups are separated by a fine mesh net, one that allows visual and chemical contact, and even permits threat displays across the barrier, they quickly learn that the would-be bullies on the other side of the tank are unable to chase and bite them. Freed from the threat of aggression by the big territorial males, the younger fish form their own communities where again, some 10% of the males escape maturational suppression and become territorial. In turn, these suppress the maturation of the remaining 90% of the males on their side of the net. Since both the older and younger communities have visual and chemical access to each other, these findings indicate that biting and nipping behaviours form some part of the suppressive signal imparted to non-territorial fish.

Our studies have shown (Muske & Fernald 1987a,b), that the territorial males differ not only in their social displays, but also in the prominence of those signals to other viewers. Becoming and remaining socially dominant produces long-term physiological changes, just as losing social dominance influences the physiological state. Given the importance of the correct production and recognition of social signals, there must be mechanisms responsible for their development and mechanisms for their transduction into physiological systems.

Social control of sex and size

In the natural environment of *H. burtoni*, there are costs and benefits associated with territoriality. The obvious benefits are that territorial males have a reliable food supply and that they are the only males that spawn. The costs are that the bright, flashy colours and active behaviours of dominant males make them conspicuous to birds of prey. Indeed, predation of territorial males occurs at a significantly higher rate than that of females or non-territorial males (Fernald & Hirata 1977b). When a territorial male is removed, the vacated space provides an opportunity for a non-territorial male to switch social state. Within a few seconds, such non-territorial males produce an eyebar and exhibit aggressive behaviours. What endogenous changes accompany this outward transformation and how are they related to one another?

Social regulation of reproduction

To understand whether social status also regulates reproduction in adult animals, adult males were converted from territorial (T) to non-territorial (NT) or vice versa and their reproductive axis examined. To do this, T males were moved into communities with larger T males, as a result of which they became NT (T→NT). Correspondingly, NT males were moved to new communities consisting of females and smaller males which they could dominate, as a result of which they became T (NT→T). In each case, the subjects remained in the altered social setting for four weeks after which the size of GnRH containing cells was measured (Francis et al 1993).

To quantify the consequences of this change in social status on reproductive competence, we measured changes in the gonad size and mean soma sizes of the POA immunoreactive GnRH-containing neurons (Fig. 4A,B). The mean value of both the soma size of POA GnRH-immunoreactive (GnRH-IR) neurons (Fig. 4A) and gonadosomatic index (GSI) (Fig. 4B) were significantly larger in both NT→T and control T males than in T→NT and NT males. In two other GnRH-IR cell groups, one located in the terminal nerve region, the other in the mesencephalon, there was no difference in mean soma sizes between T and NT males (Davis & Fernald 1990). Thus the change in POA GnRH containing neurons is not a

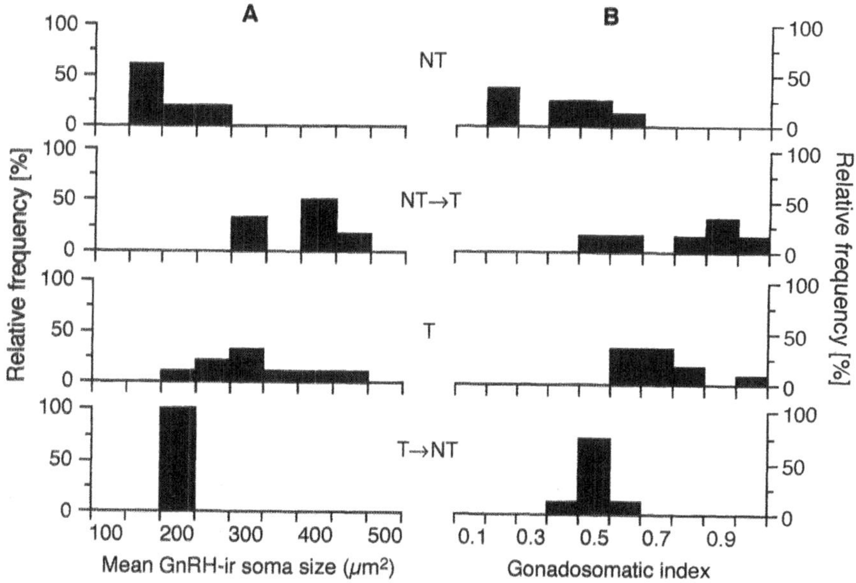

FIG. 4. The effect of social change on GnRH cell size (A; mean soma size of preoptic area GnRH-IR neurons) and gonadosomatic indices (GSI) (B) as shown in frequency histograms measured in animals from the four possible social classes. Percentage of individuals are plotted for each social category (NT, T, NT→T, T→NT). There are significant differences, in soma sizes as well as GSI, between animals that were Ts and ascended NT→Ts when compared to animals that were NTs and descended T→NTs. (Modified from Francis et al 1993.)

general property of cells expressing GnRH but rather is confined to those in the hypothalamo–pituitary–gonadal (HPG) axis.

These data show that following social change, endogenous changes occur that equip a newly dominant male for his new social and reproductive status. Conversely, animals subjected to a downgrade in social status (T→NT), lost both GnRH cell size and gonad size, in line with their new social state. Clearly, social status determines both soma size of POA GnRH-IR neurons and GSI, and both these effects are reversible. The relatively larger testes and GnRH-IR neurons characteristic of T males is a consequence of their social dominance, and when this dominance advantage is lost, both neurons and testes shrink.

Since the precipitating event in these studies was the experimentally manipulated change in social status, it is clear that in these teleosts changes in social status can initiate changes in endocrine state. However, such changes in social and endocrine systems interleave so fluently, they suggest a complex nexus of interactions rather than a linear chain of control. GnRH-containing neurons in the hypothalamus of

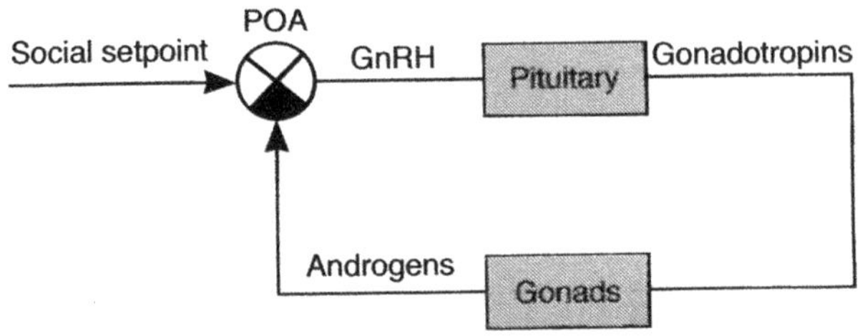

FIG. 5. Schematic illustration showing the regulation of GnRH release in male *H. burtoni* via a social setpoint. Our data show that neurons in the preoptic area integrate both social and hormonal signals to regulate GnRH release. In this model, the setpoint for the GnRH level is determined by social signals and the maintenance of the GnRH level at this setpoint is achieved by negative feedback from gonadal androgens. (Modified from Soma et al 1996.)

adult territorial males both influence and are influenced by circulating gonadal hormones. We know this because castration of territorial males caused GnRH neurons to increase in size (Soma et al 1996). This neuronal hypertrophy in castrated animals was prevented either by testosterone or by 11-ketotestosterone treatment. Oestradiol (E2) treatment did not reduce GnRH cell size in castrated animals. These results (Fig. 5) indicate that androgens reduce the size of GnRH cells through negative feedback. Since E2 had no effect, androgen influence on GnRH cell size appears to be independent of aromatization. These data are consistent with the hypothesis that the setpoint for hypothalamic GnRH cell size is determined by social cues and that this setpoint is maintained via negative feedback by gonadal androgens. Territorial males have large GnRH-containing neurons despite high circulating androgens, not because of them.

The castration experiment, above, was performed on territorial males. Enlarged GnRH neurons resulted, and though the mean soma sizes were even slightly bigger than those in control territorial males, their large size is in concert with the social dominance of the animal. To test whether GnRH neuronal cell size and social state can be dissociated, the castration experiment was replicated, this time using NT animals. Following surgery to remove gonadal tissue, the fish were returned to the social settings from whence they came, ensuring that they remained NT. Behavioural observations confirmed that these animals were indeed NT. Two weeks later, the fish were sacrificed and brains were examined for the sizes of the GnRH neurons in the POA. The number of animals that survived the surgical intervention followed by restoration to the community tank was small and thus the results are preliminary. They suggest, however, that the GnRH neurons grew to be insignificantly different from those seen in territorial males (K. Yu & R. D. Fernald, unpublished observations). Thus, through experimental manipulation it appears that GnRH neuronal soma size and social behaviour can be uncoupled.

In *H. burtoni*, the regulation of growth and development may be adaptive in their natural habitat, where territorial space is limited. In the shore pools where these animals live, only a fraction of the males can breed at any time. As noted above, these breeding males appear to be particularly vulnerable to avian predators (Fernald & Hirata 1977b), and hence territorial ownership may be relatively brief. Thus there may be a selective advantage for males to have a retarded growth rate until they have an opportunity to become territorial, whereupon they grow rapidly.

Interestingly, following our original observation, we have analysed in more detail the rate at which social interactions influence the GnRH cell size. We recently discovered that the rate of cell size change is a function of the direction of the social transition (White et al 2001). Animals moving from NT to T status achieve the changes in GnRH-containing cell size (cf. Fig. 3) in just seven days, while those animals moving from T to NT may require four weeks until completion. This result is intuitively satisfying since there is such a distinct selective advantage to being a territorial male. Preliminary analysis of the behaviour of animals that are moving in either direction is quite instructive. Many territorial males that have lost status continue to act territorial, even if only in concealed locations and at times when they are not being scrutinized by the new dominant male.

In all, these data suggest that external social signals are transduced into at least two different pathways in *H. burtoni* males. One of these is hormonal, determining the reproductive state of the animal, and the other behavioural. While in intact animals, the two pathways correspond and the hormonal cues maintain the necessary physiological state associated with social state, it is possible to dissociate the circuitry by experimental intervention, e.g. castration of NT males. Further evidence that the two systems can be dissociated comes from work in *H. burtoni* females in which the social circuit appears to be muted or missing while the endocrine circuitry shows parallel plasticity to that seen in males.

In contrast to males, female *H. burtoni* do not appear to have differences in social status. They spend most of their time at the fringes of the dominant male's territory where they school with NT males. As described above, they move into territorial waters territories only to feed or spawn. This absence of social difference amongst females prompted the question: are GnRH neurons in female *H. burtoni* similarly plastic to males and, if so, what regulates changes in cell size? As noted above, a ripe female lays her eggs and then takes them into her mouth for fertilization and brooding. The brood is carried for around two weeks prior to being released. Changes in female appearance which accompany these reproductive states are due to physiological rather than social events. Thus, differences in body colour, which in males reflects reproductive status, do not occur in females. Instead, a female that is ready to spawn will have an enlarged abdomen, due to the presence of ripe eggs.

Later, after spawning, females with distinctively large mouth cavities filled with fry are not ready to spawn and avoid males.

Since females do not engage in the aggressive social interactions which regulate male GnRH cell size, it is possible that they might not show the same plastic changes. GnRH cell size would then be sexually dimorphic, increasing in females simply as a function of development and becoming stable at maturity. This would contrast with the life-long potential for plasticity seen in males. Alternatively, since GnRH cell size in males is correlated with both social and reproductive status, cell size in females might fluctuate according to the female reproductive cycle.

To study possible changes in cell size in female *H. burtoni*, we analysed cell size as a function of reproductive state in females (White & Fernald 1993). While there is some contribution of body size to the cell size changes, body size differences do not account for all of the observed changes. Soma sizes in spawning females are typically twice as large as those in females carrying broods while post-reproductive fish have the largest neuronal soma sizes. These changes occur within the two weeks it takes to brood a clutch and the differences in GnRH neuronal soma size are comparable to those seen between dominant and subordinate males.

Taken together, these data have provided considerable insight into how social signals regulate reproductive physiology. The other major influence on the social behaviour of *H. burtoni* is changes in its physical environment.

Environmental influences on social status and size

The shorepools of Lake Tanganyika, which are the natural habitat of *H. burtoni*, are relatively unstable. Winds and the presence of large animals such as hippotomi cause considerable change in the local conditions the animals face (Fernald & Hirata 1977b). Only a fraction of the males can breed at any time and these animals appear to be particularly vulnerable to avian predators. As a consequence, reproductive opportunities may arise as frequently as they vanish because territorial ownership may be relatively brief. To untangle the causal relationship between environmental state and social status, we kept animals in stable and fluctuating habitats and assessed the consequences on the reproductive axis and body size.

In *H. burtoni*, habitat complexity influences the fraction of the male population that can sustain territories (Hofmann et al 1999). Moreover, the stability of the habitat affects duration of territorial tenure since, in a fluctuating habitat, where the three-dimensional layout changes frequently, males hold territories for a significantly shorter time period than in a stable habitat. Even a stable habitat results in a significant level of change in social status (Hofmann et al 1999). To our surprise, we found that this intrinsic instability is caused by differential

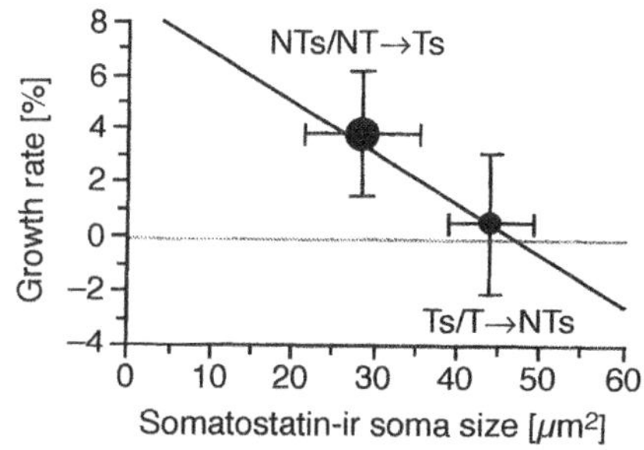

FIG. 6. Relationship between growth rates and the mean somatostatin-IR soma size in *H. burtoni*. NTs and NT→T males (filled circle; mean SDs) have smaller soma cross-sectional areas and grow faster than Ts and T→NTs (filled diamond; mean SDs). (After Hofmann & Fernald 2000.)

growth rates. Specifically, NTs and NT→Ts grow faster than Ts and T→NTs (Fig. 6). It seems likely that after territory establishment, animals allocate energy simultaneously to reproduction and growth to maintain a competitive advantage over other Ts. Indeed, animals that lose a territory slow their growth rate and may even shrink (Hofmann et al 1999).

A possible mechanism regulating differential growth is the control of somatostatin release in the pituitary. Since this neurohormone inhibits the release of growth hormone (GH) it is a likely site of control (Brazeau et al 1973, Gillies 1997). This is supported by our recent data showing that somatostatin-containing neurons in the POA change size (Fig. 6) when social status and, consequently, growth rate change (Hofmann & Fernald 2000). The somata of these neurons are significantly larger in Ts and T→NTs as compared to NTs and NT→Ts. It is unknown whether larger neurons produce more somatostatin to be released into the pituitary, or whether they represent an accumulation of somatostatin as its release is inhibited. Preliminary evidence from measurements of circulating GH (Hofmann et al 1999) suggests that the latter may be the case, thus inhibiting the release of GH from the pituitary in NTs and NT→Ts. This surprising result makes likely the social regulation of insulin-like growth factor 1 (IGF1) which mediates many of the somatic effects of GH and whose release is controlled by GH (Mommsen 1998).

Why do animals that have lost a territory (T→NTs) slow down their growth rate and even shrink? Behavioural stressors may play a role. As shown by Fox et al (1997) in *H. burtoni*, status switches in both directions can be accompanied by elevated levels of the major stress hormone cortisol with the T→NT change showing the most pronounced increase. NT→T fish with increased cortisol levels usually did not maintain territoriality. Fish descending in rank consistently showed high levels of cortisol which could, in turn cause somatic growth to be down-regulated. As has been shown in another cichlid, the tilapia *Oreochromis mossambicus*, chronic administration of cortisol leads to a reduction in body weight and reproductive parameters like gamete size and levels of sex steroids (Foo & Lam 1993). Although the regulatory interactions between GH and cortisol are very complex (Thakore & Dinan 1994, van Weerd & Komen 1998, for critical reviews), *in vivo* experiments have demonstrated an inhibitory effect of glucocorticoids on somatic growth in many vertebrates including fish (for example, Pickering 1990).

Could cortisol also be involved in the growth rate differences between established Ts and NTs? Fox et al (1997) showed that cortisol levels in Ts and NTs do not differ as long as the fish community remains unstable. However, in a situation of relatively high social stability, Ts have significantly lower levels of circulating cortisol than NTs. Under such a stable situation NTs still grow faster than Ts. Therefore, growth may not be effectively inhibited by cortisol in those animals. Rather, we hypothesize that other factors may become significant when animals maintain a particular social behaviour for many weeks (e.g. feeding habits, behavioural activity, energy expenditure).

Conclusions

In *H. burtoni* males, the brain is continually being remodelled by social behaviour throughout life. Such neural renovations make sense since there are limited resources and a clear selective advantage for males that can respond quickly to reproductive opportunities. The external phenotypic plasticity allows males to allocate physiological resources to reproduction or growth, depending on social and environmental circumstances. Our studies on this model system reveal remarkably intricate interrelationships between habitat structure, behaviour, and the brain. It seems likely that such connections exist in other species, particularly those that change sex and await discovery.

References

Arikawa K, Suyama D, Fujii D 1996 Light on butterfly mating. Nature 382:119

Borowsky R 1973 Social control of adult size in males of *Xiphophorus variatus*. Nature 245: 332–335.

Brazeau P, Vale W, Burgus R et al 1973 Hypothalamic polypeptide that inhibits the secretion of immunoreactive pituitary growth hormone. Science 179:77–79

Bushnik TL, Fernald RD 1995 The population of GnRH-containing neurons showing socially mediated size changes project to the pituitary in a teleost, *Haplochromis burtoni*. Brain Behav Evol 46:371–377

Davis MR, Fernald RD 1990 Social control of neuronal soma size. J Neurobiol 21:1180–1188

Fernald RD 1977 Quantitative observations of *Haplochromis burtoni* under semi-natural conditions. Anim Behav 25:643–653

Fernald RD 1981 Chromiatc organization of a cichlid fish retina. Vision Res 21:1749–1753
Fernald RD 1983 Neural basis of visual pattern recognition in fish. In: Ewert JP, Capranica RR, Ingle DJ (eds) Advances in vertebrate neuroethology. Plenum Press, New York, p 569–580
Fernald RD 1984 Vision and behavior in and African cichlid fish. Am Sci 72:58–65
Fernald RD 1989 Retinal rod neurogenesis. In: Finlay BL, Sengelaub DR (eds) Development of the vertebrate retina. Plenum Press, New York, p 31–42
Fernald RD, Hirata NR 1977a Field study of *Haplochromis burtoni*: quantitative behavioural observations. Anim Behav 25:964–975
Fernald RD, Hirata NR 1977b Field study of *Haplochromis burtoni*: habitats and co-habitants. Environ Biol Fishes 2:299–308
Fernald RD, Wright SE 1983 Maintenance of optical quality during crystalline lens growth. Nature 301:618–620
Foo JT, Lam TJ 1993 Retardation of ovarian growth and depression of serum steroid levels in the tilapia *Oreochromis mossambicus* by cortisol implantation. Aquaculture 115:133–143
Fox HE, White SA, Kao MH, Fernald RD 1997 Stress and dominance in a social fish. J Neurosci 17:6463–6469
Fraley NB, Fernald RD 1982 Social control of developmental rate in the African cichlid, *Haplochromis burtoni*. Z Tierpsychol 60:66–82
Francis RC, Soma K, Fernald RD 1993 Social regulation of the brain-pituitary-gonadal axis. Proc Natl Acad Sci USA 90:7794–7798
Gillies G 1997 Somatostatin: the neuroendocrine story. Trends Pharmacol Sci 18:87–95
Hofmann HA, Fernald RD 2000 Social status controls somatostatin-neuron size and growth. J Neurosci 20:4740–4744
Hofmann HA, Benson ME, Fernald RD 1999 Social status regulates growth rate: consequences for life-history strategies. Proc Natl Acad Sci USA 96:14171–14176
Johns PR, Fernald RD 1981 Genesis of rods in teleost fish retina. Nature 293:141–142
Lorenz K 1935 Der Kumpan in der Umwelt des Vogels. J Ornithol 80:2
Mommsen TP 1998 Growth and metabolism. In: Evans DH (ed) The physisology of fishes. CRC Press, Boca Raton, FL, p 65–97
Muske LE, Fernald RD 1987a Control of a teleost social signal. I. Neural basis for differential expression of a color pattern. J Comp Physiol A 160:89–97
Muske LE, Fernald RD 1987b Control of a teleost social signal. II. Anatomical and physiological specializations of chromatophores. J Comp Physiol A 160:99–107
Pickering AD 1990 Stress and the suppression of somatic growth in teleost fish. In: Epple A, Scanes CG, Stetson MH (eds) Progress in clinical and biological research, vol 342: Progress in comparative endocrinology. Wiley-Liss, New York, p 473–479
Schultz ET, Clifton LM, Warner RR 1991 Energetic constraints and size-based tactics: The adaptive significance of breeding-schedule variation in a marine fish, *Micrometrus minimus* (Embiotocidae). Am Nat 138:1408–1430
Soma KK, Francis RC, Wingfield JC, Fernald RD 1996 Androgen regulation of hypothalamic neurons containing gonadotropin-releasing hormone in a cichlid fish: integration with social cues. Horm Behav 30:216–226
Thakore JH, Dinan TG 1994 Growth hormone secretion: the role of glucocorticoids. Life Sci 55:1083–1099
van Weerd JH, Komen J 1998 The effects of chronic stress on growth in fish: a critical appraisal. Comp Biochem Physiol 120:107–112
von Uexküll J 1909 Umwelt und Innenwelt der Tiere. 2. verm. u. verb. Aufl. Springer-Verlag, Berlin
White SA, Fernald RD 1993 Gonadotropin-releasing hormone-containing neurons change size with reproductive state in female *Haplochromis burtoni*. J Neurosci 13:434–441

White SA, Nguyen T, Fernald RD 2002 Social regulation of gene expression during changes in male and female social status. Submitted

DISCUSSION

Short: Is the terminal nerve the eutherian homologue of the nerve to the vomeronasal organ?

Fernald: Yes, it is called the Zero'th cranial nerve. It gets rediscovered about every 30 years! It is only in some cold-blooded vertebrates that it projects to the retina.

Short: Is the GnRH localization around that nerve the same in eutherians?

Fernald: Yes. Indeed, the distribution we have now shown with separate gene expression patterns for each GnRH forms is conserved. In species where it has been identified there are two or three populations, each expressing a separate GnRH gene. The known GnRH receptor responds to all of them. I didn't mention this, but we have also found two receptors in this fish species that are spatially distinct, one in the pituitary and the retina, the other in the midbrain.

Short: I got the impression from what you were saying that the switching on of spermatogenesis when a fish becomes a dominant male is extremely rapid. How long does it take?

Fernald: The shortest we have looked is about 5 days.

Behringer: What happens if you place a double-sized decoy in the tank?

Fernald: That is a great suggestion, and many of my colleagues who have worked on laboratory animals imagine that with the wild animals that we can do something like this. But the fish will not go for a dummy. Videos are possible, but it is difficult to make them look realistic. We have tried to use concave mirrors, so you can have them fight with a larger version of themselves. This works well for a while, and then they catch on.

Behringer: Does fish sperm have a tail?

Fernald: Yes.

Behringer: It's remarkable that they can synthesize it so quickly. In mammals it takes weeks to produce.

Fernald: Bear in mind that these fish may leave some residue of part-made sperm behind. This may also be true for the sex-changing fish.

McLaren: You said that sex-changing fish can change from female to male or from male to female. Do different species have changes in different directions? Presumably no species changes both ways?

Fernald: That's right. There are species in which at birth some males will follow an obligate male pathway, and others will become females only to change later. Within one species there can be alternative phenotypes.

Vilain: The field of brain sexual differentiation in mammals has been heavily influenced by the theory that everything is controlled by hormones. However, in rats there is one well characterized exception to this dogma: the expression of tyrosine hydroxylase in the GnRH neurons from the mesencephalon. This is different between males and females before the apparition of the testis in the males or the ovary in the females. That is, it is independent from the fetal secretion of androgens. How do your think your model applies to other species?

Fernald: I think it is time to put the brain back in its proper perspective, as a reproductive organ in its own right. In this case my view is that social behaviour is influencing the brain, which in turn is regulating the reproductive system. The evidence for this is strong, and I don't think this is going to be a unique situation. Since we have discovered this, the same kind of process has been seen in tree shrews and musk shrews: detection of a social scene leads to a change that is clearly triggered by the brain. This is not to say that there isn't room for both the brain and endocrine system. For example, the androgen system is involved here: if you castrate these males then the GnRH cells become not eight times larger but 16 times larger and produce concomitantly more androgen. If you reimplant androgen-releasing pellets the cells shrink back down again, and the size that they attain depends on the social status. If you castrate a non-territorial male the cells will get larger, and if you then put in appropriate androgen they will come back down to a non-territorial level as long as the male is still non-territorial.

Mittwoch: Is the fluctuation in growth rate related to different amounts of food eaten? Or is it purely internally regulated?

Fernald: It is internally regulated. We controlled for food intake.

Short: Is the growth rate change due to gonadal growth, or is there also somatic growth?

Fernald: It is somatic growth, over and above any changes in gonadal size. Once we found that animal growth rate was socially regulated, we began looking at size changes. The evolutionary argument is that if you shrink a little bit upon becoming non-territorial, you won't be as obvious to the dominant individual. The mechanism of shrinking is a real puzzle.

Wilkins: From what you have described, it seems to me that in terms of social behaviour the last 400 million years of vertebrate evolution have been a waste of time! Can you estimate the amount of brain tissue in terms of cells that are devoted to this in fish versus in mammals?

Fernald: They have ~200 GnRH cells. We are now looking at immediate early genes to track the circuitry leading to these cells. It looks as if just a couple of pathways are responsible for regulating cell size change. I suspect we are going to see modifications of the connections among cells in many brain areas in response to social signals. The plasticity of these animals may give us more surprises. I am not excluding the possibility that they may have different circuits for dominance than they do for non-dominance.

Short: You gave a breakdown of the number of species of fish that show this social regulation of sexual behaviour. Of the hundreds of ciclid species that exist in Lake Tanganyika, how many show this ability?

Fernald: There is lots of evidence of other ciclids having a comparably complex social regulation. For example, in addition to the male anal fin spots, males also use fluorescent genital tassles and a number of variants on the same theme. It is in the decimation of the species in Lake Victoria that behavioural isolating mechanisms have really become well known. Lake Victoria had some 500 species. The rapid decline in species number, which has dropped by about 100 over just 20 years, has been attributed to the introduction of the nile perch. However, some nice work has shown that the decimation of species was due to eutrophication from agricultural run-off. This made the water opaque and these fabulous behavioural interactions were no longer visible. The species barriers turned out to be based on behavioural interactions and so species numbers collapsed rather quickly. All these species arose in about 12 000 years, so perhaps this shows that evolution can eliminate species as fast as they arise.

Wilkins: This is also an excellent case for sympatric speciation through sexual selection.

Short: Is there any phenotypic characteristic that distinguishes a suppressed male from a female?

Fernald: You never see a suppressed male with a mouthful of young, but that's about it. Females even have small, faint, mimic spots on their anal fins. If a dominant male loses his territory these spots will fade slowly. There is another case I should mention of female mimicry in fish. There are other examples where the suppressed males mimic female behaviour but bulk up their gonads and act as 'sneakers'. They pretend they are females, get near a male who is spawning the female and add their sperm. There are a number of variations on this theme of female mimicry.

Short: How many other mouth-breeding fish have oral sex? It seems to be a sensible strategy, conserving sperm rather than spreading them everywhere.

Fernald: Indeed, it's a rather neat evolutionary solution, and it has evolved in many species.

The battle of the sexes: opposing pathways in sex determination

Humphrey Hung-Chang Yao*, Christopher Tilmann*, Guang-Quan Zhao† and Blanche Capel*[1]

**Department of Cell Biology, Duke University Medical Center, Durham, NC 27710 and †Green Center for Reproductive Biology Sciences, Department of Pharmacology, University of Texas Southwestern Medical Center, 5323 Harry Hines Boulevard, Dallas, TX 75390-9051, USA*

Abstract. In mammals, a primordial gonad forms in XY and XX embryos that develops into a testis or an ovary depending on expression of *Sry*. *Sry* induces cell signalling pathways, including proliferation of Sertoli precursors and migration of peritubular myoid and vascular cells from the mesonephros. These events result in increased testis size and testis cord organization. Testis cord formation normally prohibits germ cells from entering meiosis. Ovarian fate is initiated in the absence of *Sry*, and has been proposed to be dependent upon the presence of meiotic germ cells in the gonad. We have shown that a developmental window exists during which testis development can be experimentally induced in XX gonads. This window closes just prior to the time that germ cells enter meiosis. Based on our work and much work that has preceded it, we suggest that the autonomous entry of germ cells into meiosis initiates the ovarian pathway and blocks testis development. *Sry* opposes this pathway by initiating testis cord formation prior to meiosis which sequesters germ cells inside cords and arrests them in mitosis. Current experiments in the lab address the hypothesis that cord formation and germ cell entry into meiosis are competing pathways in gonad development.

2002 The genetics and biology of sex determination. Wiley, Chichester (Novartis Foundation Symposium 244) p 187–202

The gonad arises as a bipotential primordium in mammals, poised in a precarious balance between male and female developmental pathways. The earliest cell types known to be present are the germ cells and the supporting cell lineage, i.e. precursors of Sertoli cells in males and follicle cells in females. The supporting cell lineage is named for its role in supporting the development of germ cells and

[1]This chapter was presented at the symposium by Blanche Capel, to whom correspondence should be addressed.

is believed to be homologous in origin and function between the sexes (McLaren 1991). Testis fate is determined by the expression of the *Sry* gene in supporting cell precursors, which specifies their development as Sertoli cells and initiates the architectural arrangement of gonadal cells to form a testis. Ovarian fate behaves as a default pathway, initiated in the absence of specification of the male pathway. Although the genes controlling ovarian fate have not been clearly identified, it is known that germ cells are required for the organization of ovarian follicles and the proper differentiation of follicle cells (McLaren 1988, 1991).

Sry is expressed between 10.5 and 12.5 days post coitum (dpc), and is required to activate the male pathway and/or repress the female pathway. When *Sry* is deleted from the Y chromosome (Gubbay et al 1992, Lovell-Badge & Robertson 1990) or carries mutations (Hawkins et al 1992, McElreavey et al 1995), an ovary forms. On the other hand, when *Sry* is expressed in the gonad of an XX embryo, a testis forms (Eicher et al 1995, Koopman et al 1991). These experiments proved that *Sry* is the only gene from the Y chromosome required to initiate testis organogenesis among the cells of the gonad primordium.

In mammals, the occurrence of ovotestes is rare. Development of ovaries or testes is strongly canalized. Once the balance is shifted in a given direction, the entire cell population in the gonad is usually recruited to whichever program is initiated. This is true even in XX↔XY mosaic gonads where XX cells are recruited to testis structures and testis development occurs normally if the proportion of XY cells is greater than 25% (Burgoyne & Palmer 1991). An exception to the usual case is the frequent formation of ovotestes in hermaphrodites where the Y^{POS} chromosome from *Mus domesticus poschiavinus* is crossed onto certain *Mus musculus musculus* strains, notably C57BL/6 (B6). The organization of ovotestes in these cases typically consists of testis cords in the centre and ovarian follicles located in the polar regions of the gonad (Bradbury 1987, Eicher et al 1995, Nagamine et al 1998, Albrecht et al 2000). To account for the formation of ovotestes in B6 XY^{POS} mice, it has been proposed that there is a narrow window of time during which *Sry* must act to initiate the male pathway and repress the female pathway (Eicher & Washburn 1986). The formation of ovotestes might result from a late-acting or lower-expressing allele of *Sry*, allowing partial induction of the female pathway (Burgoyne & Palmer 1991).

Recent evidence from organ culture experiments strongly supports the idea that the testis pathway must initiate during a narrow window of development (see below, Tilmann & Capel 1999). In addition, threshold effects relating to the timing and level of gene expression have been reported for several genes in the pathway including *Sry* itself and *Sox9*, the earliest gene known to be upregulated downstream of *Sry*. Evidence suggests that the timing and level of *Sry* expression is critical in XX mice carrying *Sry* as a transgene (Swain et al 1998) and mice carrying Y chromosome deletions that affect the level of *Sry* expression

(Capel et al 1993). In humans, heterozygosity for a mutant allele at the *Sox9* locus results in male to female sex reversal (Foster et al 1994, Wagner et al 1994).

We have defined several male-specific cell signalling pathways induced by *Sry*. Among these are pathways that control cell proliferation, Sertoli cell differentiation, and mesonephric cell migration. An increase in proliferation of supporting cell precursors appears to be involved in the specification of Sertoli cells (Karl & Capel 1998, Schmahl et al 2000). Migration of cells into the XY gonad from the adjacent mesonephros is induced by *Sry*, and is required for testis cord formation, a process that encloses germ cells inside an epithelial layer of Sertoli cells (Buehr et al 1993, Tilmann & Capel 1999).

The role of germ cells at this critical window of development is not clearly understood. Germ cells enter the gonad between 9.5 dpc and 11.0 dpc (Ginsburg et al 1990). Germ cells in XX and XY gonads proliferate similarly until 13.5 dpc (Schmahl et al 2000). At that stage, germ cells in the XY gonad are sequestered inside testis cords by Sertoli cells where they soon arrest division. Germ cells in the XX gonad then enter meiosis and arrest in prophase I (McLaren 1988). Progression of germ cells to meiosis occurs with the same timing when germ cells are located in regions other than gonads such as adrenal glands (Zamboni & Upadhyay 1983) or are assembled in lung aggregates in culture (McLaren & Southee 1997). These data suggest that entry into meiosis is an intrinsic property of germ cells that operates in a clock-like manner. Germ cells are not required for testis cord formation, although minor delays in testis cord formation have been observed in germ-cell-less mutants (H. Yao & C. Tilmann, unpublished data). However, germ cells are required for the organization of the ovary into follicles and for follicle maintenance thereafter. In sterile mutants, or in cases where germ cells are lost, the follicular structure of the ovary either never forms or rapidly degenerates (McLaren 1991).

The idea that meiotic germ cells mediate the ovarian pathway and oppose the testis pathway has been proposed previously. Burgoyne suggested that testis and ovary determination are initiated through different cell lineages. *Sry* expression in the supporting cell lineage is required to initiate the testis pathway whereas the ovarian pathway appears to be under the control of the germ cells (Burgoyne & Palmer 1991). We propose that *Sry*-mediated signalling pathways are timed to initiate cord formation before germ cells enter meiosis. If *Sry*-mediated pathways (including supporting cell proliferation, mesonephric cell migration and subsequent steps leading to cord formation) are delayed, entry of germ cells into meiosis triggers ovarian follicle formation and blocks the testis pathway (Fig. 1). It is the relative timing of these two opposing pathways that controls the fate of the bipotential gonad. In theory, this model is experimentally approachable by effectively altering either the timing of *Sry* expression or the timing of germ cell entry into meiosis.

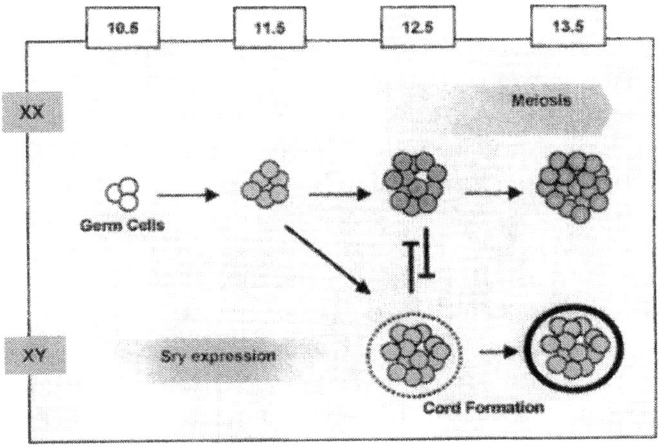

FIG. 1. Diagram illustrating our central model. Germ cells progress toward meiosis autonomously. Expression of *Sry* is timed to initiate cord formation prior to entry of germ cells into meiosis. Cord formation blocks germ cell entry into meiosis, whereas germ cell entry into meiosis blocks cord formation.

Altering *Sry* or its signalling pathways

There exist a number of reported cases where alterations of *Sry* expression lead to sex reversal. These include cases where the level of *Sry* transcript is lower (Nagamine et al 1999), the number of *Sry*-expressing cells is reduced (Burgoyne et al 1988, Schmahl et al 2000), or expression of *Sry* is delayed (Swain et al 1998). Because the usual method to detect *Sry* expression is a PCR or RNase protection assay using RNA isolated from the whole gonad, it has not been possible to distinguish a reduction in the level of *Sry* transcription from a difference in the number of *Sry*-expressing cells. Therefore, in cases were *Sry* expression is reported to be lower, either or both defects may have occurred. If a threshold level of *Sry* expression is required to trigger the testis pathway, delays in *Sry* expression may effectively reduce the level of the transcript during a critical window of development.

Signalling pathways downstream of *Sry* that execute the testis pathway also contribute to this side of the equation. For example, *Sry* expression is required to induce proliferation of pre-Sertoli cells (Karl & Capel 1998, Schmahl et al 2000). An increase in cell number in this population may be critical to initiate the testis pathway. A null mutation in *Fgf9* has been reported to lead to a dramatic reduction in the number of Sertoli and interstitial cells (Colvin et al 2001). Studies using bromodeoxyuridine to label dividing cells at early developmental stages suggest that proliferation of Sertoli precursors in *Fgf9*$^{-/-}$ XY gonads is reduced compared to *Fgf9*$^{+/-}$ or wild-type XY littermates and similar to B6 XY^{POS} gonads that develop as ovaries or ovotestes (J. Schmahl & B. Capel, unpublished results). Although many other pathways downstream are affected, these data suggest that the primary defect in *Fgf9*$^{-/-}$ gonads is a defect in Sertoli progenitor proliferation. A second male-specific pathway controlled by *Sry* is the induction of cell migration from the adjacent mesonephros into the gonad (Capel et al 1999). While this pathway is not important to build the Sertoli population or, as far as we know, in regulating the level of *Sry* expression, it is required for testis cord formation (Buehr et al 1993, Tilmann & Capel 1999). For this reason, this pathway is likely to be critical to block the entry of germ cells into meiosis. In BXD-21 XY^{POS} mice that form ovotestes, mesonephric cell migration is severely impaired, and it is always coincident with the central testicular region of the ovotestis (Albrecht et al 2000). This finding either means that late-migrating cells are excluded from polar regions of the gonad or that a failure of migrating cells to reach those regions results in 'ovarian-like' development.

The testis window

Culturing an 11.5 dpc XX gonad sandwiched between a mesonephros and an 11.5 dpc XY gonad results in the induction of cell migration from the mesonephros into the XX gonad. Examination of these sandwich gonads revealed that XX somatic and germ cells organize into cord-like structures and express *Sox9*, the earliest known Sertoli-specific marker (Tilmann & Capel 1999). In these experiments, mesonephric cell migration, cord formation, and *Sox9* expression can be induced in the XX gonad only when it is at a stage earlier than 12.5 dpc (Table 1), a timing coincident with many previous experiments suggesting that the timing of the initiation of testis development is critical (Eicher & Washburn 1986, Palmer & Burgoyne 1991). This window of development closes at ~12.5 dpc, just prior to the time at which germ cells enter meiosis.

Altering germ cell signals

To investigate the idea that XX germ cells at meiotic stages are antagonistic to the testis pathway, we first determined the earliest time that germ cell entry into meiosis can be recognized by examining the appearance of markers for meiosis in the XX gonad (Fig. 2). At 13.5 dpc, germ cells in the XX gonad were positive for SYN/COR, early synaptonemal complex proteins (Dobson et al 1994). Expression reached its peak at 14.5 dpc. Expression of phosphorylated histone 2AX (γH2AX), which is expressed during meiosis of spermatogenic cells (Mahadevaiah et al 2001), begins slightly later than SYN/COR in the XX gonad.

TABLE 1 Sandwich organ culture experiments indicate a temporal window in which the testis pathway can be induced in XX gonads

Experiments		*Migration*	*Cord formation*	*Sox9 expression*
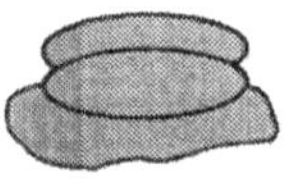	11.5 XY gonad 11.5 XX gonad 11.5 mesonephros	+	+	+
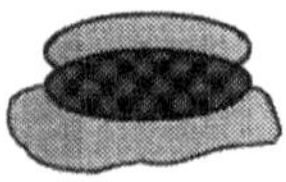	11.5 XY gonad 12.5 XX gonad 11.5 mesonephros	+/−	−	−

An 11.5 dpc or 12.5 dpc XX gonad was sandwiched between an 11.5 dpc XY gonad and a mesonephros. +, +/−, and − indicate the extent of migration, cord formation or *Sox9* expression: + indicates high extent of the occurrence of the events, +/− indicates low extent, − indicates no occurrence.

To test the possibility that germ cells at meiotic stages are responsible for the resistance of XX gonads to mesonephric cell migration and cord formation after 12.5 dpc, we compared mesonephric cell migration in organ culture assays using XX gonads with or without germ cells. Germ cells were depleted using genetic or chemical methods. XX gonads were collected at 13.5 dpc from matings between W^v/+ *and W*/+ mice. *W*/W^v mutant gonads were compared to gonads from +/+, *W*/+, and W^v/+ siblings and found to be >90% free of germ cells. These gonads were assembled with 11.5 dpc XY gonads in sandwich cultures. In cases where germ cells were severely depleted (*W*/W^v), migration occurred normally, whereas, in cases where germ cells were present, cell migration was blocked at 13.5 dpc (Fig. 3). In a second set of experiments, pregnant females were injected on day 9.5 with 10 mg/kg busulfan, a treatment that eliminates >90% of all germ cells (Merchant 1975). Gonads from these treated embryos were collected at 13.5 dpc, assembled in sandwich cultures, and compared to gonads from uninjected embryos. Cell migration occurred into 13.5 dpc XX gonads where busulfan treatment eliminated germ cells, but not into gonads where germ cells were present (Fig.3).

It has been previously reported that certain alleles of the sterile mutants c-*kit* (*W*) and *Steel* (*Sl*), exacerbate male to female sex reversal (Burgoyne & Palmer 1991, Cattanach et al 1988, Nagamine & Carlisle 1996). This finding appears to be in conflict with our hypothesis, but could be explained in several ways. First, it has been suggested that mutations in the c-*kit* pathway lead to general growth defects that may affect sex determination (Burgoyne & Palmer 1991, Cattanach et al 1988, Nagamine & Carlisle 1996). In fact, in severe *W* mutants and other cases where germ cells are lost, we have noted a 6–12 h delay in cord formation (see below in following section). This data could be interpreted to mean that germ cells at pre-

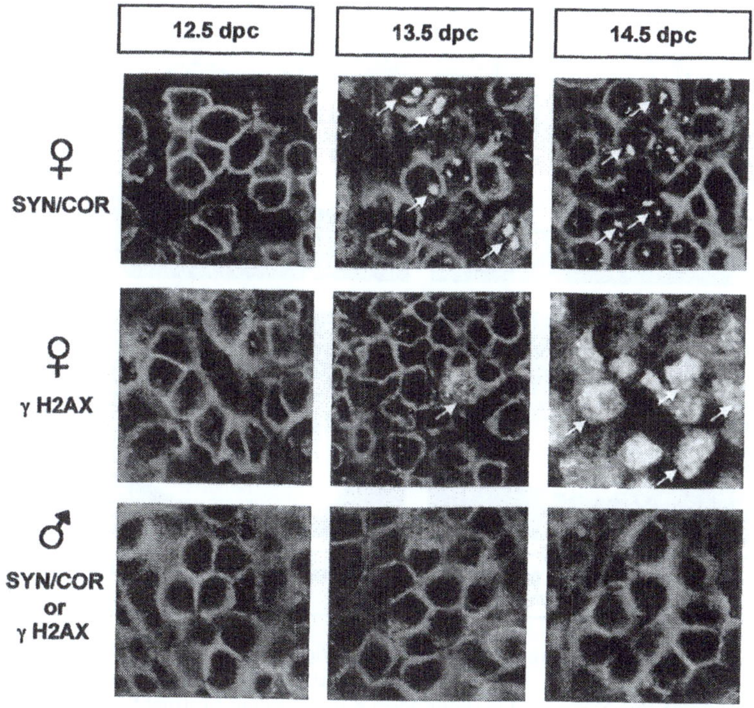

FIG. 2. Time course study of expression of meiotic markers in 12.5–14.5 dpc XX and XY gonads. Gonads were stained with antibodies against SYN/COR or γH2AX (arrows). Germ cells are outlined by an antibody that stains their surface (PECAM). Meiosis is detectable in XX, but not in XY gonads.

meiotic stages play a positive role in seeding testis cord formation. If this were true, loss of germ cells in combination with weak *Sry* signals might critically impair the cord forming process in B6 XY^{POS} gonads and contribute to failure of *W* mutants to rescue Y^{POS} (Burgoyne & Palmer 1991). Alternatively, this effect could be related to the specific allele of *W* or *Sl* and its role in germ cell development. In addition to the control of proliferation and migration of primordial germ cells, *W* and *Sl* are believed to have later roles in the control of meiosis in spermatogenesis and oogenesis. Some alleles of *W* do not affect germ cell proliferation and migration to the gonad, but instead affect adhesive interactions between germ cells and somatic cells or later stages of oogenesis or spermatogenesis (Loveland & Schlatt 1997). If abnormal germ cells arrive in the gonad in some mutant alleles of *W*, their normal signalling relationship with somatic cells may be impaired leading to disruption of testis cord formation.

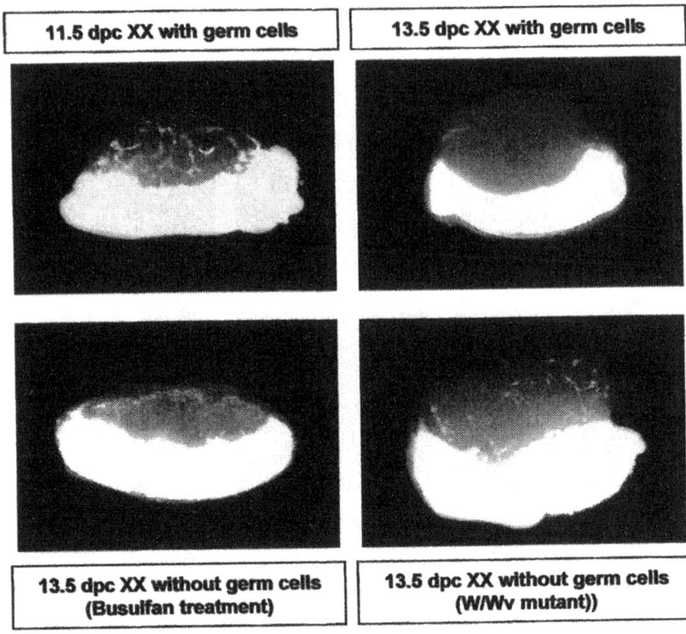

FIG. 3. Sandwich organ culture to induce mesonephric migration into 11.5 or 13.5 dpc XX gonad with or without germ cells. An 11.5 XX gonad with germ cells, a 13.5 dpc XX gonad with germ cells, or a 13.5 dpc XX gonad without germ cells (busulfan treatment or *W/Wv* mutant) was cultured between a 11.5 dpc XY gonad and a 11.5 dpc mesonephros in which GFP is ubiquitously expressed. Mesonephric migration is detected by the presence of green cells in the XX gonads.

A case in point

In *Bmp8b* null mutants, fertility defects have been characterized in the adult male (Zhao et al 1996). Based on the idea that fertility defects may actually reflect a much earlier disturbance in the organization of testis cords or the establishment of close connections between supporting and germ cell lineages, we have begun experiments to investigate the defect in homozygous *Bmp8b*^{tm1blh} gonads. We discovered that cord formation is delayed and/or incomplete at 12.5–13.5 dpc in homozygous *Bmp8b*^{tm1blb} gonads. In blue/white recombinant organ culture assays, we found no defect in the timing of mesonephric cell migration into mutant gonads. However, migrating cells often fail to organize in the gonad, suggesting that *Bmp8b* is involved in cellular interactions downstream of migration that result in proper testis cord formation or Sertoli–germ cell interactions. Immunohistochemical staining with antibodies against laminin and PECAM (a marker of germ cells, Schmahl et al 2000) revealed that cords formed normally in 39% of cases and abnormally in 61% of cases by 13.5 dpc (Table 2).

TABLE 2 Effects of germ cells on testis cord formation in *Bmp8b*^{tm1blh} XY gonads

B6 background[a]		
Stage	*Germ cells present?*	*Cord formation*
12.5	N	0/6[b]
13.5	N	9/9
B6/129 hybrid background		
Stage	*Germ cells present?*	*Cord formation*
12.5	Y	0/2
12.5	N	0/4
13.5	Y	0/8
13.5	N	4/4

[a] Germ cells are never present in gonads on a pure B6 background.
[b] Number of samples with testis cords/total number of samples.

When *Bmp8*^{btm1blb} is on the B6 genetic background, germ cells do not form (Ying et al 2000) and, therefore, never arrive in the gonad. Cord formation is delayed ~12 h, but is normal by 13.5 dpc. However, when *Bmp8b*^{tm1blb} is homozygous on a hybrid B6;129 genetic background, germ cells do arrive in the gonad in 56% of cases. In all homozygous *Bmp8b*^{tm1blb} cases where germ cells are present in the gonad, cord formation remains disrupted at 13.5 dpc. This preliminary data strongly suggests that germ cells in homozygous *Bmp8*^{btm1blb} mice are antagonistic to cord formation at 13.5 dpc. We are currently investigating whether gonadal germ cells in these mutants have prematurely entered meiosis, or are otherwise out of synchrony with the normal developmental pathway.

Summary

The unique divergence of developmental pathways in the gonad provides an ideal model to understand how regulatory genes establish cellular pathways that control the morphogenesis of organs. The discovery of *Sry* provided a clear molecular anchor point for the divergence of gonad development along the male pathway. *Sry* initiates cell signalling pathways including proliferation, cell migration, and vascular development that result in the formation of testis cords. Experiments so far are consistent with the idea that germ cell entry into meiosis is a competing pathway in the bipotential gonad that opposes *Sry* mediated pathways and initiates ovarian fate. The relative timing of these two pathways determines the fate of the gonad. This mechanism would insure that if germ cells enter meiosis, ovarian fate is specified. This makes sense in terms of reproductive fitness: once germ cells enter meiosis, their reproductive future is promoted by ovarian but not testis structure since meiotic germ cells in the embryonic testis would be rapidly depleted, leading to sterility.

On the basis of the hypothesis presented in this paper, mutations that accelerate the timing of germ cell entry into meiosis or interfere with the establishment of mitotic arrest in germ cells in XY gonads would be predicted to lead to preemption by the ovarian pathway and disruption of testis formation. Ideal experiments would involve genetic manipulation of this timing *in vivo*, avoiding possible artefacts associated with *in vitro* culture.

Acknowledgements

We are indebted to Eva Eicher, Anne McLaren, and Paul Burgoyne for their insightful experiments and many helpful discussions. Peter Moens and William Bonner generously donated antibodies against SYN/COR and γH2AX. This work is supported by grants to BC from the NIH, GM56757 and HD39963, and post-doctoral fellowship to HHY from the Lalor Foundation, Inc.

References

Albrecht KH, Capel B, Washburn LL, Eicher EM 2000 Defective mesonephric cell migration is associated with abnormal testis cord development in C57BL/6J XY (Mus domesticus) mice. Dev Biol 225:26–36

Buehr M, Gu S, McLaren A 1993 Mesonephric contribution to testis differentiation in the fetal mouse. Development 117:273–281

Bradbury M 1987 Testes of XX in equilibrium with XY chimeric mice develop from fetal ovotestes. Dev Genet 8:207–218

Burgoyne PS, Buehr M, Koopman P, Rossant J, McLaren A 1988 Cell-autonomous action of the testis-determining gene: Sertoli cells are exclusively XY in XX↔XY chimaeric mouse testes. Development 102:443–450

Burgoyne P, Palmer S 1991 The genetics of XY sex reversal in the mouse and other mammals. Semin Dev Biol 2:277–284

Capel B, Albrecht KH, Washburn LL, Eicher EM 1999 Migration of mesonephric cells into the mammalian gonad depends on *Sry*. Mech Dev 84:127–131

Capel B, Rasberry C, Dyson J et al 1993 Deletion of Y chromosome sequences located outside the testis determining region can cause XY female sex reversal. Nat Genet 5:301–307

Cattanach B, Rasberry C, Beechey C 1988 Sex reversal and retardation of embryonic development. Mouse News Letter 82:94

Colvin JS, Green RP, Schmahl J, Capel B, Ornitz DM 2001 Male-to-female sex reversal in mice lacking fibroblast growth factor 9. Cell 104:875–889

Dobson MJ, Pearlman RE, Karaiskakis A, Spyropoulos B, Moens PB 1994 Synaptonemal complex proteins: occurrence, epitope mapping and chromosome disjunction. J Cell Sci 107:2749–2760

Eicher EM, Shown EP, Washburn LL 1995 Sex reversal in C57BL/6J-YPOS mice corrected by a *Sry* transgene. Philos Trans R Soc Lond B Biol Sci 350:263–268

Eicher EM, Washburn LL 1986 Genetic control of primary sex determination in mice. Annu Rev Genet 20:327–360

Foster JW, Dominguez-Steglich MA, Guioli S et al 1994 Campomelic dysplasia and autosomal sex reversal caused by mutations in an *SRY*-related gene. Nature 372:525–530

Ginsburg M, Snow MH, McLaren A 1990 Primordial germ cells in the mouse embryo during gastrulation. Development 110:521–528

Gubbay J, Vivian N, Economou A, Jackson D, Goodfellow P, Lovell-Badge R 1992 Inverted repeat structure of the *Sry* locus in mice. Proc Natl Acad Sci USA 89:7953–7957

Hawkins JR, Taylor A, Berta P, Levilliers J, Van der Auwera B, Goodfellow PN 1992 Mutational analysis of *SRY*: Nonsense and missense mutations in XY sex reversal. Hum Genet 88:471–474

Karl J, Capel B 1998 Sertoli cells of the mouse testis originate from the coelomic epithelium. Dev Biol 203:323–333

Koopman P, Gubbay J, Vivian N, Goodfellow P, Lovell-Badge R 1991 Male development of chromosomally female mice transgenic for *Sry*. Nature 351:117–121

Loveland K, Schlatt S 1997 Stem cell factor and *c-kit* in the mammalian testis: lessons originating from Mother Nature's gene knockouts. J Endocrinol 153:337–344

Lovell-Badge R, Robertson E 1990 XY female mice resulting from a heritable mutation in the primary testis determining gene, *Tdy*. Development 109:635–646

Mahadevaiah SK, Turner JM, Baudat F et al 2001 Recombinational DNA double-strand breaks in mice precede synapsis. Nat Genet 27:271–276

McElreavey K, Barbaux S, Ion A, Fellous M 1995 The genetic basis of murine and human sex determination: a review. Heredity 75:559–611

McLaren A 1988 Somatic and germ-cell sex in mammals. Philos Trans R Soc Lond B Biol Sci 322:3–9

McLaren A 1991 Development of the mammalian gonad: the fate of the supporting cell lineage. Bioessays 13:151–156

McLaren A, Southee D 1997 Entry of mouse embryonic primordial germ cells into meiosis. Dev Biol 187:107–113

Merchant H 1975 Rat gonadal and ovarian organogenesis with and without germ cells. An ultrastructural study. Dev Biol 44:1–21

Nagamine C, Capehart J, Carlisle C, Chang D 1998 Ovotestes in B6-XXSxr sex-reversed mice. Dev Biol 196:24–32

Nagamine CM, Carlisle C 1996 The dominant white spotting oncogene allele *Kit*W-42J exacerbates XYDOM sex reversal. Development 122:3597–3605

Nagamine CM, Morohashi KI, Carlisle C, Chang DK 1999 Sex reversal caused by *Mus musculus domesticus* Y chromosomes linked to variant expression of the testis-determining gene *Sry*. Dev Biol 216:182–194

Palmer SJ, Burgoyne PS 1991 The *Mus musculus domesticus Tdy* allele acts later than the *Mus musculus musculus Tdy* allele: A basis for XY sex reversal in C57Bl/6-YPOS mice. Development 113:709–714

Schmahl J, Eicher EM, Washburn LL, Capel B 2000 *Sry* induces cell proliferation in the mouse gonad. Development 127:65–73

Swain A, Narvaez V, Burgoyne P, Camerino G, Lovell-Badge R 1998 *Dax1* antagonizes *Sry* action in mammalian sex determination. Nature 391:761–767

Tilmann K, Capel B 1999 Mesonephric cell migration induces testis cord formation and Sertoli cell differentiation in the mammalian gonad. Development 126:2883–2890

Wagner T, Wirth J, Meyer J et al 1994 Autosomal sex reversal and campomelic dysplasia are caused by mutations in and around the SRY-related gene SOX-9. Cell 79:1111–1120

Zamboni L, Upadhyay S 1983 Germ cell differentiation in mouse adrenal glands. J Exp Zool 228:173–193
Zhao GQ, Deng K, Labosky PA, Liaw L, Hogan BL 1996 The gene encoding bone morphogenetic protein 8B is required for the initiation and maintenance of spermatogenesis in the mouse. Genes Dev 10:1657–69
Ying Y, Liu XM, Marble A, Lawson KA, Zhao GQ 2000 Requirement of *Bmp8b* in the generation of primordial germ cells in the mouse. Mol Endocrinol 14:1053–1063

DISCUSSION

Short: I was always fascinated by Byskov's comparison between the rete testis and the rete ovariae, and her claim that Pflüger's cords in which the oogonia and oocytes are lining up are in fact the rete ovariae (Byskov 1986). The seminiferous cords that you are talking about presumably must ultimately anastamose with the rete testis cords. What do you think about these ovarian cords? Are they rete ovariae, and are they where the female germ cells have to be in order to induce follicular cells?

Capel: We haven't worked on ovaries enough to know that. I have never seen those cords in the sort of cultures or times that I dissect. I know they can be seen by electron microscopy, and it may be that this degree of resolution is needed to distinguish them. I haven't seen them with the laminin stain, and none of the antibodies that we have used have picked them out. The remodelling of the seminiferous cords to connect to the rete testis occurs later than the stages that we are talking about, and involves the mesonephric tubules. But I am not clear about how that happens either. It would be a good idea for someone to study this in detail.

McLaren: I was slightly confused by your first set of reaggregation experiments which concerned the inclusion of germ cells inside testis cords. Some time ago Escalante-Alcalde & Merchant Larios (1992) did some reaggregation experiments, comparing germ cells and Sertoli cells from embryos 12.5 dpc with those from 15.5 dpc in criss-cross combinations. In their study the developmental stage of the Sertoli cells determined whether or not you got nice neat cords with the germ cells inside them. It was as if the Sertoli cells were taking the initiative to shepherd the germ cells in, and the Sertoli cells didn't mind whether the germ cells were 12.5 dpc or 15.5 dpc.

Capel: Were they female germ cells at 15.5 dpc?

McLaren: No, they were all male. I don't know whether the same would have been true of female germ cells. Do your results agree or disagree with this?

Capel: If we reassociate 13.5 dpc male germ cells with 11.5 dpc male somatic cells, we get some semblance of cords forming. You might call them palisades.

McLaren: Male somatic cells as early as 11.5 dpc would not form complete cords after dissociation and reaggregation (McLaren & Southee 1997).

Capel: We culture them for two days. The somatic cells begin culture at 11.5 dpc and we look at them 48 h later. If you culture them with 13.5 dpc male germ cells they are fine, but if you try to culture them with 13.5 dpc female germ cells, it blocks cord formation in these assays.

McLaren: Did you try 13.5 dpc male somatic cells?

Capel: No.

Behringer: Do *Bmp8b* homozygous mutant male germ cells enter meiosis?

Capel: We don't know yet. The problem is that Guang-Quan Zhao has just moved to Southwestern and his mice come under a Material Transfer Agreement, which we have to navigate in order to get more. He can't send us any embryos. In our hands at least, the SYN/COR antibody only works on fresh tissue: you can fix the tissue for a couple of hours but then you must use it immediately.

Behringer: Can you tell whether the germ cells enter meiosis histologically?

Capel: We haven't really looked.

Behringer: You said that the oocytes somehow inhibit cord formation.

Capel: Yes, if they are at a stage where they have entered meiosis.

Behringer: How do they do this? Are they secreting something? If so, is there a candidate?

Capel: I don't have a candidate; I haven't really thought seriously about this. I believe there must be an active factor produced by meiotic germ cells that inhibits cell migration and cord formation. To me this is appealing from a reproductive fitness point of view, because once germ cells enter meiosis you don't want a testis to form; you want an ovary. Once germ cells have entered meiosis, if you put them in a testis they would immediately be exhausted, so you need them to promote ovarian development.

Behringer: Grant MacGregor and colleagues generated female 13.5 dpc germ cell cDNA libraries and they have done expressed sequence tag (EST) sequencing. A candidate may lurk in a database somewhere.

Capel: That is a great idea.

Swain: Do they inhibit the making of a cord or can they only destroy the cords they are in?

Capel: The cords never form in these cultures.

Swain: When you grow a testis with meiotic germ cells, are the cords destroyed?

Capel: To make that happen we might want to try to induce meiosis at some point in the male and see what happens. In theory, if this model is correct, by shifting the timing of meiosis we should be able to sex reverse a mouse.

McLaren: Did you see the converse, which is that if the cords have already formed and AMH is being produced, that this AMH kills off any meiotic germ cells that happen to be around, or at least discourages growth?

Capel: We haven't investigated this, but that is a good point.

Short: Do we accept that the disappearance of the oocytes from the free-martin ovary is a consequence of the AMH that has come across from the male co-twin?

McLaren: I would say so.

Josso: I think there is a better model for this. Richard Behringer was kind enough to give us his metallothionein/AMH mice that make a lot of AMH (Behringer et al 1990). Lionel Lyet performed a careful study of the ovaries and saw that the meiosis was retarded (Lyet at al 1995). When cells in the ovary reached meiosis they were killed off immediately.

Renfree: Our results from gonadal cultures support that. When we culture gonads in the presence of AMH the germ cells disappear and the cords form. If the culture isn't a very happy one the germ cells will disappear anyway and the cords will form. So the germ cells seem to be inhibiting cord formation in the ovary. I think the oestrogen results that I have just shown support the opposite theme, that once the female germ cells are in meiosis there is ovary formation which inhibits cords. We have thought for a long time that there is a 'conversation' between the somatic cells and the germ cells that is almost opposite in nature in the two sexes. AMH doesn't agree with germ cells, but if they disappear for some other reason the cords will form in the ovary.

Josso: Some people also say that the cords will form anyway if germ cells disappear, and that AMH is not masculinizing at all. AMH may kill germ cells off, but after that the default pathway of cord formation occurs. I tend to agree with this hypothesis.

Renfree: In our cultures, if the culture didn't go well, even without AMH, the germ cells disappear and the cords form.

McLaren: Nathalie Josso was kind enough some years back to send us some AMH. Culturing mouse female genital ridges in the presence of AMH didn't prevent the germ cells going into meiosis. Of course, the genital ridges are packed with meiotic germ cells. We didn't do any quantitation. Whether or not some of them are killed off I wouldn't know, but it certainly didn't block entry into meiosis.

Josso: What age were the genital ridges?

McLaren: They were 11.5 dpc, and we followed them for three or four days.

Josso: That finding is in contradiction with the work on metallothionein/AMH mice. There is a big difference at the same stage between a normal littermate and one with lots of AMH because of the transgene.

McLaren: Were those germ cells prevented from going into meiosis? I don't think they were. I think they went into meiosis and then they were lost later on.

Josso: The whole process of meiosis was very much delayed and there were fewer cells reaching this state. Eventually they died. AMH didn't forbid cells once and for all from entering meiosis. It was more subtle than that.

McLaren: We were interested to see whether AMH was the substance that was inhibiting entry into meiosis in the normal testis. Apparently, it wasn't.

Behringer: Blanche Capel, in your model of the gonad, how does an ovotestis fit in here?

Capel: I don't know. We have done proliferation studies on *poschiavinus*. We see proliferation dramatically reduced in Y *poschiavinus* gonads. I'm imagining that at the earliest steps, when you need to produce enough pre-Sertoli cells in order to get the pathway rolling, this is not happening, or perhaps they are only forming centrally, so that there aren't enough of them to initiate the migration at a high enough rate to block the entry into meiosis in the peripheral regions of the gonads. One of our problems is that the earliest marker that we have is SYN/COR, which is almost simultaneous with γH2AX. This is after synaptonemal complex formation has occurred. There must be an earlier decision point, which we would like to find, but we don't really have any way to identify when a germ cell has made up its mind that it is entering the meiotic pathway. In an ovotestis we need to explain why the cords are central and the non-cord regions are peripheral. Various people have called these ovarian regions, and other people say that they aren't really ovarian, they are just unorganized. I wonder if we had the right marker whether we could see that the germ cells in those regions were deciding to enter meiosis by 12.5 dpc and that we had not built up enough of a testis signal to permeate the whole gonad. The other point we need to think about is the difference between low expression of *Sry* in each cell, or not enough Sertoli cells formed. I think previous experiments with mosaics speak to this issue clearly. Paul Burgoyne, you showed that 25% of the cells needed to be XY in order to initiate testicular development.

Burgoyne: Yes, in X0/XY mosaics. A similar answer came out of studies of XX↔XY chimeras.

Capel: This is one of the reasons that I think the number of pre-Sertoli cells produced in that early proliferation step may be important and common to many testis pathways, such as alligators and chickens. It is building up enough cells in the population to produce the secondary signals. Anne McLaren, in one of your reviews of this work you mentioned that there must be paracrine signals that are important in this process.

McLaren: Yes.

Behringer: In your 5-FU- or methotrexate-treated testes, is there cord formation?

Capel: Yes, in about half of them. The way I interpret that experiment is that in any litter, there is variation in the stage of any individual embryo. If you score an individual litter for gonads that formed cords or those that didn't, you might have hit one embryo at exactly the right stage to block the critical proliferation, and the next embryo was a little later or a little earlier.

Behringer: Those testes look like they are half normal size.

Capel: They are.

Behringer: Is there still a sufficient number of Sertoli cells to make enough cords?

Capel: The whole gonad is half the size, so perhaps the proportion of Sertoli cells is sufficient. The cords are distributed throughout: there is no central localization of cords.

Mittwoch: Isn't there some old evidence that the difference between meiosis and mitosis is that meiosis has a longer premeiotic prophase? Could it be then that you don't get meiosis in developing testes because cell proliferation is faster and the prophases are shorter, in contrast to the ovary where there is less proliferation? Could there be a connection?

Capel: We didn't see a difference in proliferation in the germ cells between the male and female. We were counting proliferating cells and not looking at timing; however, if it were faster in the male we might see more BrDU labelling in the male germ cells. We didn't see this between 11.5 and 13.5 dpc.

References

Behringer RR, Cate RL, Froelick GJ, Palmiter RD, Brinster RL 1990 Abnormal sexual development in transgenic mice chronically expressing mullerian inhibiting substance. Nature 345:167–170

Byskov AG 1986 Differentiation of mammalian embryonic gonad. Physiol Rev 66:71–117

Escalante-Alcalde D, Merchant-Larios H 1992 Somatic and germ cell interactions during histogenic aggregation of mouse fetal testes. Exp Cell Res 198:150–158

Lyet L, Louis F, Forest MG, Josso N, Behringer RR, Vigier B 1995 Ontogeny of reproductive abnormalities induced by deregulation of anti-mullerian hormone expression in transgenic mice. Biol Reprod 52:444–454

McLaren A, Southee D 1997 Entry of mouse embryonic germ cells into meiosis. Dev Biol 187:107–113

General discussion III

True hermaphroditism and the formation of the ovotestis

Short: I remember reading that amazing book of van Niekerk's (1974) about the epidemic of true hermaphrodites in South Africa, who were all XX. What is our current explanation for this high familial incidence of true hermaphroditism in certain human populations?

Vilain: One of the obvious answers is recessivity. In France there are many immigrants from North Africa, and there have been studies showing a high incidence of hermaphroditism in some villages in Morocco and Algeria. The most likely cause is consanguinity, whether it is first cousins or because of insufficient outbreeding. There is also evidence of several family cases of true hermaphroditism where two siblings are true hermaphrodites. This happens when they do not carry *SRY*. Only about 10–15% of true hermaphrodites carry *SRY*. All the others do not, leaving us with a large number of patients that we can't explain. A few of them can be explained by duplication of chromosome 22q.

Josso: There are also familial cases in which one child has true hermaphroditism and another is an XX male. Both the hermaphrodite and the XX male lack a Y chromosome, but in one case there is complete virilization of the gonad by an unknown factor.

Short: In the true hermaphrodites that you have looked at, which are XX and lack *SRY*, what happens to the germ cells? Do you get oocytes in the ovarian component or any germ cells surviving in the testicular component?

Vilain: You can occasionally get female germ cells in XX true hermaphrodites. Very rarely, there are reports in the literature of fertile XX true hermaphrodites as females. Making sperm is exceedingly rare in these patients and I am not aware of any reports of fertile XX true hermaphrodites as males.

Burgoyne: It is impossible, because there are genes on the Y chromosome that are essential for the spermatogenic process.

Short: The condition of true hermaphroditism with a fertile female component is extremely common in pigs. There is a really good ovotestis that can either be bilateral or unilateral, with a sterile testicular component. If there is still a reasonable amount of uterus left you can get normal litters (Hunter 1995).

Lovell-Badge: It is also very common in moles. Most XX moles have a testicular portion to their gonad.

Burgoyne: We have to be careful about using the presence of an ovotestis in adults as an indicator of true hermaphroditism in human patients, because we are likely to miss many cases. In mice, in situations such as B6 XY^{Pos} where every gonad is affected in fetal life, only a small proportion of the ovotestes retain an ovarian component that can be recognized into adulthood.

Josso: In humans it is usually a child who comes to medical attention with external genitalia that are ambiguous.

Burgoyne: But perhaps you are already too late because you have had the AMH effect, which has wiped out the ovarian component.

McLaren: Before birth the ovotestis would have converted to either an ovary or a testis, depending on which was dominant.

Josso: So you think we diagnose too few true hermaphrodites by looking at children, and if we were to look at the fetus we would detect more.

Burgoyne: Exactly.

Short: In the intersex goat, adults almost always have testes, but if you look in the fetus they are almost always ovotestes. The ovarian component is lost before birth (Short 1972).

Josso: But then your true hermaphrodite would become what?

Burgoyne: Most usually they would have a small testis.

Josso: I have seen many slides of true hermaphrodite gonads, and the ovary is usually quite nice, with good follicles. But the testis, even in childhood, is completely dysgenetic or with very few germ cells.

Vilain: It is rare in clinical practice to see descended gonads in true hermaphroditism. When there is an ovarian component it is usually not descended.

Burgoyne: That is not surprising. To get an ovarian component there have to be very few Sertoli cells producing AMH. If there are very few Sertoli cells, the Sertoli cell factor which causes the descent will be produced in insufficient amounts.

McLaren: Nathalie, what is the chromosome situation in the patients you have studied?

Josso: Most of them are XX. Some of them are mosaic XX/XY. Very few are XY. I have seen a lot of true hermaphrodites because I work with Claire Fékété, a surgeon who is extremely good at dissecting the gonad and removing the unwanted part. Previously, surgeons used to remove the ovotestis completely.

McLaren: If the patients are XX then it is not surprising that the testicular part is totally devoid of sperm.

Vilain: There is one reported case of a mosaic mutation in *SRY* in an XY true hermaphrodite. It was mosaic at the level of the gonad. Since we are talking about true hermaphrodites, one thing that we have noticed is that in the most common form of true hermaphroditism, when there is a testis on one side (most commonly the right) and an ovary on the other, there is most often a regression of the Müllerian structures only on the testis side.

Burgoyne: Exactly the same is seen in mice.

Vilain: In this event, are we still allowed to call AMH a hormone? Of course, we detect it in the blood, but there must be some local action, which goes through the testis–blood barrier but not too far, in order to have a local action on the Müllerian structures.

Josso: The same thing occurs with testosterone. The epididymis receives much more testosterone than organs that are further away, probably through lymphatics. Organs closer to the testis are exposed to a greater concentration (Ohno et al 1971).

Short: I would have thought that it was a rule in all mammals, that when you have an ovary on one side and a testis on the other, you will have bilaterally asymmetrical Müllerian duct derivatives. They will have regressed on the testicular side but not on the ovarian side.

Burgoyne: In the experiment I mentioned earlier where there is an incompletely penetrant transgene, there are intersexes of every imaginable kind. The most mild form looks like an ovary that has just started to descend like a testis. The next stage is one where the top half of the Müllerian duct is lost and then at the bottom you can see both Müllerian duct and some vas deferens. There are all sorts of gradations, but it is always precisely matching the gonad on that side.

Behringer: We have a hypomorphic allele for the AMH ligand gene, and it expresses much lower levels. When we combine it with a null allele to reduce it a little more, the oviduct and the distal uterine horns are lost, but the body of the uterus remains. Again, this is consistent with what you are saying.

Renfree: It is even harder to explain the bilateral gonadomorph marsupials that have a hemi-pouch on one side and a hemi-scrotum on the other.

Short: I still can't hear anyone coming up with a genetic explanation for why we have an ovotestis in the first place.

Mittwoch: Apropos of the right testis and left ovary in human true hermaphrodites, we found many years ago that in human fetuses the right gonad is a little more advanced than the left, both in males and females.

Josso: In human true hermaphrodites an ovary on one side and a testis on the other is relatively rare. Usually there is an ovotestis on one side and an ovary on the other.

Vilain: Roger Short, I think the answer to your question will potentially come from linkage analysis, by grouping familial cases of XX true hermaphrodites. There are some attempts to do this. The problem is that every investigator keeps their families to themselves. There is an investigator at INRA (Institut National de la Recherche Agronomique, France), Corinne Cotinot, who is looking at pig families, trying to see if there is any linkage with true hermaphrodites. I think one of the problems with this is that the pig genome map is not very far advanced.

Behringer: There is one gene, *M33*, that as a recessive mutation in mouse that will cause true hermaphroditism.

Short: Does it always cause true hermaphroditism?

Behringer: It is variable.

McLaren: The genetic explanation would probably also be a developmental explanation. It is, after all, a delicate balance as to when *Sry* comes on relative to the progress of the default female pathway of gonadal development. If, as we know, there is a difference in the developmental stage, one could imagine an ovary on one side and an ovotestis on the other. There may be examples from mice that would illustrate this.

Koopman: Monica Bullejos has done a nice series of *in situ* hybridization experiments, looking at the timing of *Sry* expression in mouse genital ridges. We thought that there might be a difference between the left and the right that might suggest that one side is more advanced than the other. However, there is no difference that we can discern.

Capel: Eva Eicher has a strain that is a recombinant between B6 and DBA2 that always forms ovotestes on both sides. It is a stabilized phenotype.

Short: Are they fertile?

Capel: Yes.

Bullejos: In this *in situ* expression study I saw that *Sry* expression always started in the middle of the gonad and then spread to both ends (Bullejos & Koopman 2001). This could be an explanation for the ovotestes in this strain.

Short: Do you think this might blur the overall timing?

Bullejos: Yes.

References

Bullejos M, Koopman P 2001 Spatially dynamic expression of Sry in mouse genital ridges. Dev Dyn 221:201–205

Hunter RHF 1995 Sex determination, differentiation and intersexuality in placental mammals. Cambridge University Press, Cambridge

Ohno S, Tettenborn U, Dofuku R 1971 Molecular biology of sex differentiation. Hereditas 69:107–124

Short RV 1972 Germ cell sex. In: Beatty RA, Gluecksohn-Waelsch S (eds) The genetics of the spermatozoon. Bogtrykkeriet Forum, Copenhagen, p 325–345

van Niekerk WA 1974 True hermaphroditism: clinical, morphological and cytogenetic aspects. Harper & Row, New York

The evolution of chromosomal sex determination

Brian Charlesworth

Institute of Cell, Animal and Population Biology, University of Edinburgh, Edinburgh EH9 3JT, UK

Abstract. There is a great diversity of sex determination mechanisms, with evidence for numerous evolutionary transitions between different systems. For example, environmental sex determination is widespread in lower vertebrates, and genetic sex determination has probably evolved from it several times. This requires the establishment of genes that override environmental cues. Close linkage between male and female determining loci is favoured by selection, and represents the first step towards the evolution of highly differentiated sex chromosomes. Once crossing over between primitive sex chromosomes has been suppressed, the primitive Y (W) chromosome is vulnerable to the operation of forces that lead to a reduction in its effective population size. This reduces the ability of natural selection to maintain the functionality of genes on the proto-Y, so that it gradually degenerates. Primitive sex chromosome systems, and systems of neo-X and neo-Y chromosomes formed by translocations involving autosomes and sex chromosomes, provide an opportunity to test evolutionary models of the degeneration of Y chromosomes and to determine the time-scales involved. Recent data confirm that newly-evolving Y or neo-Y chromosomes experience a sharp reduction in effective population size, and indicate that degeneration can occur over a few million generations.

2002 The genetics and biology of sex determination. Wiley, Chichester (Novartis Foundation Symposium 244) p 207–224

Sexual reproduction is prevalent throughout eukaryotes, and probably represents their ancestral state. Gamete dimorphism (numerous, small, motile gametes versus few, large and immotile gametes) is the basis of the male–female distinction and is not required for sexuality: many sexual lower eukaryotes produce gametes of equal size (Hoekstra 1987). Even if there is gamete dimorphism, cosexuality (in which an individual produces both male and female gametes), is widely distributed in animals and plants (Jarne & Charlesworth 1993). As Darwin (1859) pointed out, cosexuality may well have been the ancestral state in chordates; this is certainly the case in flowering plants (Bull 1983, Charlesworth & Guttman 1999). In these groups, the distinct developmental programmes required for the production of male and female reproductive structures and gametes must have evolved before the establishment of separate males and females. Sex determination is then simply a decision to restrict an individual's development to one of two potential, pre-existing, pathways.

It is, therefore, not surprising that there is an enormous diversity of sex determining mechanisms (Table 1), as well as many evolutionary transitions between different systems (Bull 1983, 1987). For example, environmental sex determination (ESD), usually involving effects of temperature at critical stages during embryonic development, is widely distributed among cold-blooded vertebrates (Sinclair et al 2002, this volume). Phylogenetic analysis indicates that ESD may have been ancestral in the vertebrate lineage, and that several transitions from ESD to genetic sex determination (GSD) have occurred (Janzen & Paukstis 1991, Kraak & Pen 2001).

We can only speculate about the evolutionary causes of most of these systems (Bull 1983, 1987), except when within-species variation allows experimental analysis of the fitness effects of different sexual phenotypes or genotypes. Theoretical analysis shows that ESD is favoured when there is spatial heterogeneity in the environment, such that the relative fitnesses of males and females vary between different environments. For ESD in turtles, there is evidence for higher survival at a given temperature of the sex which is produced most frequently at that temperature (Janzen 1995). In contrast, GSD is commonly thought to be favoured over ESD if the environment fluctuates over a suitable time-scale, since it is disadvantageous to produce a highly skewed sex ratio over a long run of generations (Bull 1983, 1987). An alternative model has recently been proposed (Kraak & Pen 2001), but experimental evidence is currently lacking.

In other cases, the mode of transmission of the sex determinant itself generates a selective advantage. For example, a maternally transmitted cytoplasmic factor can prevent the production of male offspring, which cannot pass it on to future generations (Rigaud 1997). Similarly, systems such as haplodiploidy and female XY lemmings are associated with intrinsic transmission advantages to the genetic factors involved (Bull 1983, Fredga 1994). In other cases, such as the *M* factor of houseflies (Table 1), the change in sex determination system is in itself selectively neutral, and may have been established by genetic drift or by pleiotropic effects on fitness (Bull 1983, 1987).

The evolution of sex chromosomes

The existence of structurally and genetically highly divergent sex-determining chromosomes presents a challenge to evolutionists. In advanced systems of this kind, there is lack of crossing over between the X and Y chromosomes over all or most of their length (from now on, Z and W will be treated as equivalent to X

TABLE 1 Modes of sex determination

	Environmental sex determination
	Sex is determined by temperature during embryonic development in chelonians, crocodilians, and some lizard and fish species; by nutritional status in mermithid nematodes; by the presence or absence of female individuals in the marine echiurid worm *Bonnellia*.
	Genetic sex determination
Two factor systems	Sex is determined by a pair of Mendelian alternatives, either with male heterogamety (XX females and XY males, as in mammals, *Drosophila*, and most dioecious plants) or female heterogamety (WZ females and ZZ males, as in birds, Lepidoptera, many lower vertebrates, and strawberries). The sex-determining chromosomes may be highly distinct structurally and genetically, as in mammals, *Drosophila* and the white campion *Silene latifolia*. Alternatively, there may be a single genetic factor or small genetic region distinguishing the two, as in many dioecious plants, fishes such as the guppy (*Poecilia*), and many Dipterans such as blackflies. Intermediates with a limited amount of structural differences between the sex chromosomes are also found, as in the newt *Triturus* and the lizard *Cnemidophorus*.
Multiple factor systems	Additional factors interact with the basic sex determination system e.g. a dominant gene (M) that causes both XX and XY individuals to develop as males, so that females are always XX mm and males are either XX Mm or XY Mm. Such a gene is polymorphic in natural populations of the house fly, *Musca domestica*. In some microtine rodents, there are polymorphisms for X chromosome mutations that cause XY individuals to develop as females. Polygenic variation affecting sexual phenotype occurs in fishes such as guppies and medaka.
Haplodiploidy	In a number of arthropods, including mites and several insect taxa, diploid individuals produced by fertilized eggs develop as females, and unfertilized eggs develop as haploid males. In several species of Hymenoptera, this is underlain by a single sex determining locus with many alleles, such that heterozygotes develop as females, and homozygous diploids develop as males.
Paternal genome loss	This is genetically very similar to haplodiploidy; all offspring result from fertilized eggs, but the entire paternal genome is eliminated in males. This occurs in mites, scale insects, and sciarid flies. In some cases, elimination takes place in germ cells only, in others early in development in somatic and germ cells, so that it is then equivalent to haplodiploidy. In most cases, it is not known if sex is determined first, and chromosomes are eliminated from males, or whether elimination leads to haploidy and maleness. In *Sciara*, sex is determined by the mother, some females (Aa) producing only females, others (aa) producing only males, so that all males are aa.
Cytoplasmic sex determination	There are several species of Crustacea in which offspring sex is affected by intracellular symbionts, such as *Wolbachia* bacteria, that override the normal sex determination system of males. These are maternally transmitted, and are polymorphic in natural populations. Gynodioecy in plants (polymorphism for females and cosexuals) often involves cytoplasmic male sterility factors (probably mitochondrial), and their nuclear gene suppressors.

Sources: Bull (1983, 1987), Rigaud (1997), Charlesworth & Guttman (1999) and Kraak & Pen (2001).

and Y), as in mammals and some plant species, or a complete suppression of crossing over in the heterogametic sex, as in *Drosophila* and Lepidoptera (Bull 1983, 1987). This is often accompanied by a dearth of active genes on the Y chromosome, whereas the X usually carries a normal complement of genes for its size. The lack of active Y-linked genes is often accompanied by mechanisms which ensure approximately equal amounts of gene products at X-linked loci in males and females: dosage compensation (Bull 1983, Marín et al 2000). The Y also often contains an unusual abundance of highly repetitive DNA sequences (Bull 1983, Charlesworth et al 1994).

The Y thus presents the bizarre phenomenon of a sizeable chromosome, which often consists almost entirely of 'junk' DNA. In some groups (such as *Drosophila*), it is not required for sex determination, and in others (such as *Caenorhaditis elegans*) it has even been completely lost (Bull 1983). These are examples that represent intermediate stages between apparently single gene inheritance and fully differentiated sex chromosomes (Table 1). Even in some advanced sex chromosome systems, there may still be some genes in common between X and Y (Lahn & Page 1999). This suggests that X and Y chromosomes have diverged from a pair of ancestral, largely homologous, chromosomes. The comparative evidence shows that this must have occurred independently in many lineages (Bull 1983). But what leads to the degeneration of the Y chromosome, and the other features of advanced sex chromosome systems?

This question was first posed by H. J. Muller (1918), who also discovered dosage compensation. As he noted, a lack of crossing over between X and Y is required for them to remain genetically distinct, and must have been the precondition for the evolution of the other features of the Y chromosomes. There is an advantage to suppressing crossing over only when there are two or more polymorphic genes which interact in their effects on fitness (Barton & Charlesworth 1998). Such genes are likely to have been present from the start of the evolution of separate sexes (dioecy). If dioecy evolves from cosexuality, the simplest hypothesis is that females are created by a mutation that suppresses male function, and males by a mutation that suppresses female function (Charlesworth 1996, Charlesworth & Guttman 1999). If dioecy evolves from ESD, the simplest path involves one mutation that causes individuals to develop as females, and another that causes maleness, independently of any environmental cues (Bull 1983, Charlesworth 1996). If such mutations involve separate loci, crossing over among them would produce selectively disadvantageous neuters (Fig. 1). Invasion of the ancestral population by two successive mutations creating males and females thus requires initial close linkage of the two loci, and reduced crossing over is favoured once they are both polymorphic for the sex-determining alleles (Charlesworth & Guttman 1999). Similar principles apply to more complex multi-gene models (Charlesworth & Guttman 1999, Kraak & Pen 2001).

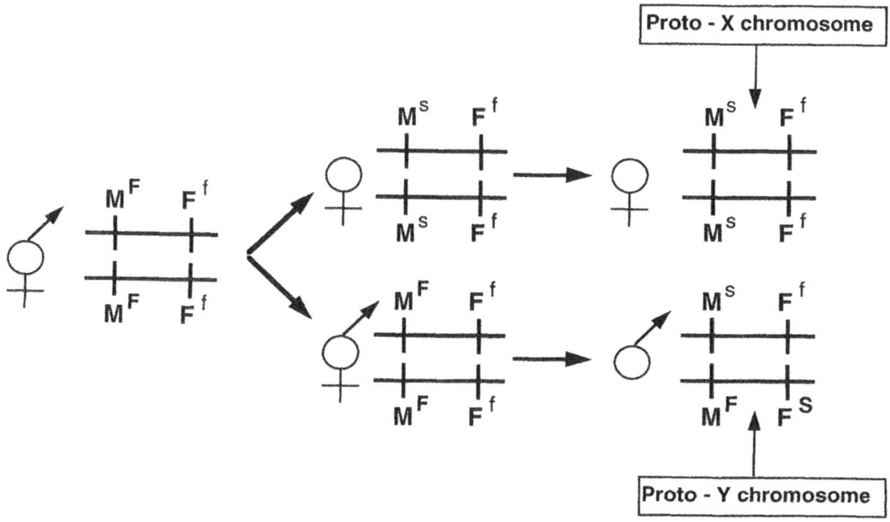

FIG. 1. The evolution of proto-X and proto-Y chromosomes from an initial cosexual state. M and F indicate loci controlling male and female fertility, respectively. Superscripts f and s indicate alleles conferring fertility and sterility, respectively. Dominant alleles are indicated by uppercase superscripts, recessive alleles by lowercase.

In addition, if there are loci with sex-dependent fitness effects, such that one allele is advantageous in males and disadvantageous in females, close linkage to the sex-determining genes is favoured by selection (Fisher 1931, Bull 1983, Rice 1996). The colour polymorphisms of the guppy, *Poecilia*, are a classic example of this: alleles conferring bright coloration are favoured by sexual selection on males, whereas alleles conferring dull colours are favoured in females, due to predation on brightly coloured individuals (Fisher 1931, Rice 1996). These genes are closely linked to the sex-determining region of the primitive sex chromosomes of this species, with alleles causing bright colours being closely associated with the male determinant. Cases where male fertility genes have apparently been transposed to the Y chromosome, such as *DAZ* in humans (Saxena et al 1996), are another possible example of this type of selection. If this process of accumulation of such 'sexually antagonistic' allelic effects is continued, it is easy to see how restricted recombination along the length of the proto-X and proto-Y chromosomes, or suppression of crossing over throughout the whole genome in the heterogametic sex, could evolve (Bull 1983, Rice 1996). Sequence comparisons of Y-linked genes in humans with their X-linked homologues do indeed provide evidence for a succession of steps towards suppressed crossing over, possibly as a result of a sequence of chromosomal inversions (Lahn & Page 1999). Of course, there is nothing inevitable about the establishment of complete crossover suppression, consistent with the numerous intermediate stages between genetic and full chromosomal sex determination (Bull 1983, 1987).

Once crossing over has been suppressed, the proto-Y chromosome has the very unusual property of constituting a large, non-recombining haploid genome, which is permanently heterozygous. A deleterious mutation can therefore become fixed on the proto-Y chromosome without becoming homozygous. This process is facilitated by the fact that the number of Y chromosomes in the population is one-third of the number of X chromosomes, so that genes on the proto-Y chromosome are more vulnerable to genetic drift than their homologues on the proto-X. Sexual selection may further reduce the effective number of breeding males in systems with male heterogamety, thus enhancing this effect (Charlesworth & Charlesworth 2000).

The absence of genetic recombination also impairs the ability of natural selection to promote the fixation of adaptively favourable mutations and resist the fixation of deleterious ones (Barton & Charlesworth 1998). A variety of specific processes can lead to the faster accumulation of deleterious mutations, or slower accumulation of favourable mutations, on the proto-Y compared with the proto-X; these have recently been discussed (Charlesworth & Charlesworth 2000) and will not be reviewed in detail here. The majority depend on the 'Hill–Robertson' effect, which involves the fact that the increase in frequency of a favourable allele due to selection at one locus may cause an increase in frequency of a deleterious allele at a closely linked locus, so that the efficacy of selection is impaired in a non-recombining genome (Fig. 2 and Table 2). Collectively, these processes can be regarded as causing a reduction in the effective population size (N_e) of an evolving Y chromosome, thereby reducing the strength of selection relative to genetic drift. Given enough time, the functionality of genes on the proto-Y chromosome is expected to decline relative to that of genes on the proto-X chromosome (Charlesworth & Charlesworth 2000). The time-scale is likely to be long; a substantial decline in the fitness of the proto-Y may take more than a million generations, depending on the magnitude of the rate of mutation to deleterious alleles, the distribution of effects on fitness of such mutations and the population size of the species (Charlesworth & Charlesworth 2000). Y-linked genes that enhance male fitness, and whose functions cannot be supplied by their X-linked homologues, are likely to resist this process of erosion, since their loss would have drastic fitness consequences (Lahn & Page 1999).

The decline in fitness of the proto-Y relative to the proto-X in the heterogametic sex promotes the evolution of dosage compensation (Charlesworth 1996). This reflects the fact that most deleterious mutations have slight effects on fitness when heterozygous; in *Drosophila*, even so-called recessive lethals have been shown to reduce the viability of their heterozygous carriers by 1–2% (Crow 1993). Any accumulation of deleterious alleles on the proto-Y chromosome will

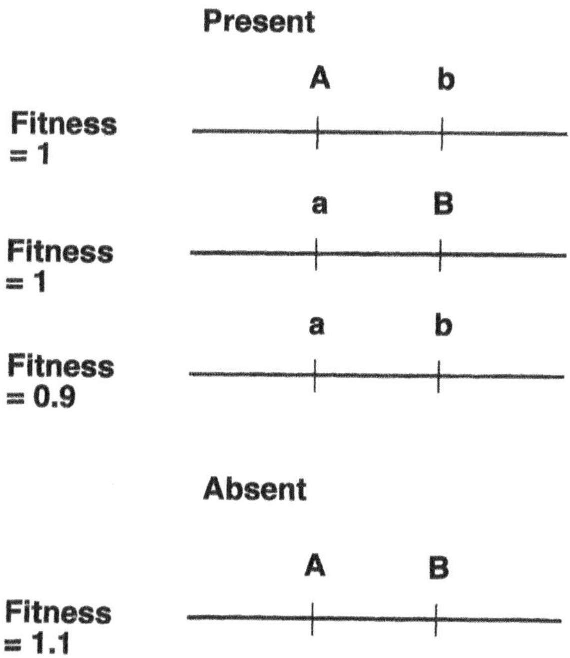

FIG. 2. The Hill–Robertson effect. A and B represent alleles at two different loci which are favoured by selection over their alternatives, a and b. If the initial state is ab, and A and B arise as independent mutations, the fittest combination AB cannot be produced in the absence of recombination.

therefore reduce the fitness of their carriers. There is thus an obvious advantage to enhancing the activity of genes transcribed from the proto-X, even at the expense of reducing the activity of their homologues on the proto-Y. Erosion of gene activity on the Y may therefore partly be an active process of down-regulation. If the up-regulation of X-linked genes were confined to the heterogametic sex, a dosage compensation system such as that of *Drosophila* would evolve, in which the end-product is a doubling of the rate of transcription of X-linked genes in males compared with females (Marín et al 2000). If, however, up-regulation is not sex-limited, X-linked activity would be doubled in both sexes, and no apparent dosage compensation would be observed, as is seemingly the case in Lepidoptera (Johnson & Turner 1979). This would in turn generate selection for restoring the level of activity in the homogametic sex to its previous, presumably optimal, level. This accounts for the systems of dosage compensation in mammals and *C. elegans*, which involve inactivation of the X and down-regulation of X-linked genes, respectively (Charlesworth 1996, Jegalian & Page 1998, Marín et al 2000).

TABLE 2 Evolutionary processes that can lead to reduced levels of adaptation and variation in a non-recombining genomic region

Hitchhiking by deleterious mutations (background selection)	A neutral or weakly selected mutation that arises in a large non-recombining population has a non-zero chance of survival only if it arises on a chromosome free of strongly deleterious mutations. The effective population size of a non-recombining chromosome can therefore be greatly reduced in a large population at equilibrium under selection and mutation. This accelerates the fixation of weakly deleterious mutations and retards the fixation of advantageous mutations.
Muller's ratchet	This involves the stochastic loss from a finite population of the class of chromosomes carrying the fewest deleterious mutations. In the absence of recombination and back mutation, this class of chromosome cannot be restored. The next best class then replaces it and is in turn lost, in a process of successive irreversible steps. Each such loss is quickly followed by fixation of a deleterious mutation on the chromosome.
Hitchhiking by favourable mutations	The spread of a favourable mutation in a non-recombining genome can drag to fixation any deleterious mutant alleles initially associated with it, so that successive adaptive substitutions on an evolving Y chromosome could lead to the fixation of deleterious mutations at many loci, contributing to its degeneration
Mutual interference among weakly selected sites	With a very large number of closely linked sites, subject to reversible mutation between favoured and disfavoured alleles and selection with a strength of the order of the reciprocal of effective population size, the mean level of adaptation is strongly reduced in non-recombining regions.

Source: Charlesworth & Charlesworth (2000).

Population genetic forces that can lead to the accumulation of repetitive DNA sequences in non-recombining genomic regions, including Y chromosomes, have been discussed elsewhere (Charlesworth et al 1994), and will not be considered further (see Table 3 for summary).

Testing the ideas

There are obvious difficulties in studying evolutionary forces that operate over very long time-scales, especially if more than one of these forces operate. In addition, advanced Y chromosomes have lost most of their active genes, so that the opportunity for detecting the signatures of Hill–Robertson effects is considerably reduced, since there is now a greatly reduced set of loci subject to selection (Charlesworth & Charlesworth 2000). Species where there are still many active genes on a predominantly non-recombining Y chromosome, as is likely to be true of some plant species (Charlesworth & Guttman 1999), are more

TABLE 3 Evolutionary processes which can lead to the accumulation of repetitive DNA in non-recombining genomic regions

Tandemly repeated non-coding sequences	Unequal crossing over among members of a tandem array can generate haplotypes with only one repeat; fixation of these by genetic drift results in loss of repeats. In consequence, long arrays can accumulate only in genomic regions with little crossing over
Transposable elements	Ectopic recombination can occur between pairs of homologous elements in different locations, generating deleterious chromosome rearrangements. This may contribute to the containment of the spread of elements; if this is less effective when meiotic recombination is infrequent, elements will accumulate in regions of reduced crossing over.
	Muller's ratchet and/or background selection can also cause the accumulation of elements in non-recombining regions, if they are associated with deleterious insertional mutations.
	Elements are also less likely to be eliminated by selection against insertional mutations in regions with low gene density, such as the Y chromosome or heterochromatin.

Source: Charlesworth et al (1994).

favourable examples for this purpose. Similarly, systems where a neo-Y/neo-X chromosome pair has been formed by fusion between an autosome and a sex chromosome (Fig. 3) offer excellent opportunities to study the processes involved in Y chromosome degeneration, especially in *Drosophila* where the absence of crossing over in males automatically ensures that a neo-Y chromosome is genetically isolated from its partner once it becomes fixed in a species (Charlesworth 1996, Charlesworth & Charlesworth 2000).

One prediction of the evolutionary models is that a newly-formed proto-Y or neo-Y chromosome which fails to cross over with its homologue X over most or all of its length will start to exhibit signs of reduced adaptation, which will become progressively more marked, the greater the age of the system. This prediction is met in the case of the neo-Y chromosomes that have evolved independently in different species of *Drosophila* (Table 4). The case of *D. miranda* shows that a high level of degeneration of Y-linked loci has been accomplished over a period of a few million generations, despite the fact that DNA sequence variation indicates an effective population size of approximately one million individuals (Yi & Charlesworth 2000a, Bachtrog & Charlesworth 2000). In birds, the rate of substitution of amino acid sequence variants at a locus with Z and W homologues is faster for the W than the Z copy, as expected if slightly deleterious variants are accumulating on the Z chromosome (Fridolfsson & Ellegren 2000). A lower rate of amino acid sequence evolution on the Y or W chromosome relative to X or Z would suggest that the rate of adaptive evolution has been slowed down

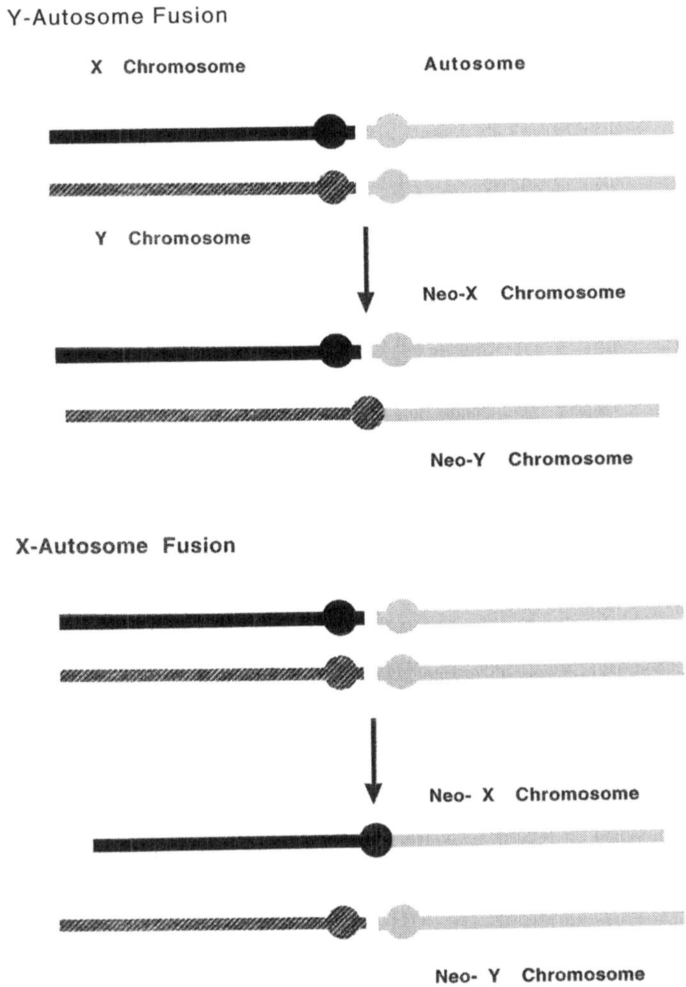

FIG. 3. Evolution of neo-X and neo-Y chromosomes by fusions between autosomes and sex chromosomes. Only males are indicated; females are homozygous for X and neo-X chromosomes.

because of its lower N_e (Orr & Kim 1998); the gene *cyclin B* in *D. miranda* shows this pattern for the neo-X and neo-Y chromosomes (D. Bachtrog & B. Charlesworth, unpublished results).

Another prediction is that the extent to which genes on the proto-X or neo-X chromosome are dosage compensated should parallel the extent of degeneration of their partner chromosomes, since dosage compensation is postulated to be an evolved response to Y chromosome degeneration. This is broadly confirmed by

TABLE 4 Properties of some neo-X/neo-Y sex chromosome systems in *Drosophila*

Species	*Age of system (Millions of years)*	*Extent of Y degeneration*	*Extent of dosage compensation*
D. pseudoobscura (X-autosome fusion)	13	Complete	Complete
D. miranda (Y-autosome fusion)	1	Partial	Partial
D. albomicans (X and Y autosome fusions)	$\ll 1$	None	Probably absent
D. americana americana	$\ll 1$	None	Absent

Sources: Bachtrog & Charlesworth (2000), Bone & Kuroda (1996), Charlesworth & Charlesworth (2000), Marín et al (2000), Mahesh et al (2001).

the *Drosophila* neo-Y chromosomes (Table 4). In particular, there is evidence that dosage compensation in *D. miranda* is patchily distributed along the neo-Y chromosome (Bone & Kuroda 1996, Marín et al 2000), as expected from the fact that only some of the neo-Y linked genes have degenerated. In at least one case, the evolutionary response to degeneration of a neo-Y-linked gene has involved duplication of the neo-X linked copy onto another chromosome (Yi & Charlesworth 2000b). In mammals, Jegalian & Page (1998) studied the inactivation status in females of a number of X-linked genes in different species of mammals, and found that it was closely associated with lack of a homologous copy on the Y chromosome, consistent with the idea that dosage compensation is an evolutionary response to a loss-of-function of Y-linked genes.

Since the standing level of neutral genetic variation is determined by the product of N_e and the mutation rate, comparisons of levels of silent polymorphism between X- and Y-linked homologues provide a test for the prediction of a reduced effective population size of the Y chromosome. In the case of *D. miranda*, there is clear evidence for such an effect from data on microsatellite loci and DNA sequence variation (Yi & Charlesworth 2000a, Bachtrog & Charlesworth 2000). Similarly, a locus on the Y chromosome of the white campion, *Silene latifolia*, shows about 20 times less variation that its homologue on the X chromosome (Filatov et al 2000). In contrast, the human Y chromosome, which has very few expressed genes, shows only a modest reduction in sequence variation (Shen et al 2000, Sachidanandam et al 2001). The observations on *D. miranda* and *S. latifolia* are consistent with the broader pattern of reduced DNA sequence variation in genomic regions with low levels of genetic recombination (Charlesworth & Charlesworth 1998).

While it is relatively easy to establish whether or not there is significantly reduced variation on evolving Y or neo-Y chromosomes, it is harder to distinguish between different possible causes of such a reduction (Table 2). The recent spread of an adaptively favourable mutation is expected to eliminate all selectively neutral or nearly-neutral variation on a non-recombining chromosome; variants arising after such an event will on average be present at much lower frequencies than in an equilibrium situation, where genetic drift has had time to raise some of them to intermediate frequencies. There should, therefore, be significant departures from the frequency distribution expected for a population at statistical equilibrium between genetic drift and mutation if variability has been reduced by a recent hitchhiking event (Charlesworth & Charlesworth 2000). There are too few variants on the *D. miranda* neo-Y chromosome to allow such a test, but the data on *S. latifolia* show no indication of such an event (Filatov et al 2000). Although the other processes listed in Table 2 can produce a departure from neutrality, their expected effects are generally smaller than that of hitchhiking events. A consistent failure to detect departures from neutral frequency distributions would therefore seem to rule out hitchhiking as a cause of Y chromosome degeneration. Other features of DNA sequence variation that might help to discriminate between the various processes are discussed by Charlesworth & Charlesworth (2000). There is plenty of scope for further work in this area.

References

Bachtrog D, Charlesworth B 2000 Reduced levels of microsatellite variability on the neo-Y chromosome of *Drosophila miranda*. Curr Biol 10:1025–1031

Barton NH, Charlesworth B 1998 Why sex and recombination? Science 281:1986–1990

Bone JR, Kuroda M 1996 Dosage compensation regulatory proteins and the evolution of sex chromosomes in *Drosophila*. Genetics 144:705–713

Bull JJ 1983 Evolution of sex determining mechanisms. Benjamin Cummings, Menlo Park, CA

Bull JJ 1987 Sex determining mechanisms: an evolutionary perspective. In: Stearns SC (ed) The evolution of sex and its consequences. Birkhäuser Verlag, Basel p 93–115

Charlesworth B 1996 The evolution of chromosomal sex determination and dosage compensation. Curr Biol 6:149–162

Charlesworth D, Charlesworth B 1998 Sequence variation: looking for effects of genetic linkage. Curr Biol 8:R658–661

Charlesworth B, Charlesworth D 2000 The degeneration of Y chromosomes. Philos Trans R Soc Lond B Biol Sci 355:1563–2572

Charlesworth D, Guttman DS 1999 The evolution of dioecy and plant sex chromosome systems. In: Ainsworth CC (ed) Sex determination in plants. BIOS Scientific Publishers, London p 25–49

Charlesworth B, Sniegowski P, Stephan W 1994 The evolutionary dynamics of repetitive DNA in eukaryotes. Nature 371:215–220

Crow JF 1993 Mutation, mean fitness, and genetic load. Oxf Surv Evol Biol 9:3–42

Darwin CR 1859 On the origin of species. John Murray, London

Filatov DA, Monéger F, Negrutiu I, Charlesworth D 2000 Low variability in a Y-linked plant gene and its implications for Y-chromosome evolution. Nature 404:388–390

Fisher RA 1931 The evolution of dominance. Biol Rev 6:345–368

Fredga K 1994 Bizarre mammalian sex-determining mechanisms. In: Short RV, Balaban E (eds) The differences between the sexes. Cambridge University Press, Cambridge p 397–418

Fridolfsson A, Ellegren H 2000 Molecular evolution of the avian CHD1 genes on the Z and W sex chromosomes. Genetics 155:1903–1912

Hoekstra RF 1987 The evolution of sexes. In: Stearns SC (ed) The evolution of sex and its consequences. Birkhäuser, Basel p 59–91

Janzen FJ 1995 Experimental evidence for the evolutionary significance of environmental sex determination. Evolution 49:864–873

Janzen FJ, Paukstis GL 1991 Environmental sex determination in reptiles: ecology, evolution and experimental design. Q Rev Biol 66:149–179

Jarne P, Charlesworth D 1993 The evolution of the selfing rate in functionally hermaphrodite plants and animals. Ann Rev Ecol Syst 24:441–466

Jegalian K, Page DC 1998 A proposed path by which genes common to mammalian X and Y chromosomes evolve to become inactivated. Nature 394:776–780

Johnson MS, Turner JRG 1979 Absence of dosage compensation for a sex-linked enzyme in butterflies (*Heliconius*). Heredity 43:71–77

Kraak SBM, Pen I 2001 Sex determining mechanisms in vertebrates. In: Hardy IC (ed) Sex ratios. Cambridge University Press, Cambridge

Lahn BT, Page DC 1999 Four evolutionary strata on the human X chromosome. Science 286:964–967

Mahesh G, Ramachandra NB, Ranganath HA 2001 Autoradiographic study of transcription and dosage compensation in the sex and neo-sex chromosomes of *Drosophila nasuta nasuta* and *Drosophila nasuta albomicans*. Genome 44:71–78

Marín I, Siegal ML, Baker BS 2000 The evolution of dosage-compensation mechanisms. Bioessays 22:1106–1114

Muller HJ 1918 Genetic variability, twin hybrids and constant hybrids in a case of balanced lethal factors. Genetics 3:422–499

Orr HA, Kim Y 1998 An adaptive hypothesis for the evolution of the Y chromosome. Genetics 150:1693–1698

Rice WR 1996 Evolution of the Y sex chromosome in animals. Biosciences 46:331–343

Rigaud T 1997 Inherited microorganisms and sex determination of arthropod hosts. In: O'Neill SL, Hoffmann AA, Werren JH (eds) Influential passengers: inherited microorganisms and arthropod reproduction. Oxford University Press, Oxford p 81–101

Sachidanandam R, Weissman D, Schmidt SC 2001 A map of human genome sequence variation containing 1.42 million single nucleotide polymorphisms. Nature 409:928–933

Saxena R, Brown LG, Hawkins T et al 1996 The DAZ gene cluster on the human Y chromosome arose from an autosomal gene that was transposed, repeatedly amplified and pruned. Nat Genet 14:292–299

Shen P, Wang F, Underhill PA et al 2000 Population genetic implications from sequence variation in four Y chromosome genes. Proc Natl Acad Sci USA 97:7354–7359

Sinclair A, Smith C, Western P, McClive P 2002 A comparative analysis of vertebrate sex determination. In: The genetics and biology of sex determination. Wiley, Chichester (Novartis Found Symp 244) p 102–114

Yi S, Charlesworth B 2000a Contrasting patterns of molecular evolution of the genes on the new and old sex chromosomes of *Drosophila miranda*. Mol Biol Evol 17:703–717

Yi S, Charlesworth B 2000b A selective sweep associated with a recent gene transposition in *Drosophila miranda*. Genetics 156:1753–1763

DISCUSSION

Graves: I am intrigued by the degeneration of the Y chromosome with the evolution of dosage compensation. Is it clear from the *Drosophila* data which comes first? In mammals we have always assumed that if you degenerate the Y then the X has to run to catch up. But if X inactivation spreads from inactive regions to non-inactive regions it could actually be quite the opposite — that you are inactivating parts of the X chromosome and the Y has to catch up.

Charlesworth: It is difficult to see how that could be a selective advantage to the evolution of dosage compensation without the decline and fall of the Y chromosome. In the standard model, as genes fall apart on the Y chromosome, it pays to up-regulate their counterparts on the X chromosome. This, of course, is the way it works in *Drosophila*. It is quite interesting that the same molecules that are involved in dosage compensation in the *D. melanogaster* X chromosome are involved in the other species, and also in cases of dosage compensation of the neo-X chromosome. There has been some nice work using antibodies to label the male-specific lethal genes to demonstrate this. The neo-sex chromosome systems are evolving dosage compensation by co-opting the same mechanisms used for the regular X chromosome. This makes sense, but exactly how this is happening is an interesting question. The mammalian and *C. elegans* systems are nuts when you start to think about it from this perspective, because what is taking place is the down-regulation of genes in females. The only sensible explanation for that, which I proposed a long time ago, is that dosage compensation initially evolved in a similar way to the situation in *Drosophila*, but that it was not male specific. Thus, the X chromosome was up-regulated in both sexes. Then, of course, it pays for you to restore X chromosome activity in females back to a regular level. This is a secondary evolutionary response to what went on in response to the decline of the Y chromosome. This is my hypothesis, but it is not easy to test. There was one example published a few years ago about a gene that has been transposed in one species of mouse from the X chromosome to an autosome, and seems to be up-regulated on the X chromosome (Adler et al 1997). This is consistent with the prediction of my model.

Graves: Are there any other examples in *Drosophila* of mismatches: genes that have degenerated on the Y but are not yet dosage compensated on the X?

Charlesworth: We don't know enough about this. The studies that have been done have been very broad, looking at dosage compensation either in terms of radioactive labelling of mRNA in polytene chromosomes, or in terms of the binding of the male-specific lethal gene products. So we can't really say that there is a one-to-one relationship.

Wilkins: I have a somewhat fuzzy and perhaps naïve question. Might the patterns of degeneration on the Y be expected to be different in systems where the key sex determinant is on the Y, as in mammals, versus where it is on the X, as in *Drosophila*? From a population genetic standpoint would you expect differences?

Charlesworth: I suspect that the *Drosophila* or *C. elegans* system, where there is this X/autosome balance control of sex determination, almost certainly has to be a secondary evolution. I find it difficult to imagine that this was a starting point for sex determination, in modelling the evolution of a sex determination system from an initial hermaphrodite or ESD ancestor. You can make some tinker toy models showing how you get from a male-determining system to an X/autosome system, but I'm not sure I believe them.

Graves: Would you expect it to be the same with a Z and W system? If you compare birds and mammals, would you expect the evolutionary forces on the W to be analogous to those on the Y?

Charlesworth: There's no reason why there should be much difference. It has been argued that there is no dosage compensation in birds or butterflies, but there has recently been some evidence suggesting bird dosage compensation. I still think there is no evidence for this in butterflies.

Mittwoch: I have never really understood why all this interference with crossing over is needed, and why sex chromosomes are needed at all. If sex is determined by a single gene all that would seem to be necessary is to have a recessive allele in homozygous form and a dominant allele in heterozygous form. This would give a permanent backcross and a 1:1 ratio. Is the evolution of sex chromosomes due to the fact that sex is determined by more than one gene?

Charlesworth: Yes, that is almost certainly the explanation. Even if it was just a single switch gene, there would also be the secondary phenomenon of the accumulation of other genes that are favourable in one sex and unfavourable in the other. I mentioned very briefly the guppy example, which was first pointed out by Fisher in 1931. In the guppy there are a number of loci that confer bright colours which are advantageous to the male for sexual selection purposes but clearly disadvantageous to females because they attract predators. Lots of experimental work has validated this, and the bright coloured genes are closely linked with the sex-determining locus. There are bright alleles in males and dull alleles in females, which is what you want. Even if you have a single factor system, it is likely that by adding on these refinements — genes that are beneficial in one sex and disadvantageous in another — you can end up with a suppression of crossing over which spreads. There is nothing inevitable about all of this, and there are plenty of examples where the sex-determining region is either very small or apparently a single gene. But my suspicion is that even in the single gene cases, if you look closely you might find a male and a female locus.

Short: Could you speculate about the relative advantages of male versus female heterogamety? This intrigues me because it seems that there is increasing conformation of Haldane's idea that the testis, not the ovary, is the prime hotspot for germline mutation (Short 1997). If the testis is the site where most of the germline mutations are occurring, is it an advantage to have a homogametic testis with a pair of like sex chromosomes as in birds, or is it an advantage to have a heterogametic testis as in mammals, where the Y-linked genes may be subjected to high rates of mutation?

Charlesworth: The higher mutation rate in the male germline has been established reasonably clearly for mammals and birds. It is not clear that it operates in lower vertebrates, and it certainly doesn't seem to apply to *Drosophila*. My guess is that the evolution of female versus male heterogamety in vertebrates is largely an accident: if you are evolving from ESD, it doesn't really matter whether your first mutation is a dominant mutation which causes you to develop as males or as females independently of the environment. In order to get to a state where you have sex chromosomes, you either have a first dominant mutation and then a second recessive mutation, or the other way round. I don't think it makes much difference. The fact that there is roughly 1:1 evolution of male versus female heterogamety in vertebrates illustrates that it is just a random happenstance of what mutation occurs first. I can't see that the mutation rate could have any influence on this.

Short: The only thing I can think of is that if you are a eutherian mammal and you have your sex-determining genes on the Y chromosome, they never escape from a testis and so are blasted by a high rate of mutagenesis. Perhaps the Y chromosome can't survive very long because it is in too much of a hotspot. In contrast, if you have the avian system, where the W is confined to the ovary, it might survive much longer.

Charlesworth: If you imagine the accumulation of deleterious mutations, it will go faster if the mutation rate is higher. I wouldn't think the difference in mutation rate is necessarily so big that it is going to have an enormous effect. It is still controversial whether in rodents there is a very large male versus female germline mutation rate difference. It has been argued that the major effect is that there is a lower mutation rate for X-linked genes compared with autosomal genes, and that it is not primarily a sex-specific difference. This may have something to do with generation time. If there is a long generation time this means more time to accumulate mutations in the male germline.

Graves: Is there any reason that you can't go directly from male heterogamety to a female heterogamety system?

Charlesworth: You can. In Bull's book there is a model of a single switch mutation that would allow you to do this (Bull 1983).

Graves: For instance, if we are thinking about mole voles and where they are going, it is assumed that they have evolved some other male heterogamety system. I can't see any reason to suppose that this is the case. It could equally well be female heterogamety.

Charlesworth: You can devise a scheme involving a modifier gene that will be essentially selectively neutral. I think there can be problems if you have YY homozygote individuals dying off unless you have reproductive compensation. This presumably exists in these mole voles, because they are producing 0/0 embryos.

Wilkins: To go back to Jenny's question: in the work that Rolf Nöthiger, Andrew Pomiankowski and I are doing on the evolution of the *Drosophila* sex determination pathway, the scheme that we have developed involves a sequence of switches from male heterogamety to female heterogemety. These are single gene changes.

Short: I was intrigued that you referred to the wood lemming and its 9:3 female:male sex ratio. I suppose one assumes that this is an adaptive advantage for an animal living in such a highly unpredictable palaearctic environment, where there can be massive die-offs and there is a need for more females to replenish the population. Is this the explanation?

Charlesworth: No, the explanation has to do with a selfish genetic advantage. The wood lemming case involves directed non-disjunction, so the modified X chromosome actually replicates itself more successfully. Bengtsson (1977) showed that if you make a simple genetic model incorporating this, you predict the observed sex ratio from the population dynamics. It is not in any sense adaptive, except from the point of view of the X chromosome itself.

Short: Is this still thought to be because of an X-linked Y suppressor?

Graves: As far as I know. I don't think anyone has found out what that suppressor is. It is not *Zfy* or *Dax1* — maybe it's *Sox3*?

Short: How about these other palaearctic lemmings, such as *Dicrostonyx*, which appear to be approaching the wood lemming in terms of a skewing of the primary sex ratio: is there not some adaptive advantage if you are living in an extreme environment in breaking Fisher's law and not having a 1:1 sex ratio?

Charlesworth: You are getting into the heresy of group selection, suggesting that selection acts at the level of the species and not the individual. That's against my religion! It is rather the odd that the cricetid rodents seem to go in for bizarre sex mechanisms. I have no idea why this should be. There is one system where there are XX females and XX males.

Graves: I suspect that they are not independent. There are six different systems in some of the old world akodont rodents, but I really wonder whether they are all that different. I think there is probably some precondition that is making it much less stable — perhaps there has been an inversion so that *Sry* is now at the mercy of a position effect.

Short: Brian, if you could wave a magic wand, would you have everyone in this room still read Fisher's *Genetical theory of natural selection* (1930)?

Charlesworth: They might want to do some skipping! It is amazing the number of insights that Fisher had. However, it has been recently pointed out that his idea on the evolution of the sex ratio was actually published by a German author in the 19th century. It is not at all clear that Fisher intended his description of the argument for the 1:1 sex ratio to be taken as original. Fisher was notorious for failing to cite other people's work, so it may be that we are wrongly assigning this to Fisher.

References

Adler DA, Rugarli EI, Lingenfelter PA et al 1997 Evidence of evolutionary up-regulation of the single active X chromosome in mammals based on Clc4 expression levels in *Mus spretus* and *Mus musculus*. Proc Natl Acad Sci USA 94:9244–9248

Bengtsson B 1977 Evolution of the sex ratio in the wood lemming. In: Christiansen FB, Fenchel TM (eds) Measuring selection in natural populations. Springer-Verlag, Berlin, p 333–343

Bull JJ 1983 Evolution of sex determining mechanisms. Benjamin Cumming, Menlo Park, CA

Fisher RA 1930 The genetical theory of natural selection. Clarendon Press, Oxford

Short RV 1997 The testis: the witness of the mating system, the site of mutation and the engine of desire. Acta Paediatr Suppl 422:3–7

The molecular genetic jigsaw puzzle of vertebrate sex determination and its missing pieces

Gerd Scherer

Institute of Human Genetics and Anthropology, University of Freiburg, Breisacherstrasse 33, D-79106 Freiburg, Germany

Abstract. Since the identification of *SRY* as the mammalian Y-chromosomal testis-determining gene a decade ago, more than a dozen additional genes essential for early gonadal development in mammals and other vertebrate classes have been identified. The location of these known pieces of the puzzle in the sex determination pathway, and how they interact, is briefly outlined. Two insights emerge: except for *SRY*, the same basic set of genes appears to operate during early gonadal development in all vertebrate classes, despite the difference in mechanisms; and vertebrate sex determination results from a complex network of regulatory interactions and not from a simple hierarchical cascade of gene actions. However, important pieces of the puzzle are still missing, such as the molecular nature of the sex switch in marsupials, monotremes and non-mammalian vertebrates; the target of SRY; the upstream regulators of *SOX9*; and the genes in the ovarian pathway. The enigma of *SRY*-positive XY gonadal dysgenesis females and *SRY*-negative XX males also indicates that the picture is still far from complete. Filling in these missing pieces is the challenge for the future.

2002 The genetics and biology of sex determination. Wiley, Chichester (Novartis Foundation Symposium 244) p 225–239

Pieces we have

The era of the molecular genetics of mammalian sex determination started off about a decade ago with the long-awaited identification of the Y-chromosomal testis-determining gene, *SRY*[1] (Sinclair et al 1990, Koopman et al 1991). With this initial switch in hand, expectations were high that progress would be rapid to unravel the gene cascade leading from the bipotential gonad to testicular organogenesis. Contrary to these early hopes, progress was rather slow. Nevertheless, positional cloning strategies, analysis of human sex reversal

[1]For consistency and simplicity, human gene nomenclature is used throughout, with gene symbols in uppercase, except where mouse genes are explicitly addressed.

TABLE 1 Genes in mammalian sex determination and early gonadal differentiation known at the year indicated

1990	*1995*	*2001*
SRY	*SRY*	*SRY*
	WT1	*WT1*
	SF1	*SF1*
	DAX1	*DAX1*
	SOX9	*SOX9*
	LHX1(LIM1)	*LHX1(LIM1)*
		EMX2
		DMRT1
		M33
		GATA4
		LHX9
		VNN1
		FGF9
		WNT4

Genes are listed chronologically, in the order of their first implication in sex determination during the time intervals 1991–1995 and 1996–2001. For references, see Koopman (2001) and text.

syndromes, and characterization of mouse knockout mutants resulted over the years in a growing list of genes and molecules implicated in vertebrate sex determination and early gonadal development (Table 1). By 1995, the list had grown to six genes, all encoding transcription factors: *WT1*, *SF1* and *LHX1(LIM1)* being essential for the formation of the bipotential gonad from the urogenital ridge, *SOX9* joining *SRY* as a testis-determining factor, and *DAX1* as an 'anti-testis' gene antagonizing SRY action. By early 2001, a further eight genes had been added to this list, which now contains over a dozen entries. These newcomers include still more genes for transcription factors, a gene (*VNN1* [*Vanin1*]) encoding a cell surface molecule, and *FGF9* and *WNT4*, the first genes encoding signalling molecules.

With our increasing knowledge of the players involved, the diagrams placing them at their respective positions in the sex determination pathway have become ever more complex. Initially only *SRY* could be drawn in splendid isolation at the root of the testicular pathway, and by the mid 1990s the picture was still comparatively simple with the then known half a dozen players (e.g. see Ramkissoon & Goodfellow 1996). But now this picture is significantly more elaborate. One attempt to put the pieces of the jigsaw puzzle together in a diagram showing the cellular and molecular interactions during gonadal induction and early differentiation is shown in Fig. 1 (for reviews, see Swain & Lovell-Badge 1999, Capel 2000, Koopman 2001). In addition to *SF1*, *WT1* and *LHX1*(*LIM1*) mentioned above, two more genes have been identified, in homozygous null mutant mice, as essential for the formation of the bipotential gonad: *EMX2* and *LHX9* (Birk et al 2000). The pathway from the supporting cell precursors to functional Sertoli cells is dependent on the proper action of the five transcription factor genes *SRY*, *SOX9*, *WT1*, *DMRT1* and *M33*. Mutation in or deletion of any of these genes results in abortive ovary development in XY individuals and in non-functional streak gonads. In addition, recent work implicates the signalling molecule FGF9 in Sertoli cell differentiation (Colvin et al 2001). The action of SRY as the most upstream regulator of the Sertolian pathway is antagonized, in an as yet unknown manner, by DAX1, since a double or higher dose of DAX1 causes XY sex reversal in humans and mice. *DAX1* expression itself is up-regulated by SF1 and, as most recently shown, by the WNT4 signalling molecule (Jordan et al 2001, Suzuki et al 2002, this volume). Interestingly, the *WNT4* locus is included in the partial 1p duplication in a human XY sex reversal case, in striking parallel to the dosage-sensitive XY sex reversal seen with *DAX1* (Jordan et al 2001).

The first sign of Sertoli cell function is the secretion of anti-Müllerian hormone (AMH, also known as Müllerian-inhibiting substance, MIS) that causes the regression of the Müllerian ducts. Five transcription factors are known to regulate *AMH* expression, four in a positive manner (SOX9, SF1, WT1 and GATA4) and one in a negative manner (DAX1). Whereas SOX9, SF1 and GATA4 act by binding to target sites in the *AMH* promoter, WT1 acts as a stimulating cofactor by protein–protein interaction with SF1. The negative action of DAX1 also occurs by protein–protein interaction with SF1, interfering with SF1–WT1 heterodimerization, and by recruitment of the co-repressor NcoR (Goodfellow & Camerino 2001). The transcription factor binding sites in the *AMH* promoter are not only defined by *in vitro* binding studies, but also by elegant *in vivo* studies in mice (Arango et al 1999). *AMH* is thus the best understood target gene in early gonadal differentiation in terms of its regulation; a success story due in part to the comfortably small size of the *AMH* promoter of only a few hundred base pairs, making the definition of functional transcription factor binding sites relatively easy.

Figure 1 reveals that several genes act at multiple steps in the pathway, serving different functions. As an example, SF1 function is needed at four steps: for formation of the bipotential gonad, for up-regulation of *DAX1*, for *AMH* expression in Sertoli cells, and for production of testosterone in Leydig cells, the second essential embryonic testicular hormone, needed for differentiation of the Wolffian ducts. In addition to their roles in Sertoli cell differentiation,

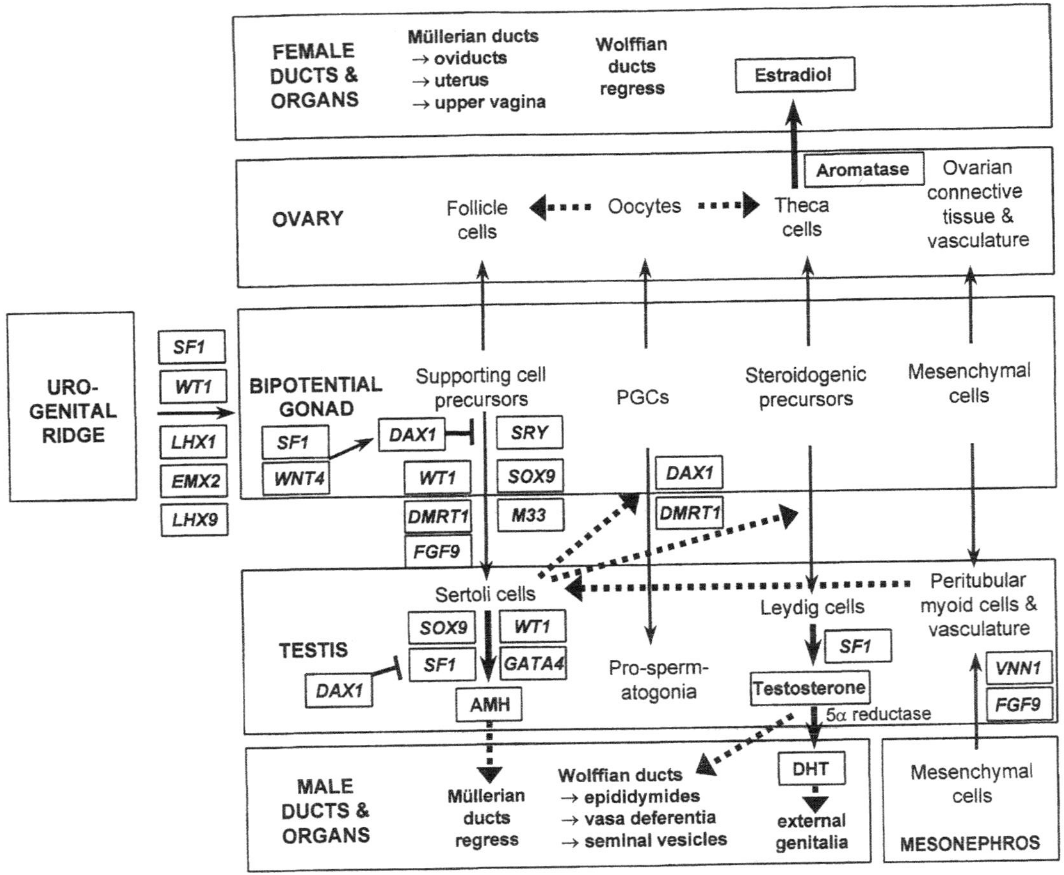
FEMALE DUCTS & ORGANS
Müllerian ducts
→ oviducts
→ uterus
→ upper vagina
Wolffian ducts regress
Estradiol
OVARY
Follicle cells
Oocytes
Theca cells
Aromatase
Ovarian connective tissue & vasculature
URO-GENITAL RIDGE
SF1
WT1
LHX1
EMX2
LHX9
BIPOTENTIAL GONAD
SF1
WNT4
DAX1
Supporting cell precursors
WT1
DMRT1
FGF9
SRY
SOX9
M33
PGCs
DAX1
DMRT1
Steroidogenic precursors
Mesenchymal cells
TESTIS
DAX1
Sertoli cells
SOX9
SF1
WT1
GATA4
AMH
Pro-sperm-atogonia
Leydig cells
SF1
Testosterone
5α reductase
Peritubular myoid cells & vasculature
VNN1
FGF9
MALE DUCTS & ORGANS
Müllerian ducts regress
Wolffian ducts
→ epididymides
→ vasa deferentia
→ seminal vesicles
DHT
external genitalia
Mesenchymal cells
MESONEPHROS

homozygous null mutant mice have revealed a role for both DAX1 and DMRT1 in spermatogenesis (Yu et al 1998, Raymond et al 2000). And, as the likely cause for its involvement in Sertoli cell development, FGF9 can directly or indirectly induce migration of mesonephric cells into XY gonads, which contribute to the interstitial cell population, including peritubular myoid cells that stimulate Sertoli cell differentiation (Colvin et al 2001). A similar role in mesonephric cell migration is attributed to VNN1 (Grimmond et al 2000, Koopman et al 2002, this volume).

The diagram in Fig. 1 summarizes gene action at early steps in gonadal development as it applies to placental mammals. The identification of these mammalian sex determination/differentiation genes has fostered comparative studies in the non-mammalian classes of vertebrates. These studies have revealed conservation of gene and protein structure and, to a large extent, of expression profiles for most of the genes such as *WT1*, *SF1*, *SOX9*, *DAX1* and *DMRT1*, indicating that the same basic set of genes operates during early gonadal development throughout the vertebrate phylum. One exception stands out: *SRY*, which is found only in placental mammals and marsupials. Another notable difference concerns the order of *SOX9* and *AMH* transcription. Whereas *SOX9* is expressed before *AMH* in mouse and human, this order is reversed in chicken and alligator, questioning the role of SOX9 as a Sertoli cell-inducing factor in birds and reptiles (Sinclair et al 2002, this volume).

Missing pieces

The molecular nature of the sex switch in marsupials, monotremes and non-mammalian vertebrates

SRY is firmly established as the switch in sex determination in placental mammals. What do we know about the molecular nature of this switch in non-placental mammals and in the other vertebrate classes? The answer is simple: almost nothing (Table 2). Although *SRY* is present in marsupials, and is even located on the Y chromosome, its sex-determining function is uncertain. A better candidate testis-determining gene for this group of mammals than the ubiquitously expressed *SRY* gene is *ATRY*, which is only expressed in developing and adult testis (Pask et al 2000). If *ATRY* could also be shown to be

FIG. 1. (*Opposite*). Gene action and cellular interactions during mammalian gonadal induction and early differentiation. Pathways of cellular differentiation and/or migration are indicated by thin black arrows, biosynthetic pathways by thick black arrows, and hormonal or unknown signalling pathways by dashed arrows. Effector genes or gene products are shown in boxes. DHT, dihydrotestosterone. Modified from Koopman (2001).

TABLE 2 The molecular nature of the sex switch in the different vertebrate classes

		Class or subclass	Molecular switch
	GSD	Placental mammals	*SRY*
	GSD	Marsupials	*SRY*? *ATRY*?
	GSD	Monotremes	?
	GSD	Birds	*DMRT1* on Z chromosome? Ovary-determining gene on W chromosome?
ESD	GSD	Reptiles	TSD: HSP–oestrogen receptor interaction?
ESD	GSD	Amphibians	?
ESD	GSD	Fish	?

GSD, genetic sex determination; ESD, environmental sex determination; TSD, temperature-dependent sex determination; HSP, heat shock protein.

present on the Y in monotremes, which lack *SRY*, it would represent a testis-determining candidate in this lineage as well.

Like mammals, birds also have a genetic sex-determining mechanism, but here it is the female that is the heterogametic sex (ZW), whereas the male is the homogametic sex (ZZ). It is still undecided whether avian sex determination is due to a dominantly acting ovary-determining gene on the W chromosome, or due to a dosage mechanism, where the interaction of an autosomal factor with a single dose (ZW) or a double dose (ZZ) of a Z-linked gene product would decide the fate of the developing gonad. No candidate gene for the dominant model has so far been identified on the W chromosome. However, *DMRT1* has emerged as an attractive 'dosage candidate', as member of a group of genes showing conservation of synteny between human chromosome 9 and the avian Z chromosome (Nanda et al 2000). What makes *DMRT1* attractive as a candidate avian sex-determining gene, besides its Z location, is the fact that it is expressed much more strongly in the male than in the female chick gonad (Sinclair 2002, this volume), and that it appears to act in a dosage-dependent manner in humans, where monosomy for the 9p region including *DMRT1* causes sex reversal. It still needs to be shown, however, that avian *DMRT1* is not subject to dosage compensation. Although it was widely accepted that dosage compensation does not occur in birds, six out of nine Z-linked genes analysed recently did show dosage compensation (McQueen et al 2001).

Reptiles, amphibians and fish show genotypic sex determination (GSD) of both the XX/XY and ZZ/ZW type as well as environmental sex determination (ESD) such as temperature-dependent sex determination (TSD). In fish, even social factors can influence gonadal differentiation (Fernald 2002, this volume). In these classes of vertebrates, as in birds, but not in mammals, gonadal development is also under the influence of sex steroids. The nature of the molecular switch in sex determination is still elusive in all of these vertebrates, be it a species with GSD or with ESD. An interesting hypothesis to explain how TSD in reptiles could work has been formulated by Pieau who speculates that temperature could be implicated in the dissociation of heat shock proteins from the oestrogen–oestrogen receptor complex, which is then activated to induce aromatase gene expression and ovary development (Pieau 1996).

The direct target(s) of SRY

SRY is a member of the HMG domain family of transcription factors that are DNA-binding and bending proteins. Its most likely mode of action is therefore that of a transcriptional regulator, binding to a recognition sequence in the promoter of a downstream target gene. Although much has been learned about the *in vitro* target sequence specificity of SRY, the physicochemistry of DNA bending by normal and mutant SRY proteins, and the three-dimensional structure of its HMG domain complexed with DNA, definitive evidence as to which gene or genes are its direct *in vivo* target(s) is still missing. *In vitro* and transfection studies that focussed on the DNA-binding specificity of SRY had implicated a number of target genes, including *AMH*, but these early reports did not stand the test of time. From our present perspective, *SOX9*, *FGF9* and *VNN1* are attractive candidates as SRY targets, as they are already expressed in the developing XY gonad at the Sertoli cell commitment stage when *SRY* is expressed. Clearly, the definitive proof for an SRY target is eagerly awaited and would fill in a major piece in the puzzle. It will then be interesting to see whether SRY acts on that target gene as an activator or as a repressor, as implicated in the double repressor model of sex determination (McElreavey et al 1993).

Upstream regulation of SOX9

The complete XY sex reversal caused by *SOX9* mutations in human, the expression profile during early gonadal development in mammalian and non-mammalian vertebrates, and the strong evolutionary conservation, assign *SOX9* a central role in vertebrate testis development. Elucidation of the upstream regulation of *SOX9* is thus a central issue. Unfortunately, the identification of regulatory elements in the *SOX9* promoter is somewhat more demanding than the definition of such elements in the *AMH* promoter, for example, because the *SOX9* 5′ control region extends over more than 1 Mb (Pfeifer et al 1999).

Present evidence indicates that gonadal *SOX9* expression is under the control of at least three regulatory elements. One is an element for low-level expression in the genital ridge of both sexes, active at E10.5 during mouse development (Morais da Silva et al 1996). Neither the location of this genital ridge enhancer, nor the factor which binds to it, are known. A regulatory element that mediates shut-off of *Sox9* expression at about E11.5 in wild-type XX fetal gonads was revealed by Bishop et al (2000) who observed up-regulated *Sox9* expression in fetal gonads of XX mice carrying a 150 kb deletion caused by a transgene insertion ~1 Mb upstream of *Sox9*, leading to dominant XX sex reversal. Under the double repressor model, an unknown repressor (DAX1?) would bind to this element; a binding which is antagonized by SRY. As a third regulatory element, a late enhancer must exist that assures up-regulation of *Sox9* expression after E11.5, when *Sry* has been switched off. This up-regulation might be brought about by SOX9 itself, acting on its own promoter in an autoregulatory loop (Swain & Lovell-Badge 1999).

My laboratory has used a comparative genomics approach to identify candidate regulatory elements for *SOX9* by way of sequence conservation during evolution, comparing the finished 2 Mb 5′ intergenic sequence of human *SOX9* with the 68 kb 5′ intergenic sequence of the puffer fish *Fugu rubripes*. This has led to the identification of five conserved sequence blocks of about 100 bp each, with 67–80% sequence identity, arranged in the same order and orientation in both species. In human, these sequence elements, labelled E1–E5, are at 28, 87, 251, 261 and 290 kb 5′ to *SOX9* (Bagheri-Fam et al 2001). A mouse line carrying a transgene construct with the distant elements E3–E5 placed in front of a 200 bp mouse *Sox9* proximal promoter fragment driving a *lacZ* reporter gene showed *lacZ* expression in E13.5 testis in a testicular cord-like fashion, besides expression in some chondrogenic areas, while a similar construct with the proximal elements E1+E2 showed no testis expression (S. Bagheri-Fam, M. Mallo & G. Scherer, unpublished work 2001). If confirmed by independent transgenic lines to rule out position effects, this result could indicate that the late gonadal *SOX9* enhancer is represented by one (or more) of the three distal elements E3–E5. Interestingly, both E3 and E5 contain one SOX consensus binding sequence.

Human sex reversal syndromes

Through positional cloning, human sex-reversal syndromes have led the way to the identification of several important genes in the sex-determination pathway such as *SRY*, *SOX9*, *DAX1* and *DMRT1*. Table 3 lists a number of syndromes with associated defects in gonadal and/or genital development where the causative gene has not yet been identified. Although complete XY sex reversal with gonadal dysgenesis is documented in only a few of these syndromes, it should be remembered that mutations in genes in the gonadal pathway such as *WT1* and

TABLE 3 Human syndromes with associated partial or complete sex reversal

Syndrome	*MIM No. or Reference*	*Location*	*Gonadal/Genital phenotype*
ATR-16	141750	16p13.3	hypospadias, CO
Fraser	219000	?	hypospadias, CO
Genitopalatocardiac	231060	?	GD, hypospadias
Male pseudohermaphroditism, Verloes type	600122	?	Severe GA
Meckel type 1	249000	17q22-q23	CO, GA
Robinow	180700	?	CO, GA
SCARF	312830	X	GA
Spastic paraplegia, optic atrophy, microcephaly, XY sex reversal	603117	?	GD
$1p^+$	Jordan et al 2001	1p22-p35	GD, hypospadias, CO, GA
10q−	Wilkie et al 1993	10q25-q26	hypospadias, CO, GA

CO, cryptorchidism; GA, genital anomalies (ambiguous genitalia, micropenis); GD, XY gonadal dysgenesis.

SOX9 can show pleiotropy, also leading to defects in genital development, and sometimes only genital development is disturbed in the absence of apparent gonadal defects. The identification of the causative gene in one or the other of the syndromes from Table 3 could thus unravel one of the missing pieces. In fact, in the case of $1p^+$ duplications, *WNT4* has just been described as a candidate for this form of dosage-sensitive sex reversal (Jordan et al 2001).

The strongest indication that important pieces of the jigsaw puzzle of sex determination are still missing comes from the unexplained cases of the following three categories of human sex reversal, where the primary defect is restricted to gonadal development: XY gonadal dysgenesis (XY GD), XX maleness, and XX or XY true hermaphroditism. Some 10–15% of XY GD females result from *SRY* open reading frame mutations, another 10–15% from *SRY* deletions due to aberrant X–Y interchange during paternal meiosis, while the remaining 70–80% are a complete mystery (see Scherer et al 1998). Several studies with large cohorts of *SRY*-positive XY GD females have failed to identify mutations in *SOX9* and *WT1* or duplication of *DAX1* (see references in Scherer et al 1998), or mutations in *DMRT1* or *DMRT2* (Raymond et al 1999). Likewise, only 80–90% of XX males are *SRY*-positive, as the mirror-image outcome of aberrant paternal X–Y interchange, while the remaining 10–20% are unexplained, except for one XX sex reversal case resulting from a partial 17q duplication that includes the *SOX9* locus (Huang et al 1999). And the true hermaphrodites? Only a handful of XX true hermaphrodites have been shown to be *SRY*-positive, and there is a single report of an XY true hermaphrodite resulting from a gonadal mosaic of cells with normal or mutant *SRY* (Braun et al 1993). The great majority of XX and XY true hermaphrodites still await a molecular explanation.

How could one find the culprit(s) in these unexplained human sex reversal cases? Mutation screens in newly described genes in the gonadal pathway such as *FGF9* or *WNT4* avail themselves, but the failures with *SOX9*, *WT1*, *DAX1* and *DMRT1* and 2 are a warning. Linkage analyses in familial XX or XY sex reversal could be performed, but such families are extremely rare, the overwhelming majority of the cases being sporadic. In view of the dosage-sensitive nature of human sex determination, submicroscopic *de novo* deletions or duplications could be sought for on a genome-wide basis, using comparative genomic hybridization on DNA microarrays that may become available in the near future. Finally, the mouse could come to rescue. The large-scale ENU (ethylnitrosurea) mutagenesis screens in mice currently underway at several centres can be used to screen for XY sex reversal phenotypes (Soewarto et al 2000) and to uncover as yet unknown genes.

The ovarian pathway

Contrary to the testicular pathway, the ovarian pathway is essentially uncharted terrain. Attempts to identify female-specific transcripts in developing mouse fetal gonads (Grimmond et al 2000, Koopman et al 2002, this volume) hold some promise to lead to some of the elusive genes involved in early ovarian organogenesis and follicle cell formation.

Concluding remark

On looking back at the state of the field about a decade ago, with *SRY* as the single piece of the jigsaw puzzle of mammalian sex determination in our hands, and at the more than a dozen pieces we now have, it becomes apparent that we have come quite some way, not only in mammalian sex determination but also in sex determination in the other vertebrate classes. However, there is still some way to go until we have all of the missing pieces and understand how they fit together. Filling in these missing pieces is the challenge for the future.

Acknowledgements

I would like to thank the Deutsche Forschungsgemeinschaft for the continuous support of our work on human sex determination over the last decade, and Ulrich Wolf and Deborah Morris-Rosendahl for comments on the manuscript.

References

Arango NA, Lovell-Badge R, Behringer RR 1999 Targeted mutagenesis of the endogenous mouse *Mis* gene promoter: in vivo definition of genetic pathways of vertebrate sexual development. Cell 99:409–419

Bagheri-Fam S, Ferraz C, Demaille J, Scherer G, Pfeifer D 2001 Comparative genomics of the *SOX9* region in human and *Fugu*: conservation of short regulatory sequence elements within large intergenic regions. Genomics 78:73–82

Birk OS, Casiano DE, Wassuf CA et al 2000 The LIM homeobox gene *Lhx9* is essential for mouse gonad formation. Nature 403:909–913

Bishop CE, Whitworth DJ, Qin Y et al 2000 A transgenic insertion upstream of *Sox9* is associated with dominant XX sex reversal in the mouse. Nat Genet 26:490–494

Braun A, Kammerer S, Cleve H, Löhrs U, Schwarz HP, Kuhnle U 1993 True hermaphroditism in a 46,XY individual, caused by a postzygotic somatic point mutation in the male gonadal sex-determining locus (SRY): molecular genetics and histological findings in a sporadic case. Am J Hum Genet 52:578–585

Capel B 2000 The battle of the sexes. Mech Dev 92:89–103

Colvin JS, Green RP, Schmahl J, Capel B, Ornitz DM 2001 Male-to-female sex reversal in mice lacking fibroblast growth factor 9. Cell 104:875–889

Fernald RD 2002 Social regulation of sex in fish: what does it tell us? In: The genetics and biology of sex determination. Wiley, Chichester (Novartis Found Symp 244) p 169–186

Goodfellow PN, Camerino G 2001 *DAX-1*, an 'antitestis' gene. In: Scherer G, Schmid M (eds) Genes and mechanisms in vertebrate sex determination. Birkhäuser Verlag, Basel, p 57–69

Grimmond S, van Hateren N, Siggers P et al 2000 Sexually dimorphic expression of protease nexin-1 and vanin-1 in the developing mouse gonad prior to overt differentiation suggests a role in mammalian sexual development. Hum Mol Genet 9:1553–1560

Huang B, Wang S, Ning Y, Lamb AN, Bartley J 1999 Autosomal XX sex reversal caused by duplication of *SOX9*. Am J Hum Genet 87:349–353

Jordan BK, Mohammed M, Ching ST et al 2001 Up-regulation of WNT-4 signaling and dosage-sensitive sex reversal in humans. Am J Hum Genet 68:1102–1109

Koopman P, Gubbay V, Vivian N, Goodfellow P, Lovell-Badge R 1991 Male development of chromosomally female mice transgenic for *Sry*. Nature 351:117–121

Koopman P 2001 *Sry*, *Sox9* and mammalian sex determination. In: Scherer G, Schmid M (eds) Genes and mechanisms in vertebrate sex determination. Birkhäuser Verlag, Basel, p 25–56

Koopman P, Bowles J, Bullejos M, Loffler K 2002 Expression-based strategies for discovery of genes involved in testis and ovary development. In: The genetics and biology of sex determination. Wiley, Chichester (Novartis Found Symp 244) p 240–253

McElreavey K, Vilain E, Abbas N, Herskowitz I, Fellous M 1993 A regulatory cascade hypothesis for mammalian sex determination: SRY represses a negative regulator of male development. Proc Natl Acad Sci USA 90:3368–3372

McQueen HA, McBride D, Miele G, Bird AP, Clinton M 2001 Dosage compensation in birds. Curr Biol 11:253–257

Morais da Silva S, Hacker A, Harley V, Goodfellow P, Swain A, Lovell-Badge R 1996 *Sox9* expression during gonadal development implies a conserved role for the gene in testis differentiation in mammals and birds. Nat Genet 14:62–68

Nanda I, Zend-Ajusch E, Shan Z et al 2000 Conserved synteny between the chicken Z sex chromosome and human chromosome 9 includes the male regulatory gene DMRT1: a comparative (re)view on avian sex determination. Cytogenet Cell Genet 89:67–78

Pask A, Renfree MB, Graves JAM 2000 The human sex-reversing ATRX gene has a homologue on the marsupial Y chromosome, ATRY: implications for the evolution of mammalian sex determination. Proc Natl Acad Sci USA 97:13198–13202

Pfeifer D, Kist R, Dewar K et al 1999 Campomelic dysplasia translocation breakpoints are scattered over 1 Mb proximal to SOX9: evidence for an extended control region. Am J Hum Genet 65:111–124

Pieau C 1996 Temperature variation and sex determination in reptiles. Bioessays 18:19–26

Ramkissoon Y, Goodfellow P 1996 Early steps in mammalian sex determination. Curr Opin Genet Dev 6:316–321

Raymond CS, Parker ED, Kettlewell JR et al 1999 A region of human chromosome 9p required for testis development contains two genes related to known sexual regulators. Hum Mol Genet 8:989–996

Raymond CS, Murphy MW, O'Sullivan MG, Bardwell VJ, Zarkower D 2000 *Dmrt1*, a gene related to worm and fly sexual regulators, is required for mammalian testis differentiation. Genes Dev 14:2587–2595

Scherer G, Held M, Erdel M et al 1998 Three novel SRY mutations in XY gonadal dysgenesis and the enigma of XY gonadal dysgenesis cases without SRY mutations. Cytogenet Cell Genet 80:188–192

Sinclair AH, Berta P, Palmer MS et al 1990 A gene from the human sex-determining region encodes a protein with homology to a conserved DNA-binding motif. Nature 346:240–244

Sinclair AH, Smith C, Western P, McClive P 2002 A comparative analysis of vertebrate sex determination. In: The Genetics and Biology of Sex Determination. Wiley, Chichester (Novartis Found Symp 244) p 102–114

Soewarto D, Fella C, Teubner A et al 2000 The large-scale Munich ENU-mouse-mutagenesis screen. Mamm Genome 11:507–510

Suzuki T, Mizusaki H, Kawabe K, Kasahara M, Yoshioka H, Morohashi KI 2002 Concerted regulation of gonad differentiation by transcription factors and growth factors. In: The genetics and biology of sex determination. Wiley, Chichester (Novartis Found Symp 244) p 68–78

Swain A, Lovell-Badge R 1999 Mammalian sex determination: a molecular drama. Genes Dev 13:755–767

Wilkie AO, Campbell FM, Daubeney P et al 1993 Complete and partial XY sex reversal associated with terminal deletion of 10q: report of 2 cases and literature review. Am J Med Genet 46:597–600

Yu RN, Ito M, Saunders TL, Camper SA, Jameson JL 1998 Role of *Ahch* in gonadal development and gametogenesis. Nat Genet 20:353–357

DISCUSSION

Harley: Did you say there was an ENU screen in Munich for XY females?

Scherer: It is a general ENU mutagenesis screen, and one of the parameters they are testing is to look specifically for XY sex reversal.

Lovell-Badge: Do you plan to look at mutations in human XX male patients for *SOX3*?

Sinclair: We tried to find some, but were unsuccessful.

Vilain: We have looked in five XX males for *SOX3* and found nothing.

Wilkins: Some fraction of the mysterious cases that Gerd Scherer mentioned of XY females and XX males that could not be traced to particular mutations may be so-called epigenetic cases: with stochastic switching off of key genes during development. Many of these genes have to be expressed with specific timing. If there are methylation events or similar phenomena that could cause temporary blockages in expression, this might produce some of these conditions. They wouldn't show up as mutations because they are developmental 'accidents' at the chromatin level. It is hard to prove this, unfortunately, but it is a possibility.

Mittwoch: Can you give an estimate of the relative proportions of unexplained XY females and XX males?

Scherer: I have searched many times for incidence rates of XY gonadal dysgenesis without success. My estimate is about 1 in 20 000, which would be the XX male figure. If this is true, then there are more unexplained XY females than XX males. The majority of XX males are *SRY*-positive.

Burgoyne: Roger Short, what is the latest on horses? Mary-Jo Kent has some pedigrees that involve X-linked mutations and sex reversal.

Short: They still seem to be a mystery. What is fascinating about horses is there are such excellent pedigrees. Clearly there are now a number of stallions that have a highly significantly skewed sex ratio in their progeny. This was how Mary-Jo picked up the first intersex cases, which were the offspring of an Arab stallion producing an excess of phenotypic female progeny. Many of these females were favoured in the show ring because they looked rather male-like, with big crests. Mary-Jo discovered that these mares had very large clitorises and ovotestes, and when she karyotyped them they were all XY. She was then sued by the owner of the stallion for defamation! We still do not know the cause of this XY sex reversal.

Harley: I have a question about your 5′ regulatory SOX9 transgenics. How does this sequence relate to Andreas Schedl's study of his yeast artificial chromosome (YAC) transgenics?

Scherer: It is discordant. We didn't expect to see expression in the testis.

Schedl: I was also surprised by our results. You have to keep in mind that we used the human sequence, and with our YAC transgenics we didn't see any expression in the gonads. We thought that this might be due to the fact that we were using human constructs in the mouse.

Scherer: This is why I was a little bit cautious when I presented these data. This is what we see; we have three other lines with the same construct and we will have to check that it is not a position effect. Unlike you, we used the mouse promoter and used the elements in complete isolation. Remember that Robin Lovell-Badge has had testis-specific expression with his *Sox9* bacterial artificial chromosome transgenics.

Lovell-Badge: We used less; only up to 70 kb. We shouldn't have seen those elements.

Harley: That suggests the existence of suppressors outside the 5′ region.

Lovell-Badge: Did you look in other tissues as well?

Scherer: Yes, there was not much expression in the prospective skeleton. We were a little surprised by this. There was much more expression with the E1/2 construct. There was no expression in the brain, which we see with E1/2. Strangely enough, there was expression in the interdigits.

Zarkower: Could you elaborate on your concern about dosage compensation and the lack of dosage compensation of *DMRT1* in birds? I am not sure I fully understand this. There is a dosage difference in the gene between the two sexes, and there is an expression difference. At face value, this means the gene is not dosage compensated.

Scherer: This has not been studied. *DMRT1* expression has not been quantified in birds yet. There was one recent paper showing that there is dosage compensation, contrary to expectations, for six of the nine genes that have been studied by quantitative PCR, but *DMRT1* could not be analysed (McQueen et al 2001).

Zarkower: So you wouldn't take the difference in strength of *in situ* hybridizations as an indication of expression strength.

I would have thought that if *DMRT1* were dosage compensated, its expression would appear equivalent in the two sexes. At any rate, *DMRT1* has been examined in the chicken by another group using northern blots, and ZZ males are reported to have twice the expression of ZW females.

Graves: It wouldn't make any sense if it were dosage compensated. You would expect it to escape.

Lovell-Badge: I think what David Zarkower is saying is that there is clearly a high level of *DMRT1* expression in males versus females, but this is not a formal argument to say that it is escaping dosage compensation.

Greenfield: That could be the action of a gonadal ridge enhancer.

Zarkower: It almost becomes a semantic point. If something escapes dosage compensation by virtue of some undefined mechanism (and we don't yet know what the mechanism of dosage compensation in the chick is), then it is not dosage compensated. The mechanism doesn't affect the argument.

Lovell-Badge: You haven't distinguished between more expression from one allele, or expression from both.

Zarkower: I don't see the point, because dosage compensation can work either by chromosomal inactivation or by differential transcription initiation off both alleles, which is what happens in flies and worms. Without knowing what the mechanism is, you can't say much.

Lovell-Badge: This would be a different mechanism from the other genes that have been looked at.

Graves: That wouldn't be clear. You are just measuring it with PCR.

Scherer: In the study I'm referring to (McQueen et al 2001), they looked at the expression of several autosomal genes as a control, and correlated this with the expression levels of Z-linked genes from male and female embryos. It was the relative expression of Z-linked genes from males versus females normalized for the expression from autosomal genes that was measured.

Graves: The ratios were between 0.7 and 1.4, but nothing like 2, except for one of the genes that was involved in chromatin packaging. One doesn't know whether this is a difference in regulation or a dosage compensation.

Behringer: Is *Fugu Sox9* expressed in testis?

Scherer: I'm not aware of any data on this. There is a report showing that *SOX9* is expressed in trout testis (Takamatsu et al 1997). In zebrafish, only one of the two *Sox9* genes, *Sox9a*, is expressed in the testis, whereas *Sox9b* is expressed in the ovary (Chiang et al 2001).

Koopman: Have you hooked up the *Fugu Sox9* upstream region to *lacZ* and put this into transgenic mice?

Scherer: We'd like to do this, but we haven't been able to yet.

Harley: How far away from *Sox9* are the conserved elements in *Fugu*?

Scherer: The compression factor in the 5′ flanking region is almost exactly a factor of 17. Each element that is separated by a particular distance in human is 17 times closer in *Fugu*. For example, the 290 kb element is around 18 kb away.

Harley: Did you pick up *Sox8* and *Sox10* in *Fugu*?

Scherer: I think we got some positives in the screen, but we are trying to concentrate on the *Sox9* region.

References

Chiang EFL, Pai CI, Wyatt M, Yan YL, Postlethwait J, Chung BC 2001 Two *Sox9* genes on duplicated zebrafish chromosomes: expression of similar transcription activators in distinct sites. Dev Biol 231:149–163

McQueen HA, McBride D, Miele G, Bird AP, Clinton M 2001 Dosage compensation in birds. Curr Biol 11:253–257

Takamatsu N, Kanda H, Ito M, Yamashita A, Yamashita S, Shiba T 1997 Rainbow trout SOX9: cDNA cloning, gene structure and expression. Gene 202:167–170

Expression-based strategies for discovery of genes involved in testis and ovary development

Peter Koopman, Monica Bullejos, Kelly Loffler and Josephine Bowles

Institute for Molecular Bioscience, The University of Queensland, Brisbane, Queensland 4072, Australia

Abstract. In recent years, strategies for gene identification based on differential gene expression have become increasingly popular, due in part to the development of microarray technology. These strategies are particularly well suited to the identification of genes involved in sex determination and gonadal development, which unlike the development of other organ systems, proceeds along two very different alternative courses, depending on the sex of the embryo. We have used a high-throughput, array-based expression screen to identify several genes expressed sex-specifically in developing mouse gonads. One of these, *vanin 1*, appears to play a role in mediating migration of mesonephric cells into the male genital ridge. Progress in characterizing other genes arising from the screen is discussed.

2002 The genetics and biology of sex determination. Wiley, Chichester (Novartis Foundation Symposium 244) p 240–252

As the previous papers in this volume have demonstrated, we now know of a number of genes that are important for the development of the gonads as testes or ovaries, and hence the development of the organism as a male or female. Considerable progress is being made in understanding how these genes fit together to form a regulatory and signalling network. However, it is abundantly clear that many pieces of the puzzle are missing.

In recent years, strategies for large-scale gene identification based on differential gene expression in vertebrate embryos have become more readily applicable, due to advances in cDNA subtraction technologies, expansion of genetic databases and improvements in their accessibility, and the advent of microarray technology. Subtraction strategies are particularly well suited to the identification of genes involved in sex determination and gonadal development, because the development of the gonads, unlike that of other organ systems, proceeds along two different courses, depending on the sex of the embryo. It is therefore possible to directly compare the transcriptional profiles of fetal testes and ovaries in order to identify genes expressed preferentially in gonads of one sex or the other — referred to in this paper for simplicity, albeit inaccurately, as 'sex-specific genes'. We and other groups have initiated high-throughput array-based expression screens aimed at identifying sex-specific genes in developing mouse gonads (Bowles et al 2000, Grimmond et al 2000, Wertz & Herrmann 2000). This paper describes the logistics of our screen, summarizes our progress to date, describes some of the pitfalls that we have encountered, and charts a course for future work in this area.

Logistics

We have based our screen on the expectation that genes important for male or female sexual development will be expressed differently between developing testes and ovaries in the fetus. Our overall strategy is to make subtracted probes enriched for genes expressed in either developing testes (male-enriched probes) or ovaries (female-enriched probes), use both types of probe to screen arrayed libraries derived from cDNA expressed in developing gonads, and identify spots that hybridize differentially with the two probes, in order to yield a large number of primary candidate genes. These are then scrutinized and prioritized by a combination of bioinformatic analysis and wholemount *in situ* hybridization screening of mouse fetal testes and ovaries. This reduces the large number of primary candidates to a much smaller number of interesting candidate genes. These are then subjected to in-depth physical, biochemical and functional analysis, in order to illuminate their exact role in sex determination and/or gonadal differentiation.

In mice, the genital ridges arise at around 10 days post coitum (dpc), and remain morphologically undifferentiated until around 12 dpc, after which testes begin to differentiate in the male. We have made subtracted probes corresponding to two time points in gonadal development. Our initial efforts involved dissection of gonads at around 13 dpc (Bowles et al 2000). At this stage, it is easy to obtain sufficient quantities of fetal gonadal tissue, since the gonads are relatively large and easy to dissect away from other tissue. It is also easy to distinguish testes from ovaries under a dissecting microscope at 13 dpc. We would expect the gonads, particularly the testes, to be transcriptionally complex at this stage, and that many transcriptional differences between testes and ovaries will exist. Subtraction at 13 dpc is most likely to yield genes involved in the differentiation, as opposed to the initial specification, of testes and ovaries.

More recently, we have made subtracted probes corresponding to a mixture of stages between 10.5 and 12.5 dpc, strongly biased to the 11.5 dpc time point. This time interval is immediately after the activation of *Sry* transcription (Koopman et al

1990); at this stage we might expect to identify male-specific genes that are close to *Sry* in the testis pathway. However, many fewer differences between the transcriptional profiles of male and female genital ridges might be expected at this stage, and hence it is likely to be more difficult to identify genes of interest.

Probes and libraries

Complex cDNA pools enriched for either male- or female-specific transcripts were made using the technique of suppression PCR (Diatchenko et al 1996, Gurskaya et al 1996). This technique not only subtracts transcripts represented in one sample from those expressed in another (by solution hybridization), but also normalizes the representation of rare versus abundant transcripts. The enriched cDNA pools are biased towards shorter (600 to 1000 bp) 3′ fragments that are ideal for use in *in situ* hybridization experiments. Rigorous controls are employed to ensure that subtraction and normalization has occurred efficiently. The cDNA pools can be labelled with radioactive or fluorescent tags to generate complex probes enriched for tissue-specific transcripts, or cloned into a plasmid vector to make gonad-specific cDNA libraries. We made both male- and female-enriched cDNA pools at 13 dpc, tested the quality of these probes by hybridization to dot blots of genes known to be expressed sex-specifically in developing gonads, and made corresponding cDNA libraries (Bowles et al 2000).

Each library was arrayed manually onto nylon filters, and screened in duplicate with male- and female-enriched probes. We also screened 2000 sequenced and gridded clones of a normalized mouse urogenital ridge (NMUR) library (Grimmond et al 2000). In both types of experiment, the intensity of hybridization of each spot gives a measure of the representation of the corresponding cDNA in the subtracted probe, while the difference in hybridization intensity of each spot with the two probes gives an indication of the sex-specificity of expression (Fig. 1). These experiments generated several hundred primary candidate genes for further study.

Evaluating primary candidate genes

Candidates derived from the suppression PCR libraries were sequenced, while sequences corresponding to NMUR clones were either retrieved from existing databases or determined *de novo*. This information was used to determine whether each gene corresponded to a known or novel gene, and in the case of known genes, what classes of molecules might be encoded. Searching of EST databases also provided information regarding the expression profile of each gene. On this basis, candidates were prioritized for further screening by wholemount *in situ* hybridization.

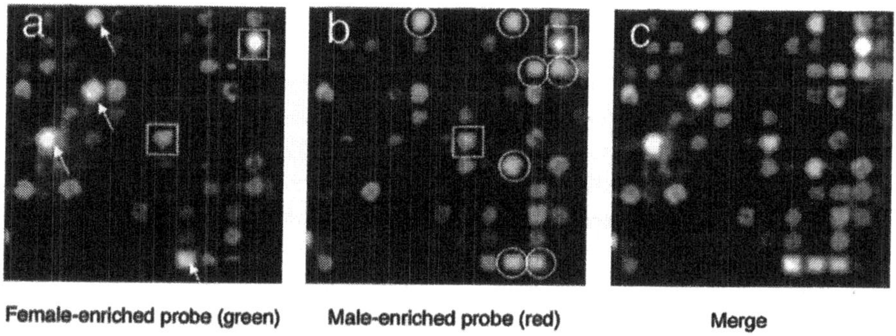

FIG. 1. Microarray screening of the NMUR cDNA library using probes enriched for male- and female-specific transcripts. Panels show detail of the microarray. (a) Signal from the green channel representing hybridization with the female-enriched probe; strongly hybridizing 'female-specific' spots are arrowed. (b) Signal from the red channel representing hybridization with the male-enriched probe; strongly hybridizing 'male-specific' spots are circled. Spots hybridizing strongly to both probes are boxed. (c) Merged image.

Labelled RNA probes were made for each gene of interest, and these were hybridized *in situ* to both male and female mouse gonads at 12.5 or 13.5 dpc, in order to verify sex-specific expression, and to determine whether each gene was expressed in the seminiferous cords, the interstitium, or the mesonephroi (Fig. 2). Expression of sex-specific genes was studied further by examining the timing of expression in the period 10.5 to 14.5 dpc, relative to morphological events in gonadal development and to the expression of known genes such as *Sry*, *Sox9* and *Amh* (*Mis*). For genes expressed in seminiferous cords, the cell type responsible for expression was determined by cutting sections of testes after wholemount *in situ* hybridization, in order to associate the hybridization signal

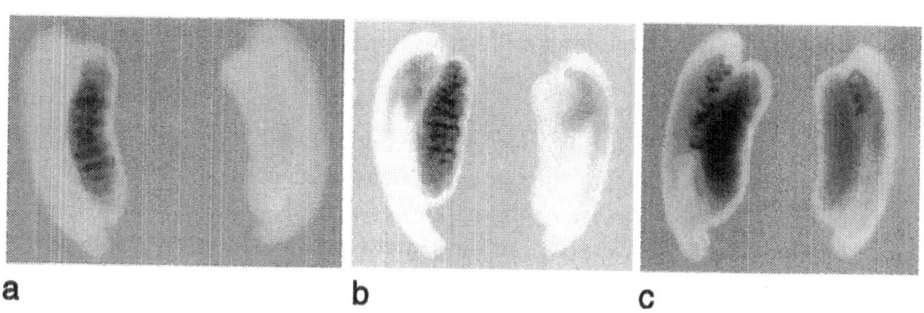

FIG. 2. Examples of *in situ* hybridization patterns seen with candidate genes arising in the screen. (a) Male-specific expression within the seminiferous cords; (b) male-specific expression in interstitial cells; (c) expression in the mesonephroi. Each panel shows a male (left) and a female (right) gonad and attached mesonephros at 12.5 dpc.

with either Sertoli cells or germ cells, and by analysing expression in the gonads of *W*[e] homozygous mutant fetuses, which lack germ cells (McLaren 1985). In some cases, gene expression was analysed during gonadal development in other species such as the chicken, to test for evolutionary conservation of sex-specific expression. Typically, gene expression was also analysed by wholemount analysis of whole embryos, to determine whether gene expression was confined to the developing gonads, or was more widespread in the embryo.

In this way, the large number of primary candidates was reduced to fewer than 10 secondary candidates for further study.

In-depth analysis of secondary candidate genes

Secondary candidate genes were prioritized for further analysis by a combination of expression data derived from wholemount *in situ* hybridization experiments, and bioinformatic data relating to the likely class of molecule encoded by each gene. We were particularly interested in genes encoding transcription factors, which might act as cell-type-specific differentiation factors, and genes encoding signalling molecules that might be considered candidates for the several signalling events known to be important for proper development of both testes and ovaries (reviewed in Capel 2000).

An example of a candidate thus prioritised for further study is the gene encoding vanin 1 (vascular non-inflammatory molecule 1). Vanin 1 is a glycosylphosphatidyl inositol (GPI)-anchored cell surface molecule expressed in perivascular epithelial and non epithelial cells, known to be involved in the migration of thymocytes from the bloodstream into the thymus (Aurrand-Lions et al 1996). It most likely plays a role in cell adhesion and/or chemoattraction. The *vanin 1* gene arose independently several times in our screen, suggesting that it might be a genuine sex-specific gene. In view of its possible role in chemoattraction in the context of thymocyte homing, and data suggesting that attraction of mesonephric cells into the male genital ridge is required for proper development of the testes (Buehr et al 1993, Martineau et al 1997, Tilmann & Capel 1999), our attention was focused on *vanin 1* as likely to be important for testis development.

Wholemount *in situ* hybridization analysis confirmed that *vanin 1* is indeed expressed male-specifically during gonadal development in mice (Fig. 3). The onset of *vanin 1* expression occurs shortly after that of *Sry*, and persists until at least 16.5 dpc. *Vanin 1* is expressed in the seminiferous cords, and section analysis showed it to be associated with the Sertoli cell lineage (Bowles et al 2000).

In order to test the role of *vanin 1* in testis development, we employed a genital ridge/mesonephros co-culture assay (Martineau et al 1997). In this assay, wild-type XY genital ridges at 11.5 dpc are cultured alongside mesonephroi from a transgenic strain of mice ubiquitously expressing green fluorescent protein

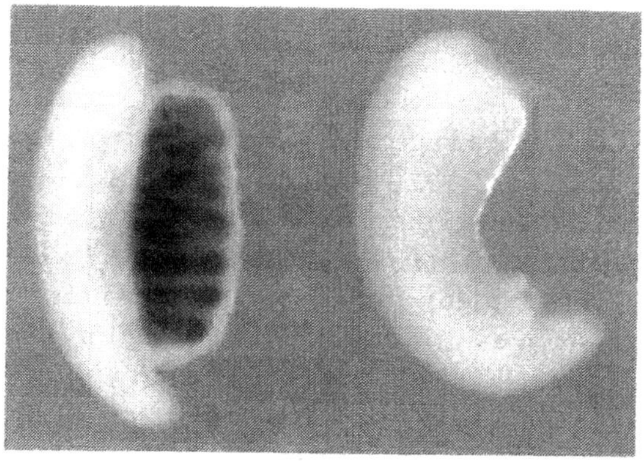

FIG. 3. Male-specific expression of vanin 1 in mouse fetal gonads. 13.5 dpc testis (left) and ovary (right) showing strong male-specific staining in the seminiferous cords.

(GFP). Migration of mesonephric cells into the genital ridge can be monitored under UV illumination. Preliminary experiments indicate that this migration is blocked in the presence of an antibody to vanin 1. These results suggest that *vanin 1* is required for the migration of cells from the mesonephros into the XY genital ridge, and is therefore important for the formation of seminiferous cords in the testis.

Progress and pitfalls

To date, we have carried out extensive screening using probes and libraries from both the 13 and 11.5 dpc stages. Several hundred primary candidate genes arising from the screen have been analysed bioinformatically, and *in situ* hybridization analysis of these candidates is continuing. A number of conclusions can be drawn from our studies to date.

First, a high proportion of primary candidates correspond to genes that show a genuine sex-specific difference in expression pattern by wholemount *in situ* hybridization of fetal gonads. For example, in one batch of 41 primary positives from a differential screen of the 13 dpc male-enriched library, a total of 10 (24%) subsequently proved to be genuinely male-specific (Bullejos et al 2001). This indicates a relatively low rate of false positives using this method, in contrast to other methods such as differential display PCR (Greenfield et al 1996, Liang & Pardee 1992, Nordqvist & Töhönen 1997).

Second, we have found a low rate of redundancy among primary positive clones at 13 dpc, supporting our prediction that the gonads are transcriptionally complex at this stage. In contrast, more redundancy is seen among primary positive clones from the 11.5 dpc libraries, suggesting lower transcriptional complexity at this stage.

Third, we have found that a high proportion of primary positive clones correspond to genes expressed in germ cells. Since germ cells are not required for sex determination, (McLaren 1985, Merchant 1975) nor for testis differentiation, we have not pursued these genes further. However, it is known that germ cells are required for the formation of follicles in the ovary (McLaren 1985, Merchant 1975), suggesting that signalling from germ cells to somatic follicular cells is important for the histogenesis of the ovary. We would therefore not discount genes expressed in germ cells during ovarian folliculogenesis.

Fourth, we have found that genes expressed differentially between testes and ovaries fall into many more classes than expected. These include genes encoding structural proteins, cytochrome P450 family members, extracellular matrix components, enzymes, membrane components and signal transduction components. Clearly, transcriptional differences between developing testes and ovaries are not confined to genes encoding transcription factors and signalling molecules.

Fifth, this diversity of classes of genes and the molecules they encode calls for a much larger repertoire of functional assays than previously envisaged. While the strategies used by molecular developmental biologists to determine the function of tissue-specific transcription factors or signalling molecules are relatively well established, strategies for investigating the role of, say, metabolic enzymes represent a path less travelled.

Sixth, the entire enterprise of developing suitable probes and libraries, testing these reagents, evaluating large numbers of primary positives bioinformatically and by expression studies, and designing and carrying out detailed functional assays has proven to be enormously labour-intensive and logistically challenging. In particular, the last step has proven to be rate limiting.

Seventh, far fewer male-specific genes have come from our screening at 11.5 dpc, compared to screening at 13 dpc. This finding is in agreement with our prediction that fewer male-specific genes will be expressed at this time point, since it is relatively soon after activation of *Sry*, the earliest-acting male-specific gene.

Finally, despite extensive searching, we have arrived at no strong leads for genes important for ovarian development. This is perhaps not surprising at 11.5 dpc, since it may well be that no genes are activated female-specifically at this time point. However, histological analyses have suggested that active organizational processes are underway in the ovary by 13.5 dpc (Fröjdman & Pelliniemi 1995, Odor & Blandau 1969, and K. Loffler & P. Koopman, unpublished data), and these may be under the control of female-specific regulatory genes. Perhaps such genes are much lower in number than those active in the developing testis at a similar time point, so that finding these genes is more difficult than finding their male counterparts. Alternatively, the set of genes involved in organizing the ovary may largely overlap with that involved in organizing the testis, and hence will not be detected in a screen based on differential gene expression.

Future challenges

As discussed in previous papers in this volume, mutations in genes such as *Sry*, *Sox9* and *Dax1* are known to affect sex determination and gonadal development in humans. However, some 80% of cases of human XY gonadal dysgenesis, 20% of cases of XX maleness and most cases of XX true hermaphroditism remain unexplained at the genetic level (Lim et al 2000). This suggests that a number of important undiscovered sex-determining genes are still at large.

Expression screens such as that described in the present paper represent an efficient, high-throughput, and unbiased means of generating large numbers of candidates for the missing links in sex determination and gonadal development. The rate-limiting step in studies such as this is the development of functional assays which are themselves efficient and high-throughput, and that can be tailored to the analysis of many different classes of molecule. For example, methods for efficient delivery of genes encoding transcription factors to cultured mouse gonads remain to be perfected, as do methods for perturbation of such genes in organ culture. In whole animal studies, suitable cell type-specific promoters need to be developed for transgenic gain-of-function assays, as do suitable *Cre* recombinase-expressing mice for tissue-specific loss-of-function (knockout) assays.

Furthermore, expression-based screens are likely to be supplemented in future by assays for non-transcriptional events, since it is naive to assume that all molecular genetic control of sex determination and gonadal development occurs at the transcriptional level. Assays that detect differences in protein expression or modification between developing testes and ovaries, or post-transcriptional events such as differential splicing, are likely to form an important part of future efforts.

In the coming decade, we are likely to see further progress in understanding one of the great black boxes in developmental biology, namely the molecular genetics and cell biology of ovarian development. Efforts to illuminate ovarian development have been overshadowed to some extent by progress in studying testis determination and differentiation. Expression-based screens are likely to yield at least a few genes that are important for the development of the ovary, and it is hoped that these will act as a springboard for discovery of other genes. At the very least, it is likely that genes will be discovered that can be used as markers of the different cell types in the ovary, so that a better understanding of the cellular events during ovarian differentiation can be gained.

Finally, important genes in many developmental processes have had a past tendency to be discovered purely by chance, often in the course of studying an unrelated process. Several genes in the sex-determining and/or gonadogenetic pathways were discovered in this way (e.g. Birk et al 2000, Katoh-Fukui et al 1998, Vainio et al 1999, Viger et al 1998). As large-scale characterization of the human, mouse and other vertebrate genomes progresses, it is likely that serendipity will continue to play an important role in filling in some of the gaps in the molecular genetic network of sexual development.

Acknowledgements

We thank Philippe Naquet for the generous gift of the vanin 1 antibody, and Andy Greenfield for permission to present microarray data. This work was supported by grant funding from the National Health and Medical Research Council of Australia, and by infrastructural funding from the Australian Research Council. PK is an Australian Senior Research Fellow of the Australian Research Council. MB is supported by the Secretaría de Estado de Universidades, Investigación y Desarrollo, Spain.

References

Aurrand-Lions M, Galland F, Bazin H, Zakharyev V, Imhof BA, Naquet P 1996 Vanin-1, a novel GPI-linked perivascular molecule involved in thymus homing. Immunity 5:391–405

Birk OS, Casiano DE, Wassif CA et al 2000 The LIM homeobox gene Lhx9 is essential for mouse gonad formation. Nature 403:909–913

Bowles J, Bullejos M, Koopman P 2000 A subtractive gene expression screen suggests a role for *vanin-1* in testis development in mice. Genesis 27:124–135

Buehr M, Gu S, McLaren A 1993 Mesonephric contribution to testis differentiation in the fetal mouse. Development 117:273–281

Bullejos M, Bowles J, Koopman P 2001 Searching for missing pieces of the sex-determination puzzle. J Exp Zool 290:517–522

Capel B 2000 The battle of the sexes. Mech Dev 92:89–103

Diatchenko L, Lau YF, Campbell AP et al 1996 Suppression subtractive hybridization: a method for generating differentially regulated or tissue-specific cDNA probes and libraries. Proc Natl Acad Sci USA 93:6025–6030

Fröjdman K, Pelliniemi LJ 1995 Alpha 6 subunit of integrins in the development and sex differentiation of the mouse ovary. Dev Dyn 202:397–404

Greenfield A, Scott D, Pennisi D et al 1996 An H-YD^b epitope is encoded by a novel mouse Y chromosome gene. Nature Genet 14:474–478

Grimmond S, Van Hateren N, Siggers P et al 2000 Sexually dimorphic expression of protease nexin-1 and vanin-1 in the developing mouse gonad prior to overt differentiation suggests a role in mammalian sexual development. Hum Mol Genet 9:1553–1560

Gurskaya NG, Diatchenko L, Chenchik A et al 1996 Equalizing cDNA subtraction based on selective suppression of polymerase chain reaction: cloning of Jurkat cell transcripts induced by phytohemaglutinin and phorbol 12-myristate 13-acetate. Anal Biochem 240:90–97

Katoh-Fukui Y, Tsuchiya R, Shiroishi T et al 1998 Male-to-female sex reversal in M33 mutant mice. Nature 393:688–692

Koopman P, Münsterberg A, Capel B, Vivian N, Lovell-Badge R 1990 Expression of a candidate sex-determining gene during mouse testis differentiation. Nature 348:450–452

Liang P, Pardee AB 1992 Differential display of eukaryotic messenger RNA by means of the polymerase chain reaction. Science 257:967–971
Lim HN, Berkovitz GD, Hughes IA, Hawkins JR 2000 Mutation analysis of subjects with 46, XX sex reversal and 46, XY gonadal dysgenesis does not support the involvement of SOX3 in testis determination. Hum Genet 107:650–652
Martineau J, Nordqvist K, Tilmann C, Lovell-Badge R, Capel B 1997 Male-specific cell migration into the developing gonad. Curr Biol 7:958–968
McLaren A 1985 Relation of germ cell sex to gonadal development. In: Halvorson HO, Monroy A (eds) The origin and evolution of sex. Alan R Liss, New York, p 289–300
Merchant H 1975 Rat gonadal and ovarian organogenesis with and without germ cells. An ultrastructural study. Dev Biol 44:1–21
Nordqvist K, Töhönen V 1997 An mRNA differential display strategy for cloning genes expressed during mouse gonad development. Int J Dev Biol 41:627–638
Odor DL, Blandau RJ 1969 Ultrastructural studies on fetal and early postnatal mouse ovaries. I. Histogenesis and organogenesis. Am J Anat 124:163–186
Tilmann C, Capel B 1999 Mesonephric cell migration induces testis cord formation and Sertoli cell differentiation in the mammalian gonad. Development 126:2883–2890
Vainio S, Heikkila M, Kispert A, Chin N, McMahon AP 1999 Female development in mammals is regulated by Wnt-4 signalling. Nature 397:405–409
Viger RS, Mertineit C, Trasler JM, Nemer M 1998 Transcription factor GATA-4 is expressed in a sexually dimorphic pattern during mouse gonadal development and is a potent activator of the Müllerian inhibiting substance promoter. Development 125:2665–2675
Wertz K, Herrmann BG 2000 Large-scale screen for genes involved in gonad development. Mech Dev 98:51–70

DISCUSSION

Short: This does seem an enormously powerful technique for pulling out the unexpected. Did you also find the expected genes such as *Sry* and *Sox9*?

Koopman: In the libraries that we made, they were not among the clones that we picked. But when we used those libraries as probes on spots of known genes, such as *Sry* and *Sox9*, we got a positive signal. So it is in those pools.

Greenfield: Sox9 also came up in the NMUR microarray screen.

Behringer: In doing these screens is chromosomal mapping an automatic thing you are putting in? And if so, how are you doing this?

Koopman: We use existing databases to find the location of known genes; for novel genes we map them as a matter of course once we have confirmed sex-specific expression.

Behringer: I didn't really see a loss-of-function approach. You mentioned knockouts, but it is apparently not a high priority. If I understand correctly, the German gene trap consortium is a public database of mouse gene traps. There is a database of sequences and these clones are freely available. Bill Skarnes at Berkeley has a gene trap public database from which you can get your knockouts pretty much for free. Terry Magnuson and John Schimenti have ENU-mutagenized Embryonic Stem (ES) cell libraries where if your gene is expressed in ES cells you can do a 96 well RT-PCR, which goes through a machine (called the WAVE machine) that then finds the mutations with which you can make mice. I think the loss-of-function approach is not unwieldy. People have created resources which you can tap into pretty easily.

Greenfield: We have several genotype-driven programmes at Harwell. We have approximately 5000 DNA samples from ENU-mutagenized mice. Several groups are now using wave machines to screen for point mutations in genes of interest. Two groups have now detected mutations causing stop codons and the mice have been recovered. This is quite a powerful approach for generating specific alleles.

Harley: Is that available to Australians?

Greenfield: Yes.

Capel: One of our next major hurdles is to find a way to introduce dominant negatives, or other ways to block in organ cultures. Adenovirus is very effective. It infects with a really high incidence. But the viruses are so unwieldy to make, it would be nice if there were ways to introduce plasmids.

Short: Peter Koopman, how early are you going to start looking?

Koopman: The earliest we have done so far is 11.5 dpc in mice. We could go earlier by tailoring the screen to answer different types of question. We have concentrated on male versus female; one could do gonad versus mesonephros or stage versus stage, for instance. The rate-limiting step in these screens is to find the needles in the haystack.

Burgoyne: Doesn't there need to be some sort of coordination between different labs? There are several of you doing this sort of thing, and you are all coming up with 800 spots each. Unless there is communication between the labs you may all start working on the same genes.

Koopman: There is some communication, but it is more a damage control mechanism at this stage. If someone could suggest a way that allows us to communicate results at an early stage, and at the same time satisfy the needs of the people doing this work—the postdocs and students whose careers are depending on being first with the breakthroughs—then I would like to hear about it.

Behringer: You get so many genes, one lab can't do them all. I wonder whether there could be some sort of consortium that just shows the *in situ* patterns. Then people could contact you for collaborations.

Greenfield: Many of us are moving away from libraries and going over to big minimal sets, but everyone is using the same big sets. We have already screened 10 000 of the NIA set, and I know that Peter Koopman has that set. We do need to coordinate, because we are going to find the same clones.

Josso: But if you had found a gene that is expressed in cartilage and then you found it was also expressed in gonad, would you have gone for it or left it to the bone people? If a gene is not only expressed in the organ of interest, would it still interest you?

Koopman: Certainly. If it is expressed in an interesting pattern in the gonads then it is of interest to us. If it is also expressed in other tissues that is fine. If it were only expressed in the gonads that would make it a bit more interesting.

McLaren: How close to saturation are you? Suppose that three people each get 800 genes, what is the overlap?

Koopman: There is no way of knowing.

Greenfield: One of the things that is quite weak at the moment is the informatics on the gene content. We can't even agree on how many genes there are. Most estimates are in the range of 38 000 to 60 000. On those large sets it is fairly certain that there isn't one probe–one gene. When the informatics is done we'll see that some of those ESTs are just distinct parts of one transcript. There will be overlap.

Schedl: I wouldn't be too concerned about redundancy. It would be good to find a way to compare the data sets, but any kind of screen needs to be repeated to make sure that clones are correctly expressed. If one takes three or four screens and somehow finds a way to compare the patterns, we can all benefit.

Greenfield: But when your postdoc has spent months finding that beautiful little red spot, to find out that they may have to give it away to someone who has had it longer is pretty tough.

Burgoyne: If you can get your RNAs that you are going to use in the screens from a situation where you are expecting very little difference, so you have a very focused question, when you put it on the screen you hopefully get six genes and not 600. Then you are immediately focused on your question of interest.

Koopman: But the haystack is the same size and there are far fewer needles.

Greenfield: The other issue here is about what the source of your target is. Peter Koopman is using a relatively simple target, the product of subtractive hybridization. I suspect that this means it is non-quantitative but very sensitive: there is better specific activity of each labelled message. But we want to move to a stage where we can start doing quantitative comparisons.

Burgoyne: The only way to do this is to use total RNA.

Wilkins: Peter Koopman, your approach potentially misses out genes that are expressed similarly but are used in slightly different ways and through different combinatorial controls. One shouldn't forget the possibility of such genes entirely, even though they shouldn't be given first priority.

Koopman: The question has been raised as to whether there are any genuinely ovarian-specific genes, or whether the same set of genes is used in the ovary as are used in testis development, but in different ways.

Short: David Zarkower, would it be a waste of time for you to use this technology in insects, if sex is expressed cell autonomously in every single cell?

Zarkower: Not at all. People are doing this. Do you mean would this be a waste of time in terms of finding things relevant to vertebrates?

Short: Yes.

Zarkower: I hope not. I am cautiously optimistic, given our experience with *mab-3*, which acts autonomously and has no obvious role in the *Caenorhabditis elegans* gonad.

Short: To put my question the other way round, Peter Koopman, are you missing out on something because you are totally sold on the hypothesis that all sexual dimorphisms stem from the gonad? Perhaps this is a bit of an untested assumption for mammals. As Marilyn Renfree was telling us, there are good examples in marsupials where there is genetic determination of secondary sexual characteristics such as the scrotum, mammary gland and pouch. Do you think we are going to come up with any cell autonomous sexual differentiation in eutherian mammals that is extra-gonadal and which you might miss?

Koopman: Yes, there are lots of things that we are likely to miss. For example, any control mechanism that is not directly transcriptional will be hard to pick up. But we still feel that we have a good chance of finding a large number of important genes for sex determination and gonadal development using screens of this type.

Zarkower: I am not sure that the autonomy or non-autonomy of tissue specificity are as much the issue as whether there is actually extensive mechanistic conservation of things involved in sex determination.

Capel: I tried one of these screens a long time ago before the techniques were so accurate. I found that anything in my library that was expressed at a low level was masked by the signal from everything else expressed at a much higher level. I wonder whether you could somehow use reassociation kinetics to eliminate the frequent probes in your mixture.

Koopman: Essentially, that's the way the suppression PCR works: it subtracts and normalizes. It enriches for rare transcripts relative to abundant ones.

Behringer: That's the theory. In practice the libraries you get still have background.

Koopman: We have used our type of probe, made by suppression PCR, versus unsubtracted male and female probes. In our experience, when you put the crude, unsubtracted male and female probes onto duplicate blots, you get a mess. However, if you put our normalized, subtracted probes onto duplicate blots, the results are very clean.

Final general discussion

Short: Perhaps we should do a bit of navel gazing and think what each of us would like to see as directions for the future. In the whole field of sex determination, what would you really like to know?

Zarkower: Partly out of effort to leave *Sry* for others, and partly because I would genuinely be interested to know the answer, I would be interested to know what determined sex in the so-called 'urbilaterian', the ancestor from which we and our model organisms are descended.

Behringer: I think there should be a big push for phenotypic screens for sex reversal and sex abnormalities, primarily in the mouse. The phenotypes are always going to be there, whereas if you come from a gene-based approach you may not get the phenotype you want.

Short: So experiments of nature are still a great inspiration.

Vilain: I would be very interested in understanding the tremendous phenotypic variability, whether it is in mice or in humans. With the same mutation there can be a whole range of sexual phenotypes. We often underestimate all manner of influences, from environment to genetic background.

Mittwoch: One of the aspects that we haven't discussed at this meeting is pregonadal sex differences. The differentiation of the genital ridge may be the pivotal act in sexual differentiation, but of course we know that this is not the first phenomenon that occurs. There are sex differences in the developmental rate of early human fetuses and mouse blastocysts, and in cleaving human and mouse embryos. Of particular significance is evidence that there is already a difference in metabolic rate in very early embryos, with males having a higher rate than females (Mittwoch 2000). This seems to be a feature at most times in life. The question is, what is the relationship between genes, metabolic rate and energy metabolism in these early embryos? Could there be more mitochondria in XY embryos? This would be particularly interesting in the developing gonad, where there is a substantial difference in developmental rate between male and female. I hope that in this new century the relationship between genotype and energy metabolism will be addressed. The male–female dichotomy is an excellent system to address this problem.

Short: This gets back to what we were discussing earlier: we mustn't be sucked into thinking that sex determination begins and ends with the gonads.

Scherer: I have two wishes. First, after 11 years of knowing about the existence of *Sry*, and 10 years since the paper that demonstrated that it was the Y-located testis determinant, it would be nice to know just one target! Second, I would like to be able to explain 30% more of the cases of XY gonadal dysgenesis.

Bujellos: I would like to know how *Sry* expression is regulated.

Harley: I think all mine have been done. I'd like to know the *Sry* targets. It is intriguing that 12 of the 14 or so players in the sex-determining pathway are transcription factors. Like Eric Vilain, I think that there is a whole range of very subtle mutations out there that are causing partial penetrance and variation in phenotypic background. I think the high-throughput and single nucleotide polymorphism (SNP) technologies will reveal subtle mutations in many sex-determining genes.

Greenfield: I agree with Richard Behringer that phenotypes are the key. I would like some kind of international consortium where we make a mouse that is primed for a mutagenesis screen. Perhaps this mouse is multiply sensitized, not just in one pathway but in many pathways that are important for sex determination. Perhaps it is heterozygous for 20 different knockouts, but doesn't have a phenotype. It might have multiple reporter genes, so you could just open up the gonads and have a quick look to see which markers it is expressing. Then you just breed this on to a mutagenized background and pick out the phenotypes.

Wilkins: I am intrigued by the possibility that the evolutionary genetics literature on sex determination evolution (which Brian Charlesworth so nicely reviewed) is coming together with the molecular genetics-type work that has been the focus of our meeting here. For this synthesis, we need more information about the molecular players in different organisms, and we also need to understand the dynamics of how genes get into pathways, and perhaps how they displace other genes. This is a problem that real evolutionary biologists have so far given relatively little attention to. A good example would be the fact that *Sry* came in some time after the monotremes. Did it capture a pre-existing pathway, or did it displace a gene? We will eventually have this information, but we will still need to understand something about the evolutionary dynamics that led to this. This is a field ripe for analysis.

Burgoyne: Leaving aside the obvious issue of the *Sry* target, I am very interested in the idea that *Sry* has to do something by a certain time in order to pre-empt a default ovarian pathway. I'd like to know what this step in the ovary is that *Sry* has to pre-empt, and in which lineage this step takes place.

Morohashi: I am a newcomer to this field, but I realize that the most interesting problem is the function of SRY for sex determination. My interests are somewhat different: how the intermediate mesoderm differentiates into a sexually indifferent gonad, and how the precursor cells differentiate into Leydig cells.

Charlesworth: I would like to know a lot more about what is going on with the degenerating genes on the Y chromosome. Is this just a passive accumulation of deleterious mutations, or are these genes actively being turned down? What is the role of transposable elements? We have some examples where transposable elements are inserting in 5′ regions and in introns: are these turning the genes down?

Swain: I would like to understand the molecular nature of the switch. Is chromatin involved? What are the components of the protein complex that are working together to make the decision?

Fernald: As an outsider, I guess my view is that this whole field reflects an interesting experiment in evolutionary discoveries. I was struck by the two experimental directions: one is to suspect that candidate genes from other species might instruct us across all animals, and the other is that there is quite a difference in sex-determining processes. For me the interesting outcome will be how much these two ideas interact. Will we find three genes that are common across sex determination in all species, and everything else is up for grabs?

Poulat: Because I did a postdoc in Gerd Scherer's lab, I was floating for two and a half years in the atmosphere of *Sox9*. Coming back to my original lab, I'm also coming back to *Sry*. It is difficult to choose between these two genes. For *Sry*, the function will be interesting to discover, in terms of its biochemical effects. For *Sox9*, there are several aspects that are puzzling, including the transcriptional regulation of this gene and why it is located in a desert with nothing 1.5 Mb upstream and 1 Mb downstream. We are also very interested in its subcellular localization. We have data showing that if it is pushed into the nucleus it can cause male sex determination even in an XX gonad.

Schedl: I have a problem here because I am interested in almost everything. Certainly, everyone wants to know what the *Sry* gene does. What Francis Poulat has touched on is important, too. We have talked a lot about regulation, but the M33 knockout has told us that chromatin and epigenetic modification is very important in regulating genes. We know very little about this. I would like to see more research done on this. I also think that the generation of gonadal anlage is very interesting. What factors are initiating the initial proliferation?

Renfree: I guess we are such an anthropocentric species that we will always be wanting to know more about humans. And the mouse has been such a fantastic model we will always know a great deal about mice and men. I would like to see us continue to study the other species and groups, and not just look at them as curiosities, but use them and incorporate them into our work as good examples of how to shed the spotlight on something from a different angle. We need to embrace the lessons that evolution can give us. I would encourage people to take a comparative evolutionary approach.

Short: Species-ism is the besetting sin of science.

McLaren: Looking to the future I think we will see more and better methods of controlling sex determination. Sperm sorting is a primitive method, and if there were other methods they could be immensely important for livestock breeding. Controlling sex determination will lead on to methods for controlling sex reversal. As far as humans go, the methods that can be used clinically today are quite unbelievably primitive. Looking even further into the future, who knows?—perhaps there will be a demand for methods of sex reversal such as we have heard of in fish, rather socially based! This would introduce new elements into society.

Koopman: I am fascinated by the pivotal role of the pre-Sertoli cell, and what makes a Sertoli cell become a Sertoli cell. What is the role of *Sox9* in this process? What is *Sox9* regulating other than the *Amh* gene?

Sinclair: My interest is in the step just below the testis switch. This brings us back to *Sry*. For me, looking at things in a comparative way may be very helpful. Are the mechanisms going be conserved or different at that point? Paul Burgoyne raised the issue of the ovarian pathway, which I think is fascinating. This is something that could be examined in birds more easily than by studying mice. Whether or not the same genes deployed in the testis pathway can be redeployed in the ovary is another fascinating question.

Graves: I really want to know how mole voles do it! I think there are general questions to be asked whenever a system changes. Although I endorse Adam Wilkins' interest in how *Sry* got its start, perhaps this was now too long ago to find out all the details. With the mole vole we have a real chance, looking at a new sex-determining system which only started a few million years ago. We may still be able to see the first stages of how a new sex-determining gene arises, and what happens at loci close to it. To me, one of the most interesting questions in the world is how genes change their function. We are seeing this all the time on the Y chromosome. We see perfectly good brain-determining genes become testis-determining genes, and ubiquitously expressed genes become spermatogenesis genes. This is a fertile field to ask the question of how genes may change their structure and their function. This will help us understand how genes function in networks.

Camerino: The main question for me is what are the driving forces in the evolution of sex determination. For example, once you have a good sex-determining system, why isn't there selective pressure to maintain this? I can understand why there is a need for a gene such as *Sry*, but why is an antitestis gene necessary? One other thing that has troubled me for a while is why we are all looking only at the level of DNA regulation. There are suggestions that RNA processing may be very important. However, we are not really in a position to be able to examine this.

Josso: It will surprise no one that my preoccupations lie with *Amh*! First, I would like to know whether it is a sex-determining gene or not. That is, whether it has an effect on the ovary in its own right. But if the answer is no, then I might not get invited to sex determination meetings, which would be a pity. I would also like to have the AMH transduction pathway straightened out so we can get down to studying other elements, such as coactivators. Finally, we may then be able to find the aetiology of all the patients with persistent Müllerian duct syndrome in whom both *Amh* and *Amh* type 2 receptor genes are normal.

Capel: My overriding interest is in how gene expression can control morphogenetic events. I am interested in how you can take a bipotential gonad primordium and express one gene, *Sry*, and trigger a series of events that reorganize the morphology of the organ. Can we understand this better for all organs by looking at this system? I agree about the proteomics approaches. I think they will be critical. Many of the genes are expressed similarly at the RNA level, and modifications will alter their function. We need to go to the protein level to understand this.

Lovell-Badge: Can I have three requests? The first is a specific one. Apart from all of the above, in terms of mammalian sex determination understanding how *Sox9* is regulated will be very important. It seems to be a central gene in the process. Perhaps the role of *Sry* will emerge if we find that it is only involved in activating a high level of expression of *Sox9*. The second thing is phenotype-based screens. We have started to do a suppressor–enhancer screen. I think this is a vital way to go, because it doesn't just rely on making null mutations: you can pick up subtle mutations. It will give you any gene that is involved, in theory. Third, and finally, in 10 years I would like to be able to come to another meeting like this.

Short: Perhaps I could also have a wish. I would go back to Wordsworth and his wonderful poem, 'Ode on the intimations of immortality', where he says, 'The Soul that rises with us, our life's Star, hath had elsewhere its setting, and cometh from afar'. To me, this is the continuity of germplasm. My wish would be that we could learn much more about the interactions between the germ cell and the soma in the gonad. Repeatedly, we have had to admit that we know almost nothing about the genetic expression of the germline. If I should be invited in my post-dotage to attend the next meeting in 10 years time, I would hope that we would begin to have some understanding of how it is that the germ cell—especially the female germ cell—is talking to the somatic tissue of the gonad, and how that somatic tissue of the gonad is responding, and talking back to the germ cell.

Reference

Mittwoch U 2000 Genetics of sex determination: exceptions that prove the rule. Mol Genet Metab 71:405–410

Index of contributors

Non-participating co-authors are indicated by asterisks. Entries in bold type indicate papers; other entries refer to discussion contributions.

Subject index

Printed and bound by CPI Group (UK) Ltd, Croydon, CR0 4YY
16/09/2026
14963471-0002